Sustainability Sciences in Asia and Africa

Sustainable Agriculture and Food Security

Series Editor

Rajeev K. Varshney, Centre for Crop and Food Innovation, Food Futures Institute
Murdoch University, Perth, WA, Australia

Sustainability as a scientific endeavor requires broad understanding of the interconnections in our global environment. It is therefore positioned at the crossroads of research in natural sciences, technology and human sciences. This book series supports the global efforts by providing timely coverage of the progress, opportunities, and challenges towards achieving sustainable development goals (SDGs) in Asia and Africa. It fosters the research in transdisciplinary academic fields spanning across agriculture and global food security, health and diseases, access to clean water and affordable energy, climate resilience and effects of global climate change on biodiversity, water and land. This series is planned to narrate the stories of scientific development and progression towards SDGs from the regions of Asia and Africa. The overall aim of the series is to provide a comprehensive publication platform and act as a knowledge engine in the growth of sustainability sciences with a special focus on developing nations. The series constitute of four sub-series each focusing on a specific theme closely linked to a set of SDGs. Each sub-series is leaders and experts in agricultural sciences, environmental sciences, biomedicine and medical sciences respectively. Books in this series are useful for environmentalists, biotechnologists, plant scientists, medical practitioners, conservation biologists, policy makers, NGOs and scientific bodies working in the areas of sustainability sciences in the developing regions.

This book series support the global efforts towards sustainability by providing timely coverage of the progress, opportunities, and challenges of sustainable food production and consumption in Asia and Africa. The series narrates the success stories and research endeavors from the regions of Africa and Asia on issues relating to SDG 2: Zero hunger. It fosters the research in transdisciplinary academic fields spanning across sustainable agriculture systems and practices, post- harvest and food supply chains. It will also focus on breeding programs for resilient crops, efficiency in crop cycle, various factors of food security, as well as improving nutrition and curbing hunger and malnutrition. The focus of the series is to provide a comprehensive publication platform and act as a knowledge engine in the growth of sustainability sciences with a special focus on developing nations. The series will publish mainly edited volumes but some authored volumes. These volumes will have chapters from eminent personalities in their area of research from different parts of the world.

Hupenyu Allan Mupambwa
Ndakalimwe Naftal Gabriel
Tamuka Nhiwatiwa
Kenneth Prudence Abasubong
Samuel K. Mafwila
Editors

Harnessing Coastal and Marine Resources for Improved Food Security in Africa

Springer

Editors
Hupenyu Allan Mupambwa
Sam Nujoma Marine and Coastal
Resources Research Centre
Sam Nujoma Campus
University of Namibia
Henties Bay, Namibia

Tamuka Nhiwatiwa
Department of Fisheries and Ocean
Sciences, Faculty of Agriculture
Engineering, and Natural Sciences
Sam Nujoma Campus
University of Namibia
Henties Bay, Namibia

Samuel K. Mafwila
Office of the Director, Katima Mulilo
Campus, University of Namibia
and the Department of Fisheries
and Ocean Sciences
University of Namibia
Katima Mulilo, Namibia

Ndakalimwe Naftal Gabriel
Department of Fisheries and Ocean
Sciences, Faculty of Agriculture
Engineering, and Natural Sciences
Sam Nujoma Campus
University of Namibia
Henties Bay, Namibia

Kenneth Prudence Abasubong
Department of Fisheries and Aquaculture
Faculty of Agriculture
Akwa Ibom State University
Akwa Ibom State, Nigeria

ISSN 2730-6771 ISSN 2730-678X (electronic)
Sustainability Sciences in Asia and Africa
ISSN 2730-6798 ISSN 2730-6801 (electronic)
Sustainable Agriculture and Food Security
ISBN 978-981-95-9724-6 ISBN 978-981-95-9725-3 (eBook)
https://doi.org/10.1007/978-981-95-9725-3

This Springer imprint is published by the registered company Springer Nature Singapore Pte Ltd.
The registered company address is: 152 Beach Road, #21-01/04 Gateway East, Singapore 189721, Singapore

Preface to the Book Series

The UN's second Sustainable Development Goal (SDG) aims to end hunger, achieve food security and improved nutrition, and promote sustainable agriculture by 2030. Yet over 700 million people still suffer from undernourishment, meaning that global food production systems will need to expand their capacities and produce almost twice their existing levels to feed the world's ever-growing population. This poses a significant challenge, and while farmers, researchers, and policymakers the world over have been tireless in their efforts to address the global food security gap, we're a long way off from where we need to be to meet SDG 2: Zero Hunger.

In recent years, the adoption of new research approaches has led to improvements in food access, farmer autonomy, and the stabilisation of agriculture's global carbon footprint. The advent of these approaches, such as genomics-assisted breeding, has dramatically improved crop breeding programmes worldwide, yet the challenge to reduce hunger remains especially pronounced in the Global South.

Furthermore, the agri-food systems in Asia and Africa are severely constrained by the subsistence nature of farming, limited access to new research tools and technologies, and the increasing impacts of environmental pollution, soil degradation, and biodiversity loss.

This book series, "Sustainable Agriculture and Food Security", aims to support global efforts towards sustainability by providing timely coverage of the progress, opportunities, and challenges of sustainable food production and consumption in Asia and Africa. The series narrates the success stories and research endeavours from Africa and Asia on issues related to SDG 2: Zero Hunger. It fosters research in transdisciplinary academic fields spanning sustainable agriculture systems and practices, pre-seed and post-harvest innovations, and food supply chains. The series aims to provide a comprehensive publication platform and serve as a repository of knowledge in sustainability sciences, with particular focus on developing nations.

Our seas, rivers and lakes are a cornerstone of life on Earth, with fisheries and aquaculture providing livelihoods for an estimated 60 million people. Aquaculture is now the fastest-growing food production sector, providing 20 per cent of animal protein to over 3 billion people globally. It is therefore essential that forward-looking research into our aquatic ecosystems is part of the conversation on global food security and sustainability.

This book, "Harnessing Coastal and Marine Resources for Improved Food Security in Africa", edited by Dr Hupenyu A. Mupambwa, Prof Naftal Gabriel, Prof Tamuka Nhiwatiwa, Dr Kenneth P Abasubong and Prof Samuel K. Mafwila, serves as a comprehensive tome detailing Africa's extraordinary and vibrant marine and coastal ecosystems. From the utilisation of seaweed as a renewable resource in Namibia to detailing the impact of marine pollution in Nigeria, this book offers unprecedented insights into coastal and marine resource management in Africa.

I commend the Editors for including multidisciplinary and continent-wide perspectives and congratulate all the authors for their valuable contributions.

I wish to extend my sincere thanks to the Springer staff, particularly Dr. Sanchi Bhimrajka, Editor (Books, Medicine and Life Sciences), and Mr. Nick Melchior, Editorial Director (Books, Medicine and Life Sciences, Asia), for their constant support in the completion of this compendium.

The cooperation received from my supervisors and research leaders, such as Prof Andrew Deeks (Vice Chancellor), Prof Peter Eastwood (Deputy Vice Chancellor-Research and Innovation), Prof Peter Davies (Pro Vice Chancellor), and my colleagues, Anu Chitikineni and Nic Ingram from Murdoch University (Australia), is also gratefully acknowledged. I want to thank my family members – Monika Varshney, Prakhar Varshney and Preksha Varshney for their love and support in discharging my duties as Series Editor.

Rajeev K. Varshney
Director – WA State Agricultural Biotechnology Centre
Director – Centre for Crop & Food Innovation
International Chair – Agriculture & Food Security
Murdoch University, Perth, WA, Australia

Preface

Africa's coastal and marine zones are endowed with some of the world's most diverse and productive ecosystems. From rich fisheries and unique desert–coast interfaces to expansive seaweed beds, tidal wetlands, and saline-tolerant plant species, these environments provide vital food, income, and livelihood opportunities for millions of Africans. Yet, the potential of these resources remains unevenly harnessed, constrained by ecological pressures, climate change, pollution, limited technological innovation, and socioeconomic disparities.

Harnessing Coastal and Marine Resources for Food Security in Africa emerges against this backdrop with the aim of consolidating contemporary knowledge, highlighting innovative solutions, and providing research-driven insights that support sustainable food systems across the continent. This edited volume brings together multidisciplinary perspectives from leading African researchers. Collectively, the chapters explore the ecological, socioeconomic, technological, and policy dimensions of coastal and marine resource management.

The book opens with a focus on marine biomass resources, including seaweed utilization in animal nutrition and the transformative role of marine-derived organic amendments in horticultural production. These chapters demonstrate the growing relevance of coastal bio-resources for both terrestrial and aquatic food systems. Subsequent chapters address water security, pollution, and ecosystem health, core determinants of sustainable food production. Topics include solar-driven desalination technologies suited to arid coastal regions, the impacts of aquatic pollution on food organisms, and the implications of climate change–driven parasitic diseases in fisheries.

A strong emphasis is placed on marine and coastal fisheries, featuring historical analyses, socioeconomic assessments, and practical examples of sustainable management strategies from Namibia and other African regions. Discussions on bycatch utilization, small-scale fisheries, and the contribution of marine aquaculture to nutritional security further highlight opportunities for expanding Africa's blue economy. The volume also offers forward-looking perspectives on renewable energy integration, aquafeed innovations, value chain expansion, and economic analyses such as the fisheries–economic growth nexus. Notably, case studies on

Salicornia species and seaweed valorization underscore the promise of nontraditional marine crops for enhancing food and raw material security.

By bringing these diverse themes together, this book provides a holistic understanding of how coastal and marine resources can be sustainably leveraged to strengthen Africa's food systems. It is intended for policymakers, researchers, development practitioners, coastal planners, students, and anyone seeking evidence-based pathways for advancing food and nutritional security on the continent.

Henties Bay, Namibia — Hupenyu Allan Mupambwa
Henties Bay, Namibia — Ndakalimwe Naftal Gabriel
Henties Bay, Namibia — Tamuka Nhiwatiwa
Akwa Ibom State, Nigeria — Kenneth P. Abasubong
Katima Mulilo, Namibia — Samuel K. Mafwila

Acknowledgments

The editors of *Harnessing Marine and Coastal Resources for Food Security in Africa* extend their profound gratitude to all contributing authors whose research, insights, and dedicated scholarship made this volume possible. Your commitment to advancing knowledge in marine and coastal resource management is deeply appreciated.

We also sincerely thank the reviewers for their valuable and constructive feedback, which greatly improved the quality, clarity, and academic rigor of the chapters. Your time, expertise, and critical perspectives strengthened this book in meaningful ways.

Our appreciation further goes to the respective institutions of the authors for granting them the time and support needed to contribute to and refine this work. Institutional backing remains an essential pillar in fostering collaborative scholarship across borders, disciplines, and research platforms.

Finally, we acknowledge the ongoing efforts of researchers, practitioners, and policymakers across Africa who continue to champion sustainable utilization of marine and coastal resources for the continent's food security. This book is a testament to your vision and commitment.

Hupenyu Allan Mupambwa
Ndakalimwe Naftal Gabriel
Tamuka Nhiwatiwa
Kenneth P. Abasubong
Samuel K. Mafwila

Contents

Series Editor and Editorial Board Members

About the Series Editor

Rajeev K. Varshney FRS FAA
WA State Agricultural Biotechnology Centre;
Centre for Crop and Food Innovation;
International Chair – Agriculture and Food Security,
Food Futures Institute, Murdoch University,
WA, Australia

Professor Rajeev K. Varshney FRS FAA is a globally recognised agricultural scientist with over two decades of leadership in genomics and molecular breeding. He has made a significant contribution to global food and nutrition security by developing genomic resources for "orphan" crops and leveraging these resources to identify genetic loci and candidate genes associated with drought tolerance, pest resistance, and yield improvement in crops widely grown by smallholder farmers in sub-Saharan Africa and India.

He has led major international research programs that have delivered improved, climate-resilient crop varieties to some of the world's most resource-constrained farming systems, contributing to increased farm productivity, enhanced livelihoods, and improved nutritional outcomes.

Currently, Professor Varshney leads several national and international initiatives in wheat, pulse, and horticultural crops in Australia, with a strong focus on enhancing farmer profitability, environmental sustainability, and global food security.

With more than 600 peer-reviewed publications, over 90,000 citations, and an h-index exceeding 150, he is consistently ranked among the world's most highly cited researchers in plant science.

Editorial Board Members

Chike Mba, Deputy Director, Plant Production and Protection Division, Food and Agricultural Organization (FAO), Rome, Italy.

Trilochan Mohapatra, Chairman, Protection of Plant Varieties and Farmers' Rights Authority, Ministry of Agriculture and Farmers Welfare, Government of India, New Delhi, India.

Soumya Swaminathan, Chairperson, MS Swaminathan Research Foundation, Chennai, India.

Wendy Umberger, Chief Executive Officer, Australian Council for International Agricultural Research (ACIAR), Canberra, ACT, Australia.

About the Editors

Hupenyu Allan Mupambwa is currently the Head of the Sam Nujoma Marine and Coastal Resources Research Centre (SANUMARC) and also a senior research fellow in desert and coastal agriculture under the Sam Nujoma Marine and Coastal Resources Research Centre at the University of Namibia. He holds a PhD in soil science from the University of Fort Hare in South Africa, specializing in organic soil fertility management, an MSc focused on soil physical properties in conservation agriculture, and a BSc in crop science from the University of Zimbabwe. With 17+ years of experience in government agriculture research, university teaching, and research, he has published 30+ journal articles, 27+ book chapters, and edited 4 books. His research emphasizes waste beneficiation, vermicomposting, and organic soil fertility for resource-poor farmers.

Ndakalimwe Naftal Gabriel is currently an Associate Professor in the Department of Fisheries and Ocean Sciences, School of Fisheries and Agriculture, University of Namibia (UNAM). He holds a PhD in Fisheries and Aquatic Sciences from UNAM, and his postdoctoral studies were conducted at Kentucky State University's Aquaculture Research Center in the United States, as well as at the Asian Institute of Technology (AIT) in Thailand. He earned his MSc in Aquaculture from Nanjing Agricultural University in Jiangsu, China, from 2013 to 2015, and his bachelor's degree in Biological Sciences from UNAM. He has been awarded several prestigious scholarships, including a Fulbright scholarship for his postdoctoral studies, The World Academy of Science-TWAS fellowship, Nanjing Municipal People's Government International Student Scholarship, Chinese Government Scholarship, and recognition as one of the top postgraduate students at Nanjing Agricultural University in 2014 and 2015. He has over 10 years of experience working in the field of aquaculture as a researcher and 9 years as a lecturer at UNAM. With more than 40 published research articles, including over 15 book chapters and 3 edited books, Gabriel has also reviewed numerous peer-reviewed articles for respected journals and supervised a significant number of graduate students in the fields of aquaculture and fisheries sciences.

Tamuka Nhiwatiwa is Associate Professor of Fisheries Biology at the University of Namibia. He has a PhD in Biology from KU Leuven (Belgium), MPhil degree in Limnology from University of Zimbabwe, and BSc Honors degree in Biological Sciences from the University of Zimbabwe. He has 20 years' experience as an academic and researcher. His research focuses on freshwater and marine systems ecology, fisheries, aquaculture, and limnology. He has also vast experience in other disciplines of aquatic ecology that include water resources management, pollution control and biological monitoring, river and lake ecology, fish biodiversity, and invertebrate biology. He has published 125 research contributions (articles, book chapters) and is a highly rated researcher according to academic research metrics.

He reviews for several international journals and publishers. As an academic and researcher, he has accrued vast exposure and knowledge of Africa's aquatic systems, and also gained knowledge through regional and international collaborations of aquatic systems within Southern Africa and beyond. He has supervised and trained many undergraduates, master's, and PhD students from various countries over the years. In terms of reach, he has done studies in South Africa, Mozambique, Tanzania, Zambia, DRC, and Malawi; he has many established links with many institutions across the world (e.g., Germany, Belgium, the UK, Finland, and Norway) where there are ongoing collaborations. He also has a wealth of experience in initiation, coordination, and implementation of international collaborative projects, which require ability to integrate with peoples of different cultures and backgrounds.

Kenneth Prudence Abasubong from Nigeria holds an MSc and a PhD from Nanjing Agricultural University (NAU), China, specializing in aquaculture nutrition, particularly fish nutrient metabolism, excretion, immune pathways, feed science, and the eco-intensification of fish farming using recycled waste, fish health, and sustainable aquaculture development. He studied at the University of Uyo, Nigeria (PGD in Aquatic Environmental Management) from 2012 to 2013 and at the University of Calabar for his bachelor's degree in Animal Sciences. He has attended several international conferences and symposiums in aquaculture and fisheries sciences and has published over 35 articles, 11 book chapters, and 2 edited books. He has received several awards, including the Best International Student at NAU in 2018. Additionally, he has reviewed numerous peer-reviewed articles for respected journals, including Elsevier, Springer, Wiley, and Frontier, and has supervised many undergraduate and graduate students in the fields of Aquaculture and Fisheries Sciences. Currently, he is undergoing postdoctoral research training in the Faculty of Fisheries and Protection of Waters at the South Bohemian Research Center of Aquaculture and Biodiversity of Hydrocenoses, Institute of Aquaculture and Protection of Waters, Czech Republic. He has over ten years of experience as an animal and aquaculture nutrition researcher.

Samuel K. Mafwila is Director for the Katima Mulilo Campus, University of Namibia. He is an Associate Professor in the Department of Fisheries and Ocean Sciences, University of Namibia. He has a PhD in Marine Science from the Zoology Department, University of Cape Town, an MSc in Applied Marine Science from the Zoology Department, University of Cape Town, a BSc (Honors) in Marine Biology from the Entomology and Zoology Department, Rhodes University, and a Postgraduate Diploma in Education (Biology and Physical Science) and a Bachelor of Science (Zoology and Geography), both from the University of Namibia. He has more than 25 years of working experience in both public and private sector where he served in various capacities. He is the National Chairperson of Scientific Committee on Oceanic Research (SCOR), and Coordinator of the FishFORCE Academy. He has served as a Course Director for the Regional Graduate Network in Oceanography in Namibia, and former Manager for the I&J and NAMDEB

Mariculture Venture. He is a certified project management professional, and he has managed project portfolios including various EU-funded Projects and UK Research Funded projects. He serves on the Boards of several Corporate Companies and is an Advisor for various scientific bodies. He has published more than 30 scientific papers and specialist technical reports. His research covers a wide range of thematic areas: Marine Ecosystem Health, Marine Biology, Biodiversity, Aquatic and Benthic Ecology, Oceanography, Climate Change, Aquaculture, Fish and Fisheries Management, Blue Economy, Integrated Coastal Zone Management, and Marine Spatial Planning.

Contributors

James Abah Department of Applied Educational Sciences, School of Education, University of Namibia, Katima Mulilo Campus, Katima Mulilo, Namibia

Kenneth Prudence Abasubong Department of Fisheries and Aquaculture, Faculty of Agriculture, Akwa Ibom State University (AKSU), PMB 1167, Obio Akpa Campus, Oruk Anam, Akwa Ibom State, Nigeria

Maama Kwao Akita Institute for Environment and Sanitation Studies, University of Ghana, Legon, Accra, Ghana

Peter Chukwudi Department of Biology and Forensic Sciences, Admiralty University of Nigeria, Ibusa, Delta State, Nigeria

Canicio Dzingirai Department of Economics, University of Namibia, Windhoek, Namibia

Ekaette Godwin Edelduok Department of Zoology, Akwa Ibom State University, Mkpat Enin, Akwa Ibom, Nigeria

Eyo Idorenyin Edem Department of Fisheries and Aquaculture, Faculty of Agriculture, Akwa Ibom State University (AKSU), PMB 1167, Obio Akpa Campus, Oruk Anam, Akwa Ibom State, Nigeria

Umoh Uwana Emmanson Department of Fisheries and Aquaculture, Faculty of Agriculture, Akwa Ibom State University (AKSU), PMB 1167, Obio Akpa Campus, Oruk Anam, Akwa Ibom State, Nigeria

Ndakalimwe Naftal Gabriel Department of Fisheries and Ocean Sciences, Faculty of Agriculture, Engineering, and Natural Sciences, Sam Nujoma Campus, University of Namibia, Henties Bay, Namibia

Ubong Uwem George Department of Fisheries and Aquaculture, Akwa Ibom State University, Mkpat Enin, Akwa Ibom, Nigeria

Martha K. Hausiku-Ikechukwu Sam Nujoma Marine and Coastal Resources Research Centre, Sam Nujoma Campus, University of Namibia, Henties Bay, Namibia

Shikangalah Rosemary Department of Environmental Science, University of Namibia, Windhoek, Namibia

Elia N. M. Ruben Sam Nujoma Marine and Coastal Resources Research Centre, Sam Nujoma Campus, University of Namibia, Henties Bay, Namibia

Department of Crop Production and Agricultural Technologies, Ogongo Campus, University of Namibia, Oshakati, Namibia

Al-Mas Sendegeya Department of Electrical and Mechanical Engineering, Namibia University of Science and Technology, Windhoek, Namibia

Arisha September Sam Nujoma Marine and Coastal Resources Research Centre, Sam Nujoma Campus, University of Namibia, Henties Bay, Namibia

Vistorina Shithigona Social Security Commission, Windhoek, Namibia

Fimanekeni Ndaitavela Shivute Science and Technology Division, Multidisciplinary Research (MR), Centre for Research Services (CRS), University of Namibia, Windhoek, Namibia

Uugulu Shoopala Department of Geo-Science, University of Namibia, Keetmanshoop, Namibia

Evans Simasiku Department of Wildlife and Tourism, Katima Mulilo Campus, University of Namibia, Katima Mulilo, Namibia

Simeon S. Hamukoshi Namibia Green Hydrogen Research Institute, University of Namibia, Windhoek, Namibia

National Commission on Research, Science and Technology, Olympia, Windhoek, Namibia

Martin Tjipute Department of Fisheries and Ocean Sciences, Faculty of Agriculture, Engineering, and Natural Sciences, Sam Nujoma Campus, University of Namibia, Henties Bay, Namibia

Endjambi Tobias Department of Environmental Studies, The International University of Management, Windhoek, Namibia

Uwem George Ubong Department of Fisheries and Aquaculture, Faculty of Agriculture, Akwa Ibom State University (AKSU), PMB 1167, Obio Akpa Campus, Oruk Anam, Akwa Ibom State, Nigeria

Mfon Timothy Udo Department of Fisheries and Aquaculture, Faculty of Agriculture, Akwa Ibom State University (AKSU), PMB 1167, Obio Akpa Campus, Oruk Anam, Akwa Ibom State, Nigeria

Department of Fisheries and Environmental Management, University of Uyo, Uyo, Akwa Ibom State, Nigeria

Michael Chukwuemeka Umehai Department of Agricultural Extension, Delta State University, DELSU, Abraka, Nigeria

Chapter 1
Promoting Solar-Driven Desalination for Water and Food Security in an Arid Environment: The Case of the Namibian Coastal Regions

Shikangalah Rosemary, Endjambi Tobias, and Uugulu Shoopala

Abstract Namibia, the driest country in sub-Saharan Africa, faces severe water scarcity and food insecurity due to its arid climate, erratic rainfall, and high evaporation rates, which threaten the availability of water for domestic use, rain-fed agricultural activities, livestock, and fisheries, which are critical to the livelihoods. Despite having a 1572 km coastline, Namibia underutilises desalination, relying instead on depleting groundwater and ephemeral rivers. This study assessed the feasibility of scaling solar-driven desalination to enhance water and food security, leveraging Namibia's abundant solar resources (Global Horizontal Irradiance >6.6 kWh/m^2/day). This research employed a qualitative methodology, analysing several relevant documents, existing initiatives (e.g. the Orano plant), pilot solar-powered projects, and reviewing relevant policies. The findings reveal that Namibia can mitigate water stress by upscaling desalination for water production for domestic and agricultural purposes, which is highly plausible, especially when the process is combined with solar power to reduce energy costs. Conventional desalination faces barriers, including high energy costs (3–4 kWh/m^3), infrastructure gaps, and environmental concerns. Solar-integrated initiatives, such as graphene oxide membranes and zero liquid discharge, have the potential to reduce energy expenses and provide an environmentally friendly water production process while producing a substantial amount of water. Successful case studies from Israel and Saudi Arabia have demonstrated scalable models. Simultaneously, local pilots (for example University of

S. Rosemary (✉)
University of Namibia, Department of Environmental Science, Windhoek, Namibia
e-mail: rshikangalah@unam.na

E. Tobias
The International University of Management, Department of Environmental Studies, Windhoek, Namibia

U. Shoopala
University of Namibia, Department of Geo-Science, Keetmanshoop, Namibia
e-mail: suugulu@unam.na

H. A. Mupambwa et al. (eds.), *Harnessing Coastal and Marine Resources for Improved Food Security in Africa*, Sustainability Sciences in Asia and Africa,
https://doi.org/10.1007/978-981-95-9725-3_1

Namibia's solar-powered desalination) confirm the technical viability of the approach, producing 2300 litres per hour of high-quality water. Upscaling desalination could also benefit ecological restoration and ecotourism, reinforcing the climate-resilient system and ensuring sustainability. The study recommends policy reforms to streamline water governance, further development of public-private partnerships for financing, and enhanced community engagement to ensure the sustainable use of water resources.

Keywords Climate adaptation · Desalination · Solar energy · Namibia

1.1 Introduction

Namibia is widely known as the driest country in sub-Saharan Africa, with approximately 92% of its land classified as arid, semi-arid, and hyper-arid (World Bank Group 2021; Frankl and Madeira 2024). The annual average rainfall varies significantly between regions, ranging from 600 mm in the northeast region (Zambezi) to 10 mm in the coastal areas, such as the Erongo, Hardap, and Karas regions (Mupambwa et al. 2021; Atlas of Namibia Team 2022). The arid to semi-arid areas are also characterised by mean annual temperatures ranging from 14.3 °C to 24.2 °C, with daily maximum temperatures ranging between 25 °C and 35 °C (reaching 40 °C in some areas), and high evaporation rates exceeding 3000 mm/year in some areas (World Bank Group 2021; Atlas of Namibia Team 2022). These conditions are exacerbated by climate change, which has led to increased temperatures and prolonged dry spells, aggravating water scarcity and food insecurity in the country (Spear et al. 2018; IPCC 2021).

Namibia has experienced recurrent droughts over the past decade, with the period from 2012 to 2019 being one of the most severe on record (Shikangalah 2020). This has resulted in vast economic losses for farmers, as both livestock and crops have died due to water scarcity (Mupambwa et al. 2021). The 2024 drought was equally devastating, affecting nearly 1.26 million people, or 41% of the population (Hempapu 2024, October 25; ACAPS 2025). The country's economy and food security are heavily reliant on climate-sensitive sectors, including rain-fed agriculture, fisheries, and tourism. Nearly 70% of the country's population relies on subsistence farming, which is highly susceptible to rainfall variability, and approximately 90% of the country's agriculture is livestock farming (MAWF 2015; Murray and Ruppel 2025). This dependency underscores the urgent need for adaptive strategies to mitigate climate risks.

Namibia's 1572 km coastline along the Atlantic Ocean holds immense potential to address the country's water scarcity and food insecurity; however, it remains a critically underutilised resource. While desalination has proven effective in other water-stressed regions, Namibia's current capacity is mainly limited to mining operations, leaving urban and agricultural sectors without reliable access to freshwater (Namibia Water Corporation (NamWater) 2025). Small-scale solar-powered systems are being implemented across the country, primarily for electricity consumption and pumping water from aquifers through boreholes (Ministry of Environment and Tourism 2019). This study assessed the feasibility of scaling up solar-driven desalination to improve water security for domestic and agricultural use, while exploring its potential to stimulate ecotourism and promote ecological restoration in the region. By assessing the challenges and opportunities of large-scale implementation, this study aims to inform strategies that capitalise on Namibia's abundant solar energy and coastal proximity to foster long-term resilience and sustainable development.

1.2 Methods

This study employed qualitative analysis to assess the potential of solar-powered desalination in enhancing water and food security in Namibia's coastal regions. This study reviewed secondary sources, including climate and hydrological data from the Atlas of Namibia Team (2022), IPCC reports, NamWater publications, and case studies of existing desalination projects, both locally (e.g. the Orano plant) and internationally (e.g. Israel and Saudi Arabia). The study further examined documents such as Namibia's Harambee Prosperity Plan II and the Namibian water and energy policies to identify gaps in supporting desalination initiatives and potential benefits. In addition, this study analysed the feasibility of green hydrogen production. Finally, the study proposes scalable desalination initiatives that apply to the arid and water-stressed Namibian region. Thematic analysis was used to analyse several documents and identify challenges and solutions. This triangulation strengthened the validity of the study and its outcomes (Meydan and Akkaş 2024).

1.3 Climate Change and Its Impacts on Water and Food Security in Namibia

Namibia's coastal regions, which are already arid, are experiencing worsening climatic conditions due to global warming. Historical climate data have revealed a consistent decline in rainfall over the past century, particularly in the Namib Desert and transitional zones near the coast (Engelbrecht et al. 2015). Under the high-emission scenario (RCP8.5), projections indicate a further 10–20% reduction in mean annual rainfall by 2050, which will exacerbate water scarcity (IPCC 2021). Rising temperatures, increasing by 1.7 to 5.4 °C by mid-century (World Bank Group 2021), are likely to intensify evaporation rates, further reducing water availability in already stressed aquifers and ephemeral rivers. These trends suggest that droughts, which are already frequent, will become more prolonged and severe, threatening both freshwater supplies and the productivity of coastal fisheries and agriculture

(Shikangalah and Mapani 2021; Engelhard et al. 2024). Water demand in the country is expected to reach up to 772 million cubic meters per year by 2030, while at the same time, water shortages are expected to be more severe within the next 4 to 7 years if no viable long-term solutions are implemented (Murray and Ruppel 2025).

Namibia's coastal regions rely on a fragile mix of water supply systems that are increasingly strained by climate change and growing demand. Groundwater is one of the primary sources of water in coastal regions, with the Omaruru, Kuiseb, and Omdel aquifers supplying critical water to towns such as Arandis, Henties Bay, Swakopmund, and Walvis Bay (von Oertzen et al. 2024). However, overextraction has led to declining water tables and seawater intrusion (Matengu 2020). Surface water systems are also vulnerable. The Swakop and Omaruru Rivers, which feed into coastal dams, are ephemeral and often dry before reaching their endpoints (NPC 2017; Shikangalah and Mapani 2021). Water distribution inefficiencies have compounded challenges, including water scarcity in various schemes, unstable electricity, aging infrastructure, and frequent breakdowns, as well as increased demand from the mining sector and reliance on expensive water trucking during emergencies (NamWater 2024). This further demonstrates the need to increase the water supply capacity.

Agriculture in Namibia's coastal regions is highly vulnerable to climate change owing to its dependence on erratic rainfall, limited irrigation, and high temperatures that increase evapotranspiration. Rain-fed crop production, already marginal, decreased significantly by 43% in 2024 (Namibian Agronomic Board 2024) and is expected to experience yield reductions of 20–30% for staples, such as maize and millet, by 2030 (World Bank Group 2021). Livestock farmers along the coast face similar challenges, with heat stress reducing cattle productivity and drought diminishing grazing land (NPC 2017). The fisheries sector faces threats from warming ocean temperatures of 0.4 °C per decade and Benguela Niños, which result in low oxygen and poor nutrient levels in the central to northern parts of the Namibian Exclusive Economic Zone (Hyder et al. 2024). These factors have led to a decline in fish stocks, such as sardines, horse mackerel, and hake, which are crucial for food security and export revenues (MFMR 2023). Moreover, the country's tourism industry, which is centred around its unique desert landscapes and wildlife, is vulnerable to climate-induced ecosystem shifts (Goverment of the Republic of Namibia 2018). Without urgent adaptation measures, particularly those reliant on subsistence farming and fishing, they will face escalating risks to their livelihoods and food security in the future. Climate change can impact the distribution and abundance of desert wildlife, which is a key attraction for tourism.

1.4 Desalination Current Efforts and Challenges

Desalination, although a potential solution, remains underdeveloped beyond its use in mining, leaving municipalities struggling to meet their water demands (NamWater 2025). Seawater Reverse Osmosis (SWRO) technology has been successfully implemented in Namibia's mining sector, particularly through the Orano desalination plant that supplies water to uranium operations near Swakopmund (Christelis and Struckmeier 2018). The Orano Desalination Plant produces 20 million m^3/year, primarily for uranium mines, but also sells approximately 12 million m^3/year to NamWater for municipal use to meet coastal water demand (NamWater 2025). Desalination facilities face three key barriers to their broader adoption. First, energy costs are prohibitive; desalination consumes 3–4 kWh/m^3 (Moumni et al. 2024), and Namibia is highly reliant on importing power (60–70%), at approximately USD 4 billion in 2019 to USD 5 billion in 2023 (IEA 2024). This results in the country having the highest electricity prices in Southern Africa (IEA 2024) and makes operations expensive (Dall and Hoffmann 2017). Second, infrastructure gaps persist owing to the complex nature of the landscapes. Namibian coastal landscapes vary significantly (Atlas of Namibia Team 2022), which can make it more difficult to implement proper piping and increase energy use for pumping. Third, environmental concerns are significant regarding the annual discharge of brine, which has the potential to increase salinity levels in nearshore ecosystems (Hutson and Taganova 2023; IEA 2024). Despite these challenges, there are opportunities. Hybrid solar desalination pilots were conducted on small-scale desalination units in Henties Bay by the University of Namibia, utilising 100% solar power for SWRO technology and producing 2300 L/h of fresh water per hour with an excellent mineral composition (Mupambwa et al. 2021). Wind energy is also utilised along the coast, but not for desalination. It is practised particularly in the Lüderitz area, where average wind speeds exceed 6–12 m/s, with a potential capacity factor of 30–40% (Ministry of Mines and Energy 2022; Iileka et al. 2024). Between Hanties Bay and Walvis Bay, wind power can produce up to 20 MW, whereas between Oranjemund and Lüderitz, up to 400–600 MW can be generated (Ministry of Mines and Energy 2022). A more recent effort in water and energy production in Namibia involves the use of a green hydrogen desalination process, primarily through reverse osmosis (von Oertzen 2021; Altenburg and Kantel 2024; IEA 2024). While there are very high expectations from this approach, the end production of brine and its impact on marine water bodies and ecosystems are also anticipated. The lack of a Strategic Environmental Impact Assessment for combined discharges (Schmidt and Frank 2023) poses a risk of unmonitored damage, as highlighted in a feasibility study.

Namibia's policymakers have developed a series of policy documents to guide the country toward sustainable development. Namibia Vision 2030 aims "to improve the quality of life of the people of Namibia" (Government of the Republic of Namibia 2004). Furthermore, the National Development Plan (NDP5) focuses on diversifying the economy and promoting industrial development, which includes investing in infrastructure development, particularly energy infrastructure, to ensure energy security and enhance value-added processes for natural resources (IEA 2024). Building on Vision 2030 and the NDP5, Harambee Prosperity Plan I, introduced in 2016, focused on reducing poverty and improving the quality of life. Harambee Prosperity Plan II (2021–2025) focused on strategic actions (IEA 2024). According to von Oertzen et al. (2024), Namibia's policy landscape for water and food security appears robust on paper, but its implementation is uneven along the coast. Many internal water management policies are not enforced by law but are practised in the general administration of water matters (Goverment of the Republic of Namibia 2013; von Oertzen et al. 2024). This is likely to pose challenges for groundwater extraction, potentially leading to unchecked depletion.

Furthermore, institutional fragmentation is likely to exacerbate these delays. For instance, although the Ministry of Agriculture, Fisheries, Water, and Land Reform (MAFWLR) holds primary responsibility for water management and regulation, a combination of four ministries oversees water resource management. This includes the MAFWLR, Ministry of Health and Social Services, Ministry of Environment, Forestry, and Tourism, and Ministry of Works, Transport, and Communication (Govermenet of the Republic of Namibia 2000), which could potentially lead to bureaucratic bottlenecks if not handled correctly. Closing these gaps requires urgent action, including consolidating water governance under a single authority, mandating climate risk assessments for coastal development, and aligning budgets with adaptation priorities.

1.5 The Potential of Desalination for Water and Food Security in Coastal Namibia

In Namibia, water stress is expected to worsen (40–80%) by 2040. However, the country receives Global Horizontal Irradiance of at least 6.6 kWh/m^2 per day and over 2400 kWh/m^2 per year, which is not fully utilised (IEA 2024). However, solar energy is one of the least utilised sources in the country (Frankl and Madeira 2024; IEA 2024). Scaling this initiative could transform desalination from a mining sector novelty into a cornerstone of coastal water and food security.

The technological feasibility of desalination in Namibia's coastal regions presents significant opportunities and notable challenges. According to NamWater (2025), Orano's production capacity has the potential to increase to a medium- to long-term capacity of up to 45 million cubic meters per year. Technological challenges persist in membrane durability and brine management; however, recent advances in graphene oxide membranes (GOM) and zero-liquid discharge (ZLD) systems show potential for addressing these issues in Namibia's context. GOM (polymeric and inorganic) has emerged as a crucial material that can filter ions and molecules with high precision, blocking all substances except water vapour (Joshi et al. 2015; Ma et al. 2017; Jia et al. 2022). Instead of using a heating source to achieve water evaporation, condensation recycling, and salt crystallisation, GOM utilises a permeable membrane as a transitional medium, facilitating the extraction of freshwater resources while simultaneously concentrating brine through the force of transmembrane pressure differentials (Wang et al. 2024).

Furthermore, graphene is an excellent opportunity for photothermal-photochemical synergistic water treatment because of its exceptional medium for photon-to-thermal energy conversion, which includes broadband light absorption, ultrafast hot carrier dynamics, tuneable electronic structure, and low evaporation enthalpy (Li et al. 2023; Zhang et al. 2025). ZLD is an engineering approach to water treatment in which all water and precious salts are recovered (up to 95% of a plant's wastewater), and contaminants are reduced to solid waste, eradicating liquid waste (Modi et al. 2022; Panagopoulos 2022; Elewa 2024). According to Saltworks (2017), ZLD reduces waste volumes and decreases the costs associated with waste management. Recycling water on-site lowers water acquisition costs and risks and can also result in fewer treatment needs versus treating to meet stringent environmental discharge standards, reducing trucks associated with off-site wastewater disposal and their associated greenhouse gas impact and community road incident risk, improving environmental performance and regulatory risk profile for future permitting, and recovering valuable resources, for example, ammonium sulphate fertiliser or sodium chloride salt for ice melting (Saltworks 2017; Modi et al. 2022; Panagopoulos and Michailidis 2025). The ZLD can be powered by solar energy through a scheme comprising an electrolyser powered by photovoltaic (PV) panels and an evaporator whose energy is supplied by solar panels. The advantages of the electrodialysis process are that it can concentrate brine up to 100,000 ppm (Loganathan et al. 2016; Gadalla et al. 2023) and has up to 79–98% water recovery through the reverse osmosis-electrodialysis reversal system (Korngold et al. 2009; Gadalla et al. 2023). Therefore, the two systems, GOM and ZLD, although with some limitations of their own, are more suitable for Namibia, as they are more efficient, have low energy consumption, and have a reduced environmental impact. Current desalination methods pose risks to marine ecosystems and fisheries and threaten long-term water security. Therefore, switching to GOMs and ZLD ensures sustainable hydrogen production while protecting Namibia's natural resources. Resource recovery and avoidance of environmental damage justify the higher upfront cost of ZLD.

1.6 The Need for Green Hydrogen to Be Combined with GOM and ZLD Approaches for Brine Treatment

This section utilises a green hydrogen feasibility study in Namibia, conducted by Schmidt and Frank (2023), to highlight the need for combining approaches. Green hydrogen poses several environmental and economic challenges. When green hydrogen is produced from desalinated seawater, the ecological risks include the effects of brine discharge on marine ecosystems, such as increased salinity and reduced dissolved oxygen levels (Ihsanullah et al. 2021; Linden 2024), as well as the introduction of contaminants that may exacerbate H_2S plumes. This can create high-salinity zones, which can lead to fish mortality and ecosystem disruption (Schmidt and Frank 2023). Additionally, there are economic and operational risks associated with the high-water demand for hydrogen production, as producing 1 kg of H_2 requires 9–14.5 kg of water (Schmidt and Frank 2023). Table 1.1, derived

Table 1.1 Projected brine discharge vs. water demand from 2030 to 2050

Scenario	Brine discharge (million m^3/a)	% of Namibia's water demand	Fishing impact risk
Current (2023)	106.3	0.6% (H_2 only)	Low (localised)
Scenario 1 (HYPHEN Project)	150.2	1%	Moderate (Lüderitz zone)
Scenario 2 (2035)	235.7	4.9%	High (multiple plants)
Scenario 3 (2050)	398.7	19.4%	Severe (coastal collapse risk)

Data sourced from Schmidt and Frank (2023)

Table 1.2 Comparison of brine disposal methods for surface and evaporation discharge (data sourced from Tables 1, 6, and 7 of Schmidt and Frank 2023) to GOM and ZLD

Method	Cost (US$/$m^3$)	Energy use (kWh/m^3)	Environmental impact	Feasibility for Namibia
Surface discharge (current)	0.05–0.30	Low (pumping only)	High (salinity rise, H_2S plumes, fish mortality)	Risky for fisheries and ecosystems
Evaporation ponds	3.28–10.04	Low (solar-dependent)	Moderate (land use, salt recovery possible)	Limited by land availability
ZLD	3–27	6–45 (high efficiency)	Low (no brine discharge, resource recovery)	Best for sustainability and circular economy
GOM and ZLD	Higher CAPEX	Lower than RO	Minimal (reduced fouling, no chemicals)	Ideal for long-term efficiency

from the reports of Schmidt and Frank (2023), illustrates the exponential growth of brine. By 2050, brine discharge will nearly quadruple (from 106.3 to 398.7 million m^3/a), and green H_2 production will consume 1% of water in 2023, which will translate to 19.4% of Namibia's total water demand by 2050 (Fig. 3 of Schmidt and Frank 2023), adding more pressure to the water problem. Lastly, the report highlighted the energy and cost inefficiencies, as reverse osmosis desalination consumes 2–6 kWh/m^3, despite brine discharge being relatively inexpensive due to low operational costs ($0.05–0.30/$m^3$) (Table 1.2). Furthermore, Lüderitz (a critical fishing zone) will face a moderate risk by 2035. Although it currently faces a low risk (localised discharge), it will face a severe risk by 2050 (Table 1.1), highlighting the potential for coastal collapse.

Table 1.3 presents a comparison of brine management, highlighting the critical differences between Namibia's current desalination practices and the proposed advanced system utilising GOM and ZLD. Currently, Namibia relies on RO with surface discharge, which has low operational costs ($0.05–0.30/$m^3$) but severe environmental consequences. The brine produced (with a TDS concentration exceeding 55,000 mg/L) is discharged into the ocean, increasing salinity, depleting oxygen,

Table 1.3 Brine management comparison

Factor	Current RO, surface discharge (Schmidt and Frank 2023)	Proposed GOM and ZLD system
Brine impact	High (marine damage)	None (full recovery)
Energy use	2–6 kWh/m^3 (RO) and pumping	Lower (GOM efficiency)
Chemical use	High (anti-scaling, antifouling)	Reduced (GOM resists fouling)
Resource recovery	Grey water/ammonia	Salts, lithium, magnesium
Long-term cost	Low OPEX, high environmental risk	Higher CAPEX, but offsets via mineral sales

Data sourced from Tables 1, 2, 6, 7; Figs. 10, 12, 13 of Schmidt and Frank (2023)

Table 1.4 Economic and environmental benefits of ZLD and GOM over the current methods

Stage	Current system (Schmidt and Frank 2023)	GOM + ZLD system
Pretreatment	High chemical use (Table 1.2)	Reduced chemicals (GOM advantage)
Desalination	RO (2–6 kWh/m^3)	GOM (lower energy, external data)
Brine Output	141.5M m^3/d global (p. 7)	Zero liquid waste
Environmental risk	High (fishing harm, p. 19–24)	None
Economic value	Energy, grey water (ammonia)	Energy and salt/mineral sales

and posing a risk of fish mortality, particularly in critical fishing zones such as Lüderitz. This method also requires heavy chemical pretreatment (antiscaling and antifouling agents), further contaminating marine ecosystems (Table 2 of Schmidt and Frank 2023). In contrast, the GOM and ZLD systems have higher upfront costs than RO ($ 3-$27/m^3) (Table 1.2), but they offer sustainable and environmentally friendly alternatives for coastal regions.

Table 1.4 shows that the current methods pose risks to fisheries and ecosystems, whereas ZLD prevents this. In contrast, GOMs have higher efficiency, offering better salt retention and lower energy consumption than traditional RO (as noted in the report, supported by external research), and reduce fouling, scaling resistance, and the need for pretreatment chemicals (Han et al. 2021; Ren et al. 2025). ZLD recovers >95% of water and crystallises salts, is a resource recovery process, as it extracts valuable minerals (e.g. sodium, magnesium, calcium, lithium, and boron) from brine and produces marketable salt via evaporation ponds, which are feasible (Panagopoulos and Giannika 2022; Sabour and Ghorashi 2024). Both GOM and ZLD comply with Namibia's goals for carbon neutrality, and ZLD aligns with the marine spatial plan goals for sustainable water use by preventing coastal ecosystem collapse and creating a circular economy from desalination byproducts.

1.7 Synergies with Agriculture, Ecological and Ecotourism Benefits

The potential applications of desalinated water extend far beyond municipal drinking supplies in Namibia's coastal region. Agricultural irrigation using desalinated seawater, while energy-intensive, could revolutionise food production in the region, particularly along ephemeral rivers where soil quality is high, but freshwater availability limits cultivation (Shikangalah and Mapani 2021). Mupambwa et al. (2019) have highlighted several potential climate-smart agriculture ideas, including hydroponics, aquaponics, and mushroom farming activities. Such farming activities could benefit from utilising a hybrid renewable desalination-powered system in coastal areas at the regional and household levels. Aquaculture represents another promising application, with the Namibia Mariculture Project demonstrating that desalinated water can support the farming of freshwater fish (MFMR 2023). Tilapia aquaculture, particularly using species like *Oreochromis mossambicus* and hybrids, has the potential to thrive in desalinated water due to their tolerance of low to moderate salinity (Kamal and Mair 2005; Sarker et al. 2023), making them ideal for arid coastal regions of Namibia. Initiatives such as Ubuntu Aquaponics Farming in Swakopmund, though on a small scale, have successfully combined tilapia farming with hydroponic vegetable production in recirculating systems. This approach maximises water use while contributing to food security and supporting local entrepreneurship in Namibian coastal towns. Namibia can also learn from other desert agriculture initiatives in other parts of the world, such as Israel's Negev Desert, where drip irrigation and aquaponics support high-value crops like tomatoes, cucumbers, and olives (Katzir 2014; Tal 2021). Countries like Algeria and Egypt also provide good models for integrated aquaculture–agriculture systems in arid environments (Corner et al. 2020). Olive farming has also been successfully established in the Namibian coastal town of Swakopmund, where groundwater from the lower Swakop River has supported both asparagus and olive production, though challenges such as mica deposits clogging dripper lines and variations in water quality between plots persist (Marx 2009). Desalinated water could provide an alternative or supplementary source to strengthen the reliability of irrigation projects along the Swakop River and in other coastal towns, thereby contributing to improved food security and employment creation in these areas. These diversified applications could significantly improve the economic viability of desalination investments while addressing multiple dimensions of food insecurity in coastal communities. To fully realise this potential, proactive government policy is essential to allocate and prepare suitable land for agricultural development. The government should leverage its existing land inventory to identify and designate specific coastal zones for climate-resilient agriculture, prioritising areas with suitable soil conditions near desalination infrastructure. This must be coupled with transparent and affordable land-lease agreements to incentivise investment from both local entrepreneurs and commercial farmers. As emphasised by the Food and Agriculture Organisation (2021), national food security strategies must integrate land-use planning that facilitates sustainable intensification, ensuring that newly irrigable land is used efficiently for high-value food production rather than low-employment, water-intensive crops. Such targeted land allocation will be the critical link between the production of desalinated water and the creation of viable, job-creating agricultural enterprises.

The strategic use of desalinated water could also support ecological restoration efforts along Namibia's degraded coastline. The rehabilitation of Namibian coastal ephemeral river estuaries, where freshwater flow is not limited to the rainfall season but is highly impacted by climate change, could benefit from the controlled release of desalinated water to maintain critical ecosystems, especially those affected by climate change (Ndjaula et al. 2023). Inland, artificial water points supplied by desalinated water could help stabilise populations of endangered species, such as the Damara tern, which nests in coastal deserts (Braby et al. 2009). The replenishment of near-coastal aquifers through managed aquifer recharge with excess desalinated water could create buffers against seawater intrusion while improving the vegetation cover (Steinmuller and Chambers 2018; Bennett 2022). These ecological benefits would extend to combating desertification. Pilot projects using drip-irrigated desalinated water have shown the potential to restore arid degrading ecosystems and avoid further risks and further degradation in National Parks along the coast, such as Namib Naukluft Park (Bertelmann et al. 2024; Henschel and Jürgens 2024) and can be used for desert agricultural activities in such water stressed regions (Gamboa et al. 2025). Such ecological restoration efforts would not only preserve biodiversity but also enhance the natural capital that underpins Namibia's tourism economy and its coastal resilience to climate change.

Reliable water access through desalination could significantly enhance Namibia's coastal-ecotourism sector. Water-secure resorts along the coast could expand their operations while maintaining strict environmental standards, addressing the current limitations that have led to temporary closures due to water rationing during droughts. The unique desert-adapted wildlife that forms the cornerstone of Namibia's tourism appeal, including the famous desert elephants and oryx herds, depends on reliable water points that desalination could maintain without depleting natural aquifers (Bhushan and Bhushan 2020; Huntley 2023; Joshipura 2025). Furthermore, sustainable water management could become a marketing point for eco-conscious travellers, with certification programs such as the Namibian Ecotourism Certification Scheme potentially incorporating water sustainability criteria (Manhera 2025). Coastal lagoons are important environments that make up 13% of the Namibian coastline and are affected by climate change (Ndjaula et al. 2023). The expansion of activities such as kayaking and birdwatching in coastal lagoons (de Lange and Bricker 2025), currently limited by freshwater inflows, could diversify tourism offerings if supplemented with carefully managed desalinated water. This synergy between water security and tourism development can create a virtuous cycle in which tourism revenues help fund water infrastructure, while reliable water access enhances the tourism experience.

1.8 Future Outlook for Coastal Namibia

To overcome the challenges of water and food insecurity and harness the full potential of desalination, Namibia must adopt a comprehensive, multi-pronged strategy that addresses the financial, technical, and governance aspects. Agricultural applications warrant particular attention to the use of blended desalinated water for various crops, which is a viable option that needs to be explored. Chile is one of the arid countries investing in seawater desalination, producing up to 8200 L/S (Molinos-Senante et al. 2025). Lessons, both as advantages and challenges to work on, can be used as a starting point for improvement in systems such as those in Namibia. Research priorities should include the long-term ecological monitoring of brine discharge impacts (Feng et al. 2025) and the development of Namibia-specific membrane technologies, such as reverse osmosis, that can handle the unique mineral composition of the Benguela Current waters (Elewa 2024; Wang et al. 2024). These coordinated measures could position Namibia as a regional leader in sustainable desalination, while addressing its pressing water and food security challenges.

Examining successful desalination projects in other arid regions provides valuable lessons for Namibia's coastal water security strategy. Israel's national water system, which sources 80% of its domestic water from desalination, demonstrates the potential for large-scale implementation (Water Resources Podcast 2023). However, Namibia would need to adapt this model to its smaller population and more dispersed settlements. The Al Khafji solar-powered desalination plant in Saudi Arabia, which is expected to produce 60,000 m^3/day using photovoltaic technology (Engineered Design Insider (EDI) 2023), offers a model for utilising Namibia's abundant solar resources. Closer to home, the Mossel Bay desalination plant in South Africa's Western Cape provides insights into managing environmental impacts while supplying water to a coastal town of a similar size to Walvis Bay (Dodd and Rishworth 2023). These case studies collectively suggest that Namibia would benefit from a hybrid approach, combining medium-scale municipal plants with smaller decentralised units for specific agricultural or industrial uses while prioritising renewable energy integration and environmental safeguards.

The success of desalination initiatives in coastal Namibia fundamentally depends on meaningful community engagement and capacity building. The Topnaar communities along the Kuiseb River have demonstrated how indigenous knowledge can inform sustainable water management when integrated with modern technologies (Sullivan and Ganuses 2024; Thornton 2024). Community water committees in coastal towns, such as Henties Bay, have proven to be effective in managing local water resources when given proper authority and resources (von Oertzen et al. 2024). These participatory approaches must be complemented by robust awareness campaigns on water conservation to ensure that increased availability through desalination does not lead to wasteful practices in the long run. Furthermore, by embedding desalination projects within broader community development frameworks, Namibia can ensure that its coastal water solutions are technically sound and socially sustainable. Policy reforms should prioritise the establishment of a dedicated desalination development under the Ministry of Agriculture, Water, and Land Reform, capitalised through a combination of government allocations, climate adaptation financing, and private sector contributions. Technological investments should focus on hybrid renewable energy desalination systems, building on the current small-scale desalination pilot projects (Fellows 2025). Building on this foundation of community-led governance and technological investment, the strategic deployment of potable water from mixed-mode desalination presents a transformative opportunity to address food security in Namibia directly. By harnessing Namibia's abundant solar and wind resources to power hybrid desalination systems, the country can produce a consistent, climate-resilient supply of freshwater without exacerbating carbon emissions (IRENA 2022). This water can be strategically allocated to fuel sustainable agriculture in coastal regions, enabling the establishment of climate-resilient agro-hubs. These hubs would support drip-irrigated farms cultivating high-value, nutrient-dense crops and facilitate the development of protected agriculture, such as greenhouses, which drastically reduce water loss and protect crops from extreme weather (UN Water 2021). For Namibians, particularly those in arid coastal communities whose livelihoods are increasingly threatened by climate-induced droughts and unpredictable rainfall, this initiative is a lifeline. It directly benefits vulnerable populations by creating local jobs in agriculture and water management, reducing reliance on expensive and volatile food imports, and improving nutritional outcomes through access to fresh, locally grown produce. By creating a decoupled agricultural system that is no longer solely dependent on erratic rainfall, this model secures a food production pathway that can withstand the escalating impacts of climate change, turning a region of scarcity into one of sustainable abundance. This shift would particularly benefit communities most affected by climate change by ensuring that potable water does more than meet domestic needs; it becomes the backbone of resilient, sustainable agriculture for current and future generations.

1.9 Conclusion

The analysis of desalination's potential for Namibia's coastal regions reveals both the urgency of action and the feasibility of solutions for the country's water and food security. With 41% of Namibia's population affected by drought in 2024 and coastal aquifers facing irreversible salinisation, conventional water management approaches are insufficient. The current RO of green hydrogen and discharge options are cheap, but they have a high potential risk of being ecologically destructive, threatening marine ecosystems and fisheries. The initial costs of GOM and ZLD are higher, but they mitigate long-term environmental costs and generate revenue from mineral recovery, which aligns with Namibia's sustainability goals. Therefore, transitioning to these advanced approaches is critical for sustainable water production and the protection of Namibia's marine-dependent economy.

Furthermore, realising this potential requires not only substantial initial financial resources, but also technical support and minimising governance barriers through coordinated national and international efforts. The window for action is narrowing, as climate projections indicate further aridity along Namibia's coast, with rainfall expected to decrease by as much as 19% by the 2080s (World Bank Group 2021). The government must act decisively to implement the recommended strategies. Investors should recognise the long-term value of sustainable water infrastructure for solar-powered desalination processes, and researchers should focus on context-specific innovations for different communities or urban centres along the Namibian coast. Combining GOM and ZLD with green hydrogen would work even better for brine and brackish water treatment, and the energy produced from green hydrogen can be further utilised for producing water.

References

Assessment Capacities Project (ACAPS) (2025) Country analysis: Namibia. https://www.acaps.org/en/countries/namibia#

Altenburg & Kantel (2024) Green hydrogen in Namibia. Oppor Risks (Vol. 6, No. INIS-DE--4720). German Institute of Development and Sustainability (IDOS) gGmbH, Bonn (Germany)

Atlas of Namibia Team (2022) Atlas of Namibia: its land, water and life. Namib Nat Found. Windhoek

Bennett N (2022) Impacts of saltwater intrusion on drinking water. Doctoral dissertation, Worcester Polytechnic Institute

Bertelmann P, Wyss D, Pius ET, Kappas M (2024) Spectral imaging of grass species in arid ecosystems of Namibia. Front Remote Sens 5:1368551

Bhushan B, Bhushan B (2020) Overview of arid desert conditions, water sources, and desert plants and animals. In: Bioinspired water harvesting, purification, and oil-water separation. Springer, pp 11–46

Braby J, Braby RJ, Braby N, Simmons RE (2009) Protecting Damara Terns Sterna balaenarum from recreational disturbance in the Namib Desert increases breeding density and overall success. Ostrich 80(2):71–75

Corner R, Fersoy H, Crespi V (2020) Integrated agri-aquaculture in desert and arid lands-learning from case studies from Algeria, Egypt and Oman. Food & Agriculture Org

Christelis G, Struckmeier W (2018) Groundwater in Namibia: An explanation of the hydrogeological map. Minist Agric, Water For

de Lange R, Bricker KS (2025) Setting the stage: highlights of former Wttc tourism for tomorrow awards. In: Ecotourism and Sustainable Tourism, pp 3–20

Dall EP, Hoffmann JE (2017) The techno-economic optimisation of a 100MWe CSP-desalination plant in Arandis, Namibia. AIP Conference Proceedings 1850 (SolarPaces 2016), p 170002

Dodd C, Rishworth GM (2023) Coastal urban reliance on groundwater during drought cycles: opportunities, threats and state of knowledge. Cambridge Prisms: Coastal Futures 1:e11

EDI (2023) Green industry. https://www.ediweekly.com/saudi-solar-powered-desalination-plant-will-worlds-largest/#:~:text=The%20plant%20will%20desalinate%2060%2C000,supply%20about%208.5%20million%20people

Elewa MM (2024) Emerging and conventional water desalination technologies powered by renewable energy and energy storage systems toward zero liquid discharge. Separations 11(10):291

Engelhard GH, Bova C, Gusha MNC, Harrod OL, Kadhila T, Kanyimba A et al (2024) Climate risk assessment of the fisheries in Namibia. Rev Fish Biol Fish 34(4):1231–1260

Engelbrecht F, Adegoke J, Bopape MJ, Naidoo M, Garland R, Thatcher M, Gatebe C (2015) Projections of rapidly rising surface temperatures over Africa under low mitigation. Environ Res Lett 10(8):085004

Fellows CM (2025) What does the seawater desalination industry need from polymer scientists? Polym Int 74(2):87–94

Feng X, Zhang X, Cao J, Zhang L, Song H, Kou X et al (2025) Case studies on the identification of environmental impact indicators in brine and distinguishing the direct and indirect contributions of the desalination process. Desalination 608:118860

Food and Agriculture Organization (FAO) (2021) The state of the world's land and water resources for food and agriculture- systems at breaking point. FAO, Rome. https://doi.org/10.4060/cb7654en

Frankl P, Madeira R (2024) Renewable energy opportunities for Namibia. In: Renewable energy opportunities for Namibia. International Energy Agency, Paris. https://hdl.handle.net/11159/703245

Gadalla MA, Fatah AA, Elazab HA (2023) A novel renewable energy powered zero liquid discharge scheme for RO desalination applications. Case Stud Chem Environ Eng 8:100407

Gamboa G, Palenzuela P, Ktori R, Alarcón-Padilla DC, Zaragoza G, Fayad S et al (2025) Thermal seawater desalination for irrigation purposes in a water-stressed region: emerging value tensions in full-scale implementation. Desalination 593:118213

Government of the Republic of Namibia (2000) - Ministry of Agriculture, Water and Rural Development (2000) National water policy white paper: Policy framework for equitable, efficient, and sustainable water resources management and water services [White paper], Windhoek, Namibia.

Government of the Republic of Namibia (GRN) (2004) Namibia Vision 2030: Policy Framework for Long-Term National Development. Office of the President. Windhoek, Namibia

Government of the Republic of Namibia (2013) Integrated Water Resources Management, Namibia (IWRM). Windhoek, Namibia

Government of the Republic of Namibia (GRN) (2018) National climate change policy. Ministry of Environment and Tourism

Han ZY, Huang LJ, Qu HJ, Wang YX, Zhang ZJ, Rong QL et al (2021) A review of performance improvement strategies for graphene oxide-based and graphene-based membranes in water treatment. J Mater Sci 56:9545–9574

Hempapu O (2024) Namibia's N$482m drought relief deficit. New Era. https://neweralive.na/namibias-n482m-drought-relief-deficit

Henschel J, Jürgens N (2024) Ecology of psammophily in the Namib dunes. Southern African Dunes. Springer, Windhoek, pp 1–50

Huntley BJ (2023) Adaptations to life in the Namib Desert. In: Ecology of Angola: terrestrial biomes and ecoregions. Springer International Publishing, Cham, pp 249–274

Hutson N, Taganova J (2023) A new strategy for merging the environmental and commerce challenges of the Caspian. Central Asian J Water Res 9(2):76–102

Hyder K, Wilhelm M, Kanyimba A, Bova C, Gusha MN, Harrod OL, Kadhila T, Khan U, Kreiner A, Nghipangelwa S, Olwage E, Pinnegar JK, Potts W, Rivers N, Shakalela E, Snow B, Tshiningayamwe S, Unengu U, Veii A, Warikandwa T, Engelhard GH (2024) Climate risk and adaptation for fisheries in Namibia. One Ocean Hub Report:8. https://doi.org/10.14465/2024.ooh02.pol

Ihsanullah I, Atieh MA, Sajid M, Nazal MK (2021) Desalination and environment: a critical analysis of impacts, mitigation strategies, and greener desalination technologies. Sci Total Environ 780:146585

International Renewable Energy Agency (IRENA) (2022) Renewable energy and water: opportunities and challenges. IRENA, Abu Dhabi. Retrieved from https://www.irena.org/publications/2022/Jul/Renewable-Energy-and-Water

International Energy Agency (IEA) (2024) Renew Energy Oppor Namib. https://hdl.handle.net/11159/703245

Intergovernmental Panel on Climate Change (IPCC) (2021) Climate change 2021: The physical science basis. Cambridge University Press
Iileka S, Likius DS, Tomas P, Chembe DK & MuAshekele H (2024) Viability of Green Hydrogen Production at Benguela Wind Energy Community Project in Lüderitz, Namibia. J Sustain Dev Energy, Water Environ Syst, 12(3):1–13
Jia F, Xiao X, Nashalian A, Shen S, Yang L, Han Z et al (2022) Advances in graphene oxide membranes for water treatment. Nano Res 15(7):6636–6654
Joshi RK, Alwarappan S, Yoshimura M, Sahajwalla V, Nishina Y (2015) Graphene oxide: the new membrane material. Appl Mater Today 1(1):1–12
Joshipura V (2025) Geomorphology of desert environments. Educohack Press
Kamal AHMM, Mair GC (2005) Salinity tolerance in superior genotypes of tilapia, Oreochromis niloticus, Oreochromis mossambicus and their hybrids. Aquaculture 247(1–4):189–201
Katzir R (2014) Advance farming in the desert-the Israeli experience. https://www.cabidigitallibrary.org/doi/pdf/10.5555/20163289129
Korngold E, Aronov L, Daltrophe N (2009) Electrodialysis of brine solutions discharged from an RO plant. Desalination 242(1–3):215–227
Li W, Zhang X, Shi Y, Hu X, Wang X, Liang N et al (2023) A dual-modal biosensor coupling a cooperative catalysis strategy for sensitive detection of AFB1 in agri-products. Food Chem 426:136553
Linden MVD (2024) Environmental effects of brine disposal and seawater usage for offshore green hydrogen production and storage in the Dutch North Sea. Master's thesis
Loganathan K, Chelme-Ayala P, El-Din MG (2016) Pilot-scale study on the treatment of basal aquifer water using ultrafiltration, reverse osmosis and evaporation/crystallisation to achieve zero-liquid discharge. J Environ Manag 165:213–223
Ma J, Ping D, Dong X (2017) Recent developments of graphene oxide-based membranes: a review. Membranes 7(3):52
Manhera R (2025) The role of green marketing in promoting sustainable supply chain in Namibia. In: Eco-logistics and sustainable supply chain innovations. IGI Global Scientific Publishing, pp 425–458
Marx V (2009) Impacts of upstream uses on the alluvial aquifer of the Swakop River, Namibia. Diplomarbeit, Albert Ludwig University of Freiburg
Matengu BM (2020) Groundwater assessment and sustainable management of the coastal alluvial aquifers in Namib Desert, Namibia: Omdel Aquifer as case study. Doctoral thesis, University of the Western Cape, South Africa
Meydan CH, Akkaş H (2024) The role of triangulation in qualitative research: converging perspectives. In: Principles of conducting qualitative research in multicultural settings. IGI Global, pp 98–129
Ministry of Agriculture, Fisheries, Water and Forestry (MAFWF) (2015) Namibia agriculture policy. Ministry of Agriculture, Fisheries, Water and Forestry, Windhoek
Ministry of Environment and Tourism (2019) Delivering transformative impacts on the ground: contribution of projects mobilised through the Ministry of Environment and Tourism from 2014-2019. Government of Namibia. https://www.eif.org.na/uploads/files/Delivering%20transformative%20impacts%20on%20the%20groundLOWRES.pdf
Ministry of Mines and Energy, Government of the Republic of Namibia (2022) The 2022 National Integrated Resource Plan (NIRP 2022) update
Modi R, Kavaiya AR, Vanzara PB, Raval HD (2022) Membranes in zero-liquid-discharge systems for efficient processes toward a sustainable environment: a review. J Environ Eng 148(11):03122004
Molinos-Senante M, Rivera D, Donoso G, Melo Ó (2025) Desalination in times of drought. In: Droughts in Chile: impacts, monitoring, and adaptation (management) policies. Springer Nature Switzerland, Cham, pp 387–407

Moumni M, Mohamed MEA, Zahra MF (2024) Solving the energy consumption barrier in brackish water reverse osmosis desalination plants: a genetic algorithm and energy recovery approach. Adv Syst Sci Appl 24(02):203–218

Mupambwa HA, Hausiku MK, Namwoonde AS, Liswaniso GM, Haulofu M, Mafwila SK (2021) Climate change implications and mitigation in a hyperarid country: a case of Namibia. In: African handbook of climate change adaptation. Springer International Publishing, Cham, pp 2247–2268

Murray R, Ruppel OC (2025, March) Country report for Namibia. In: Legal pathways to sustainable soil management in Africa. Nomos Verlagsgesellschaft mbH & Co. KG, pp 467–548

Ministry of Fisheries and Marine Resources (MFMR) (2023) Namib Maric Dev Plan. MFMR

Mupambwa HA, Hausiku MK, Nciizah AD & Dube E (2019) The unique Namib desert-coastal region and its opportunities for climate smart agriculture: A review. Cogent Food & Agric, 5(1):1645258

Namibian Agronomic Board (2024) Harvesting/marketing season 2024. https://www.nab.com.na/wp-content/uploads/2024/10/White-maize-production-forecats-report-2024-final.pdf

NamWater (2024). NamWater Integrated Annual Report (2023/2024), Namibia Water Corporation Ltd, Windhoek. https://www.namwater.com.na/images/docs/2024_Annual_report.pdf

Namibia Water Corporation (NamWater), 2025. Hydrological Services. https://www.namwater.com.na/index.php/services/56-hydrological-services?start=7

National Planning Commission (2017) Namibia's 5th National Development Plan: NDP5 at a Glance 2017/18–2021/22. National Planning Commission

Ndjaula HO, Hausiku-Ikechukwu MK, Haulofu M, Mupambwa HA (2023) Climate change research perspectives in marine and coastal environments of Namibia. In: Handbook of nature-based solutions to mitigation and adaptation to climate change. Springer, pp 1–23

Panagopoulos A (2022) Brine management (saline water & wastewater effluents): sustainable utilization and resource recovery strategy through Minimal and Zero Liquid Discharge (MLD & ZLD) desalination systems. Chem Eng Process-Process Intensif 176:108944

Panagopoulos A, Giannika V (2022) Decarbonised and circular brine management/valorisation for water & valuable resource recovery via minimal/zero liquid discharge (MLD/ZLD) strategies. J Environ Manag 324:116239

Panagopoulos A, Michailidis P (2025) Membrane technologies for sustainable wastewater treatment: advances, challenges, and applications in zero liquid discharge (zld) and minimal liquid discharge (mld) systems. Membranes 15(2):64

Ren RY, An YC, He ZH, Han JL, Wang GZ, Ajibade FO et al (2025) Fabrication of graphene oxide membranes for industrial applications: toward efficient separation of target organic compounds and inorganic salts. Sep Purif Technol 373:133586

Sabour SMJS, Ghorashi B (2024) A comprehensive review of major water desalination techniques and mineral extraction from saline water. Sep Purif Technol 349:127913

Saltworks (2017) What is zero liquid discharge & why is it important? https://www.saltworkstech.com/articles/what-iszero-liquid-discharge-why-is-it-important/

Sarker BS, Sabuz MSR, Azom MG, Easmin N, Ali A, Alam MS, Islam MS (2023) Genotype responsive growth performance and salinity tolerance of tilapia hybrid (Oreochromis niloticus× O. mossambicus). Aquaculture 575:739729

Schmidt R, Frank D (2023) Feasibility Study for Green Hydrogen in Namibia: Report on desalination, brine treatment options, disposal options and its potential impacts on maritime life. GreeN-H2 Namibia.

Shikangalah RN (2020) The 2019 drought in Namibia: an overview. J Namibian Stud 27:37–58

Shikangalah RN, Mapani BS (2021) Ephemeral river systems and their ecosystem provisions to the local populations: a review of the Huab and Ugab Rivers, Namibia. Int Sci Technol J Namibia 14:46–62

Spear D, Zaroug M, Daron J Ziervogel G, Angula M Haimbili E et al (2018) Vulnerability and responses to climate change in drylands: the case of Namibia

Steinmuller HE, Chambers LG (2018) Can saltwater intrusion accelerate nutrient export from freshwater wetland soils? An experimental approach. Soil Sci Soc Am J 82(1):283–292

Sullivan S, Ganuses WS (2024) Cultural heritage and histories of the Northern Namib/Skeleton Coast National Park. In: Etosha pan to the skeleton coast. Open Book Publishers, pp 307–342

Tal A (2021) Israeli agriculture: innovation and advancement. In: From food scarcity to surplus: innovations in Indian, Chinese and Israeli agriculture. Springer Singapore, Singapore, pp 299–358

Thornton JL (2024) The elemental feast: a taste of nature. African Books Collective

UN Water (2021) Water and climate change. The United Nations World Water Development Report 2021. Paris: UNESCO. Retrieved from https://www.unwater.org/publications/un-world-water-development-report-2021/

Von Oertzen D (2021) Issues, challenges and opportunities to develop green hydrogen in Namibia. Konrad-Adenauer-Stiftung.

von Oertzen D, Schneider M, Heyns P (2024) Namibia's water–energy–food nexus: National Development in Uncertain Times, vol 7. Konrad-Adenauer-Stiftung, pp 1–223. https://www.kas.de/namibia

World Bank Group, 2021. Namibia climate risk profile: Vulnerabilities and adaptation responses. World Bank Group Climate Change Knowledge Portal. https://climateknowledgeportal.worldbank.org/country/namibia/climate-datahistorical

Wang X, Mao YF, Shen X, Zhao J, Zhou J, Liu Z (2024) Advances and prospects in graphene oxide membranes for seawater salt ion sieving and rejection. Sep Purif Technol 329:125216

Water Resources Podcast (2023) Importance of seawater desalination and wastewater reuse in Israel. https://wrp.beg.utexas.edu/node/31#:~:text=Seawater%20desalination%20now%20provides%20~%2080,desalination%20plants%20are%20being%20developed

Zhang Y, Wang H, Zhang J (2025) Application of graphene-based solar driven interfacial evaporation-coupled photocatalysis in water treatment. Catalysts 15(4):336

Chapter 2
Sustainable Mariculture Practices in Africa's Coastal Desert Zones: A Pathway to Food Security and Economic Stability

Arisha September, Ndakalimwe Naftal Gabriel, and Tamuka Nhiwatiwa

Abstract With the world population projected to reach an astonishing 9.7 billion by 2050, an immense increase of pressure on finite food resources such as traditional land-based agriculture and wild capture of fish stocks is anticipated. Mariculture has globally become the solution that various individuals, communities, businesses, and countries have turned to due to it being able to supply the demand of food, alleviating the pressure on wild fisheries stock, preventing the continuation of habitat degradation and providing socioeconomic benefits thereby stimulating local economies. Although Africa's coastal desert zones present formidable challenges for food production systems, it provides the potential for the development of sustainable mariculture through the strategic development and implementation of mariculture practices, production systems, and policies. The development of mariculture in Africa's coastal deserts countries, despite extreme aridity, limited fresh water, and poor terrestrial agricultural potential including Namibia, South Africa, Eritrea, Djibouti, Egypt, and Morocco, is a reflection of the strategic adaptation to the evolving environmental reality. While rearing marine organisms such as fin fish, crustaceans, shellfish, and algae, the use of these coastal deserts offers a transformative approach to food security, simultaneously, aiding towards the conservation of natural marine environments and boosting local economies. Various methods of mariculture are observed among the countries, including open sea systems, sheltered bay systems using cages or tanks or land-based systems such as recirculating aquaculture systems, polyculture tank systems, ponds, and multi-trophic mariculture approach. This ensures the optimization of the utilization of the resources while ensuring positive yields. However, the successful sustainable

A. September (✉)
Sam Nujoma Marine and Coastal Resources Research Centre, Sam Nujoma Campus, University of Namibia, Henties Bay, Namibia

N. N. Gabriel · T. Nhiwatiwa
Department of Fisheries and Ocean Sciences, Faculty of Agriculture, Engineering, and Natural Sciences, Sam Nujoma Campus, University of Namibia, Henties Bay, Namibia

H. A. Mupambwa et al. (eds.), *Harnessing Coastal and Marine Resources for Improved Food Security in Africa*, Sustainability Sciences in Asia and Africa,
https://doi.org/10.1007/978-981-95-9725-3_2

implementation of mariculture heavily relies on the adoption of appropriate systems, innovative tools and technologies, compliance to rigorous standards of operations, community empowerment, and the establishment of essential governance and policy frameworks. In this chapter, we analyze what steps these African countries have taken, and need to further implement to ensure that they sustainably contribute to food security and economic well-being. Mariculture provides a pathway to enhanced food security through the provision of marine protein sources and is a holistic development tool capable of fostering resilient food systems, generating diverse employment, and driving economic growth in some of the continent's most challenging environments. This chapter provides a framework for scaling mariculture as a strategic tool for sustainable development across Africa's coastal desert zones in accordance to the African Union Blue Economy strategy (2020) and the United Nations Sustainable Development Goals.

Keywords Mariculture · Coastal deserts · Sustainability · Food security · Economy · Fin-fish · Shell-fish

2.1 Introduction

With the world population projected to reach an astonishing 9.7 billion by 2050, an increase in the demand of food security, and employment is observed globally (Molotoks et al. 2021; Falcon et al. 2022; Valavanidis 2023). The total population of Africa of 1.448 billion as estimated in 2023 forms 18.14% of the world's current total population estimated at 7.979 billion (Kaba 2024a, b), thereby prompting an increase in the need for food security and economic stability (Mwaniki 2006; Sasson 2012). With the effects of global climate change such as unpredictable weather patterns, desertification, and the decrease of wild fish stocks due to their overexploitation, land-based agriculture and marine wild-capture fisheries are unable to meet such demands sustainably (Kolding et al. 2016; Blanchard et al. 2017; Clavelle et al. 2019). Although coastal zones constitute about 5% of the world's surface, approximately 50–70% of the global human population lives in these areas (Steyl and Dennis 2010). Despite Africa's total coastline of 40,188 km only forming 11.3% of the world total coastline of 356,000 km (Kaba 2024a, b), it supports a diversity of habitats and resources including mangroves, coastal wetlands, rocky and sandy shores, lagoons, and coastal arid and semiarid desert zones (Steyl and Dennis 2010). With Africa being no different to the rest of the world, most of its population is observed to be clustered within major coastal towns thereby increasing the demand of food resources and economic stability. Thereby, challenges arise socially, politically, environmentally, and economically to sustain such demands particularly for coastal arid and semiarid desert zones. The implementation of sustainable mariculture in these coastal desert zones plays a transformative role. Instead of being hindered by the challenges and constraints within such zones, when coupled with strategic management, mariculture within coastal desert zones creates resilient food systems, supports livelihoods, boosts economies, and may contribute to the restoration of ecosystems. This chapter aims to determine the sustainable mariculture practices used within Northern African, Middle East African, and Southern African coastal desert countries that aid toward food security and economic stability through assessing various case studies within these regions

2.2 Regional Overview of Africa's Coastal Desert Regions

2.2.1 Northern Africa (Egypt and Morocco)

The Northern African coastal strip borders the Mediterranean Sea, the Red Sea, and the Atlantic Ocean, thus resulting in dynamic oceanic and coastal differences among arid coastal countries within this region (Diop et al. 2014; Agoubi 2021; Hzami et al. 2021; Ennabili et al. 2025). Additionally, the hyperarid Sahara Desert exerts a powerful influence on these coastal arid zones, by bringing hot, dry continental air to the shores (Shahin 2007; Eckardt et al. 2022). In Egypt and Morocco, mariculture activity is found mainly within sheltered lagoons, estuaries, and onshore systems in which finfish, shellfish, and seaweed are cultured (Cataudella et al. 2015; Dickson 2022; Ennabili et al. 2025). The typical mariculture production systems include longlines and rope systems used for bivalves such as the Pacific oyster (*Crassostrea gigas)* and Mediterranean mussels (*Mytilus galloprovincialis)* in Morocco (Idhalla et al. 2017; Ortega et al. 2021), and Pearl oysters (*Pinctada radiate)* along with a variety of clams in Egypt (Sadek 2010; Dickson 2022; Mandour et al. 2025). While onshore recirculating aquaculture systems (RAS) are used for high-value finfish such as the European seabass (*Dicentrarchus labrax)*, which in some cases, are coupled with small tank hatchery systems (Dickson 2022; Elhetawy et al. 2025). However, integrated multi-trophic approaches are seen in pilot projects using finfish and shellfish (Freeman and Angel 2019; Corner et al. 2020; Ennabili et al. 2025) and are increasingly incorporated within Northern African mariculture (Ortega et al. 2021; Elhetawy et al. 2025). These mariculture activities provide seasonal and year--round employment opportunities for its coastal communities (Cataudella et al. 2015; Corner et al. 2020); however, conflict with artisanal fishers does arise due to the use and access of nearshore areas (Dickson 2022; Farouk et al. 2024; Parug et al. 2024). While small-scale farmers target local markets, the larger commercial enterprises produce for export to larger European markets (Sippel 2015; El-Sayad 2017). This contributes to local food stability and economic growth; however, the economic viability depends on farmers' access to sustainably producing hatcheries, efficient logistics systems, and meeting market certification for local and international sale (El-Sayad 2017; Dickson 2022; Parug et al. 2024).

2.2.2 *Middle East Africa (Djibouti and Eritrea)*

Despite mariculture being nascent in the Middle East African region due to the arid coastal conditions and limited freshwater sources within this region, it is steadily growing as observed within recent years (El-Sayad 2017; Billi 2022; Dickson 2022). The Red Sea's unique features including its coral reef biodiversity (Bruckner 2011;) provide suitable conditions for onshore and nearshore mariculture activities when production activities are managed effectively (Cziesielski et al. 2021; Diop and Scheren 2016). The nature of this coastal arid region provides extreme challenges to local livelihoods dependent on traditional agricultural food supply. However, coastal or saltwater aquaculture provides a suitable alternative for food supply and contribution to employment creation (Araya 2012; Habte-Tsion 2016). In the region, various mariculture systems dependent on the scale of production and species cultured are used. However, land-based recirculating aquaculture systems with small-to-medium hatcheries are more commonly employed (Mapfumo 2022). The Eritrean coastline, with gradually sloping beaches and shallow bays, provides a conducive environment for fish farming (Araya 2012; Habte-Tsion 2016; Tesfamichael and Mohamud 2016). Eritrean freshwater aquaculture is faced with challenges such as cold temperature, feed availability, and salinity levels of water sources, thereby proving a unique opportunity for mariculture production to sustainably provide food sources and support economic growth and resilience (Habte--Tsion 2016; Eyayu et al. 2023; Mirera and Okemwa 2023). Along the northern coast of Eritrea, mullet (*Mugil cephalus*), milkfish (*Chanos chanos*), and artemia are cultured in earthen ponds (Mirera 2011; Araya 2012; Darias and Mirera 2023); however, the biggest seawater farm is focused on shrimp and finfish production. The farm comprises grow-out ponds for white-leg shrimp (*Penaeus vannamei*), a hatchery, a laboratory, and interlinked earthen lakes for milk-fish, and Nile tilapia (*Oreochromis niloticus*) culture (Habte-Tsion 2016; Mirera and Okemwa 2023; Darias and Mirera 2023). More than 90% of Diboutji is desert area characterized by hot and dry climatic conditions (Harsh 2018), where groundwater is often the only water source with acceptable quality. Thereby presenting limitations within traditional agricultural practices and freshwater aquaculture (Jalludin 2014; ElMountassir et al. 2024). Despite not being a particularly logical area for mariculture and aquaculture, the small-scale, noncommercial culture of clams, oysters, and seaweed presents a good alternative for traditional fin-fish culture (Abdullah 2002; Poynton and Lovatelli 2006; Van der Knaap 2020). Its marine territory in the Gulf of Aden and the Red sea, however, presents a biologically rich but underutilized environment (Abdullah 2002; Poynton and Lovatelli 2006; Chirdon et al. 2025), suitable for the establishment of mariculture (Bruckner 2011; Van der Knaap 2020). However, within the Middle Eastern coastal arid African region, limited aquaculture and mariculture regulation and national strategies are in place to support such sustainable development despite forming a part of development strategies (Mwandja and Nyandat 2013; Elsayed 2017; Dickson 2022).

2.2.3 *Southern Africa (Namibia and South Africa)*

Within the Southern region of Africa, Namibia and South Africa are seen to have better developed mariculture production initiatives in place along their coastal desert towns (Iitembu 2005; Mapfumo 2011; Erasmus et al. 2023). The private--government mariculture partnership initiatives observed within the region, although currently having lower input than optimally possible, are promising examples of arid coastal mariculture activities focusing mainly on oysters, mussels, and abalone for with the recent reintroduction of seaweed farming (Mapfumo 2011; Tolley et al. 2019; Erasmus et al. 2023). Despite the arid conditions of the south-western African region, the nutrient-rich, cold Benguela current provides a conducive environment for the establishment of mariculture productivity. The Namibian coastline is classified as hyperarid, low rain (~ 15 mm along Walvis Bay and Lüderitz), frequent occurrence of fog and strong predominantly southerly or south-westerly winds (Mupambwa et al. 2019; Pulfrich and Kemper 2022). In Namibia, longlines and floating raft systems are used for the production of pacific oysters (*Crassostrea gigas*) and mussels (*Mytilus galloprovincialis & Perna perna*) in shallow lagoon waters in Lüderitz and Walvis Bay, supported by land-based hatchery and nursery systems (Iitembu 2005; Erasmus et al. 2023). Oysters and mussels are sold within local markets; however, oysters may be exported regionally. However, the land--based production of high-value abalone (*Haliotis midae*) species incorporates onshore flow through hatcheries, nurseries, and grow-out ponds/tanks which are exported to international markets (Iitembu 2005; Iitembu et al. 2023; Pulfrich and Kemper 2022; Erasmus et al. 2023). Despite the benefits provided by the cold Benguela current, both nearshore and onshore culture systems face risks from extreme upwelling events, hypoxia, harmful algal blooms, and infrastructure damage (Brown et al. 2020; Pitcher and Louw 2021). In remote coastal arid regions such as Lüderitz, mariculture plays a vital role in the implementation of upstream employment contributing to various livelihoods within the region. Furthermore, the potential of integrated multitrophic aquaculture of mussels within oyster or abalone farms (IMTA) and coexistence of mariculture activities, particularly those in Walvis Bay, with marine eco-tourism can be incorporated to promote circular blue economy activities (Iitembu 2005; Iitembu et al. 2023; Pulfrich and Kemper 2022; Erasmus et al. 2023). The west coast of South Africa, from the Orange River to as far as Saldanha Bay, lies within the arid coastal belt of the Namaqualand Desert and the Karoo, with little rainfall (100 mm per year in Namaqualand) and warm temperatures (Mapfumo 2011; Olivier et al. 2013). The sheltered bays and rocky shores, supported by the cold Benguela current, allow for the culture of mussels (*Mytilus galloprovincialis*) on longline systems, oysters (*Crassostrea gigas & Ostrea edulis*) in rafts and intertidal racks, as well as abalones (*Haliotis midae*) in land-based systems, as seen in Namibia (Olivier et al. 2013, Erasmus et al. 2023). Additionally, despite not being successful within Namibia, the culture of seaweed (*Gracilaria gracilis* and *Ulva lactuca*) takes place in land based tanks and pond cultures, and is often integrated with abalone and oyster culture (Olivier et al. 2013; Rothman et al. 2020).

2.3 Challenges and Successes

2.3.1 *Northern Africa (Egypt and Morocco)*

Northern African countries, particularly Egypt, have a significantly well-advanced aquaculture sector due to its historical advantages of fresh water backed by national development frameworks and policies (El-Sayed 2017; Corner et al. 2020; Dickson 2022). However, the mariculture sector within these countries faces various challenges. The weak enforcement of marine spatial planning (MSP) and licensing procedures remains a key challenge, thereby resulting in the competition for land and coastal water (Khalil et al. 2021; Rockmann et al. 2022). This places the coastal zone under threats from aquatic pollution through urbanization and inadequate reinforcement of waste management and biosecurity standards (Khalil et al. 2021; Rockmann et al. 2022; Farouk et al. 2024). Furthermore, the culturing of high-value species like sea bass or sea bream in Egypt faces delayed development due to limited hatchery capacities and a lack of technical support within the mariculture sector (Megahed and Aly 2009; Dickson 2022; Elhetawy et al. 2025). In addition, the lack of financing for high operational costs impacted by the cost of feed and access to cold chain logistics introduces challenges for the export of products, thereby impacting the economic viability of mariculture (Megahed and Aly 2009; El-Sayed 2017; El Mahrad et al. 2020; Dickson 2022). Despite this, the success of mariculture in the region is measured by its contribution and commercialization of protein products for local and international markets. Technological advancements, growing regional policy alignment efforts, and technical cooperation to better sanitary standards successfully contribute to closing the national marine protein supply deficits (El-Sayed 2017; El Mahrad et al. 2020; Dickson 2022), therefore positively impacting employment and food security.

2.3.2 *Middle East Africa (Djibouti and Eritrea)*

Regardless of the extreme aridity this coastal desert region faces, large-scale commercial mariculture investments along with the use of advanced technologies are observed Eritrea. Thereby, stimulating high operational costs due to increased demand for energy, feed, husbandry, and management (Habte-Tsion 2016; Mirera and Okemwa 2023; Darias and Mirera 2023). However, due to this, Eritrea has seen significant successes with highly intensive recirculating systems for highly nutritional and valued species such as sea bream (Araya 2012; Habte-Tsion 2016; Tesfamichael and Mohamud 2016). Leveraging the arid and coastal conditions, mariculture successfully utilizes saline water and salt-affected land for earthen ponds for shrimp farming which is otherwise unsuitable for traditional agriculture

(Habte-Tsion 2016; Mirera and Okemwa 2023; Darias and Mirera 2023). Successes in the region can be further stimulated through strong governmental support and significant capital investments, promoting management that includes strict monitoring of recirculating systems, maintaining biosecurity and water quality (Dickson 2022; Mapfumo 2022; Weldetsion 2023).

2.3.3 *Southern Africa (Namibia and South Africa)*

Although coastal desert mariculture along the west coast of South Africa and along the Namib desert in Namibia benefits from the Benguela current, it is faced with governmental and financial challenges within its operations (Iitembu 2005; Finke et al. 2020; Erasmus et al. 2023). This includes high operational costs including the costs of feed, energy, land utilization, and logistics as well as the capacity building of workforce due to its remoteness (Iitembu 2005; Finke et al. 2020; Erasmus et al. 2023). Despite the benefits provided by the Benguela current, the environmental variability leads to fluctuations in water quality and temperature, and could introduce harmful algal species throughout blooms (Hutchings et al. 2009; Van der Lingen et al. 2016). This brings forth a detrimental challenge as high-value bivalve, filter-feeding organisms such as oysters and abalones are the main mariculture products for both local and international markets (Brown et al. 2020; Fernandes-Salvador et al. 2021; Erasmus et al. 2023). Additionally, the complexity of obtaining licensing and permitting such as sea use rights, within the region presents a significant hurdle for first-time investors and the expansion of operational enterprise activities. This delays the implementation of mariculture projects thereby discouraging investments and economic growth through employment (Iitembu 2005; Mouton 2019; Brown-Webb 2023; Iitembu et al. 2023; Erasmus et al. 2023). Furthermore, poaching, theft, and vandalism of abalones from onshore commercial farms and intertidal farms are another key challenges, commonly observed within the South African mariculture sector which negatively impacts operational profits and deters investments (De Greef 2014; Potgieter 2021; Pay 2023). While these challenges may be detrimental to the successful outcomes of mariculture within region, its positive impacts such as job creation and economic diversification are particularly felt within historically isolated coastal towns such as Lüderitz (Erasmus et al. 2023). The Namibian and South African bivalve mariculture sectors' economic successes are linked primarily, to the export of premium products, cultured according to international standards of management, to international markets (Iitembu 2005; Olivier 2013; Brown-Webb 2023; Erasmus et al. 2023). This is supported by strong research and industry linkages to improve hatchery systems, husbandry practices, and disease management.

2.4 Strategies for Sustainable Mariculture Development in Coastal Desert Zones

Despite the challenges mariculture faces in these coastal desert zones, its proximity to the nutrient-rich marine ecosystems presents an opportunity for sustainable mariculture (Crespi and Lovatelli 2011; Chithambaran 2016; Lounas et al. 2020). However, the sustainability of mariculture is only observed when strategic measurements in production systems, management, and policy implementation are incorporated thereby integrating ecological understanding, technological advancements, and community empowerment (Chithambaran 2016; Abdelzaher et al. 2023; Fisher et al. 2023). The adoption of such strategies enables mariculture in coastal desert zones to align with the United Nations Sustainable Development Goals, particularly Goal 2 (Zero Hunger) and Goal 14 (Life Below Water) (Hambrey 2017; Abdelzaher et al. 2023; Abeysinghe et al. 2025). While complementing the African Unions Blue Economy strategy (2020) which emphasizes that mariculture indeed contributes to job creation, food security, environmental resilience, and economic development (Odunlade and Isikwei 2025). Therefore, to ensure sustainable mariculture within these coastal desert zones, countries within these regions should adopt and integrate ecological, production, socioeconomic and governmental strategies. These strategies directly address core challenges of water scarcity, competition for coastal space, high operational costs, and institutional limitations and ensures its successful contribution to food security and economic stability (Abdelzaher et al. 2023; Fisher et al. 2023; Odunlade and Isikwei 2025).

2.4.1 Northern Africa (Egypt and Morocco)

Northern Africa has made great advances such as the improved hatchery capacities, improved seed strains, and the adoption of land-based intensive recirculating systems in freshwater aquaculture which is seen within the its mariculture sector as well (Corner et al. 2020; Dickson 2022). However, challenges within this region are primarily linked to governance, competition for coastal resources and their dependence on the importation of feed (El-Sayed 2017; El Mahrad et al. 2020; Khalil et al. 2021; Dickson 2022; Rockmann et al. 2022). Therefore, strategies within this region should address strengthening regulations and advancing existing systems and feed supply (Table 2.1).

2.4.2 Middle Eastern Africa (Djibouti and Eritrea)

Mariculture has been observed to be high capital and high technology based within the Middle East African region; therefore, strategies toward sustainable mariculture should focus on ensuring long-term economic viability, energy efficiency, and

Table 2.1 Strategic measurements for Sustainable Mariculture in Coastal Desert zones: Northern Africa (Egypt and Morocco)

Strategy	Challenges addressed
Policy development and implementation	
Integrated Coastal Zone Management (ICZM): develop and implement mariculture zones through marine spatial planning within National ICZM policies to reduce land and water utilization conflicts with urbanization and Industrial sectors.	Land/water utilization conflicts Reduces vulnerability of coastal sites
Improved permitting systems and enforcement: establish efficient, and advance licensing systems and enforcement standards within management and monitoring policies. This may positively impact investment for mariculture and improve biosecurity and compliance due to strict enforcement.	Improves slow and tedious processes for licensing Inadequate enforcement standards Boosts investments
Feed production targeted incentives or subsidies: government to provide incentives for local sustainable production of feed. Alternatively, subsidize feed import through reduction of taxes or proportion of financial assistance for purchase of feed.	Reduces high operation costs Reduces the reliance on imported food containing fishmeal
Mariculture production systems	
Use of recirculating aquaculture systems (RAS): promote the investment in to the use of RAS for high-value species for minimized water utilization and environmental conservation.	Reduces aquatic pollution and environmental impacts Sustainable use of water resources Reduces high operating costs
Integrated Muti-Trophic Aquaculture (ITMA): through promoting co-culture of different symbiotic species such as finfish with biofilters such as seaweed or shellfish to improve sustainability. This may also provide multiple streams of income from singular culture systems.	Reduces high operation costs Reduces aquatic pollution and environmental impacts Increases profits Promotes sustainable use of water resources and culture space
Management	
Establish training programs: Develop develop and implement mariculture training programs focusing on all components of the various culturing systems (hatchery management, water quality monitoring, biosecurity standards).	Improves technical skills of workers Increases skilled workforce Increases production efficacy and efficiency Boosts investments
Establish the value-chain of cultured products: establish and invest in the different methods to add value to cultured products through enhancing the cold-chain logistics, processing facilities, meeting quality standards, skilled workforce in order to increase export and product value.	Low-value mariculture products Market demand may be increased Increases profits Increases employment of skilled workforce Promotes food security

circular integration within the utilization of the environment (Bruckner 2011; Habte-Tsion 2016; Mirera and Okemwa 2023) (Table 2.2).

2.4.3 Southern Africa (Namibia and South Africa)

Although the Namibian coast and the Western South African coast of the Southern African region benefit from the Benguela current, institutional complexity, environmental variability, and access to markets impact the successful outcomes of its mariculture industries (Iitembu 2005; Hutchings et al. 2009; Finke et al. 2020; Erasmus et al. 2023). Therefore, strategies should focus on providing an enabling environment while maximizing the utilization of this pristine environment. Through integrating environmental stewardship, technological resilience and socioeconomic factors. Sustainability of mariculture in the Southern African region requires adaptive governance, circular production models, and inclusive regional cooperation (Brown-Webb 2023) (Table 2.3).

2.5 Mariculture Lessons Learned

2.5.1 Northern Africa (Egypt and Morocco)

The primary lessons for sustainable mariculture within the Northern African region are the need for transition towards sustainable and intensive systems such as recirculating aquaculture systems (RAS) for land-based operations, as seen in freshwater aquaculture (Lounas et al. 2020; Dickson 2022; Paruğ et al. 2024). This significantly reduces water consumption and effluent discharge. Furthermore, promoting government and private investments in local and resilient hatcheries, disease surveillance and management, and feed production (Khalil et al. 2021; Farouk et al. 2024). Lastly, the need to use participatory coastal zoning to reduce user conflicts, and to raise farm productivity and food-safety compliance (Khalil et al. 2021; Rockmann et al. 2022).

2.5.2 Middle Eastern Africa (Eritrea and Djibouti)

In the Middle Eastern Africa, the main lesson drawn from the Djibouti and Eritrean mariculture sector is the viability of high-tech energy-efficient technologies, intensive systems in resource-limited desert environments, and otherwise unproductive land (Habte-Tsion 2016; Mirera and Okemwa 2023; Darias and Mirera 2023). Provided there is sufficient investment capital and a commitment to robust water

Table 2.2 Strategic measurements for Sustainable Mariculture in Coastal Desert zones: Middle Eastern Africa (Djibouti and Eritrea)

Strategy	Challenges addressed
Policy development and implementation	
Energy subsidies or incentives for renewable energy integration: regulatory authorities or governments to establish and offer preferential tariffs or direct subsidies for mariculture or coastal aquaculture operations. Alternatively, establish incentives for the integration of renewable energy, i.e., solar or wind energy for energy-intensive recirculating aquaculture systems and operations.	Reduced high operation energy costs Reduces high capital for startups promoting investments Reduces water resources utilization and waste Reduces aquatic pollution Indirectly reduces environmental impacts
Research and development (R&D) and intellectual property (IP) protection: fund and establish national research and development (R&D) centers dedicated to understand local mariculture challenges such as disease resistance and management, selective breeding of local and adaptive strains and to develop desert-adapted technologies while protecting such findings of the center, small-scale farmers or commercial operators.	Secures product demand in markets Reduces the reliance of foreign expertise and technologies Informs and strengthens local policies and managing frameworks
Mandatory effluent standards: establish effluent management protocols and standards which are mandatory across the sector. Enforce and monitor these policies. Provide incentives for the integration of technologies which enhance filtration or the transition to Integrated Multi-Trophic Aquaculture (IMTA) approach.	Reduces aquatic or land pollution Reduces environmental impacts Improves overall standards for local and international certification Increases profits
Mariculture production systems	
Modular and mobile recirculating aquaculture system (RAS): promote the use of modular and climate-controlled Recirculating Aquaculture systems through community education and local training and incentivizing the incorporation of such systems.	Sustainable use of water resources Reduces high operation costs Reduces aquatic pollution

(continued)

Table 2.2 (continued)

Strategy	Challenges addressed
Species diversification: through research and development (R&D) programs promote emphasis on high-value, climate-resilient local mariculture species to reduce risks associated with the culturing of imported species.	Improves local strain selection Promotes species resilience Promotes sustainability of culture species Reduces high operation costs through lowering import costs Boots investment
Management	
Public-private partnerships (PPP): implement management frameworks which structure government investments to reduce initial risk for private small-scale farmers and commercial enterprises such as providing serviced land, leasing government infrastructure at lower costs or establishing specialized hatcheries to supply culture activities.	Reduces high capital for startups promoting investments Strengthens government-private relationships Reduces high operational costs Boosts investment
Integration of circular economy: develop market linkages to transact aquaculture byproducts to promote integrated economic activities across sectors, i.e., utilization of fish waste as organic fertilizers used in such desert agricultural fields.	Promotes effective and sustainable waste management Reduces pollution Increases economic viability Boosts investment Reduces independence on foreign imports Enhances local and global certification

and waste management. The need to prioritize rigorous environmental impact assessment with attention to reef connectivity and its contribution to the success of other sectors (Abdullah 2002; Poynton and Lovatelli 2006). Lastly, the need to build regional capacity in disease surveillance and harmonized sanitary standards within the sector to enable local and international trade (Faour- Klingbeil and Todd 2020; Mapfumo 2022).

Table 2.3 Strategic measurements for Sustainable Mariculture in Coastal Desert zones: Southern Africa (Namibia and South Africa)

Strategy	Challenges addressed
Policy development and implementation	
Streamline Marine Spatial Planning (MSP): a clear legal framework dedicated to MSP should be developed to formally allocate offshore and nearshore mariculture sites which fast-track mariculture operations startups and assist in faster permitting and impact assessment processes while providing greater security for investors.	Reduces industry conflicts Improves licensing and permitting processes Promotes better enforcement Initiates a controlled environment and reduces impacts Boosts investment
Establish coastal development funds: establish government-backed funding programs dedicated to providing capital, grants, or loans to current and emerging small and medium enterprises (SME's) in mariculture.	Stimulates industry expansion Promotes increased employment Reduces high capital for startups promoting investments Provides better access to industry for small-scale projects
Mariculture production systems	
Specialization of cold-water species: capitalize on the unique environmental advantages of the Cold Benguela current through prioritizing high-value, cold-water species such as Atlantic salmon, Abalones, and oysters while using robust systems.	Promotes sustainable mariculture Promotes species and system resilience Reduces operation costs through eliminating the need to optimize systems with technologies
Incorporate biosecurity best practices: set up and enforce national biosecurity frameworks and working plans along with mandatory health certification of seed stock, and strict disposal protocols.	Protects water quality Reduces the risk of disease transmission Reduces environmental impacts Promotes positive local and global certification of products
Management	
Capacity building and extension: promote investment in accredited practical training and mentoring programs, for technical skills, financial literacy, and cooperative management for community members, small-scale farmers, and emerging enterprises.	Improves technical skills of workers Increases skilled workforce Increases production efficacy and efficiency Boosts investments

(continued)

Table 2.3 (continued)

Strategy	Challenges addressed
Market linkage and certification: establish and apply standards which support export certification which will assist in accessing high-value markets including international and local tourist markets.	Increased product value Reduced lack of market access Increased profits
Risk mitigation: establish robust security measures against theft, vandalism, and poaching that is supported by legal policies to deter such activities and promote community engagement and ownership to safeguard such coastal development.	Reduces poaching, vandalism, and theft Increases community involvement Lowers investment risk Promotes local development and economies

2.5.3 Southern Africa (Namibia and South Africa)

The critical lessons for sustainable mariculture within the Southern Africa region include the are the urgent need for streamlined, coherent, and enabling legal and regulatory frameworks that facilitate international investment and boost economic growth (Brown-Webb 2023; Agiotis 2024; Tran 2025). Furthermore, the success of mariculture requires a strong focus on biosecurity and good site selection to minimize environmental risks in highly productive, yet fragile, cold-water ecosystems, paired with the use of onshore intensive production systems. Through adopting early warning systems for environmental fluctuations, while having insurance or contingency plans in place for risk management (Brown et al. 2020; Brown-Webb 2023; Falconer et al. 2023; Erasmus et al. 2023). Additionally, workforce capacity building is required to enhance employment and operation outcomes, which in turn positively impact economic growth and food security (Iitembu et al. 2023; Erasmus et al. 2023). Lastly, the need to secure market access for upcoming small-scale and expanding commercial ventures is essential for achieving equitable social benefits through establishing flexible business models within the mariculture sector (Brown-Webb 2023; Erasmus et al. 2023).

2.6 Conclusion

Sustainable mariculture in coastal desert zones is feasible (Crespi and Lovatelli 2011; Corner et al. 2020), but requires aligning supportive policy and regional harmonization, production systems, and robust management & economic instruments simultaneously (Chithambaran 2016; Dickson 2022; Fisher et al. 2023). The regionally specific lessons are practical and promote the use of shared services, stimulation of technical capacity, and the use of modular energy-efficient technologies in parallel with participatory coastal zoning and regional biosecurity measures (Brown et al. 2020; Mapfumo 2022; Brown-Webb 2023). Government and private investment should target hatcheries, monitoring systems, and logistics for coastal desert mariculture to become environmentally responsible, economically viable and able to contribute to food security (Chithambaran 2016; Dickson 2022). Mariculture in coastal desert zones is not possible through a single solution but a set of context-specific opportunities. When strategically sited, technically supported and governed by inclusive management frameworks, shellfish, finfish, and seaweed culture provide options that create livelihoods in otherwise marginal coastal economies (Fisher et al. 2023; Brown-Webb 2023; Bunting 2024). Thereby, promoting food security and boosting economies sustainably.

References

Abdelzaher MA, Farahat EM, Abdel-Ghafar HM, Balboul BA, Awad MM (2023) Environmental policy to develop a conceptual design for the water–energy–food nexus: a case study in Wadi-Dara on the Red Sea Coast, Egypt. Water 15(4):780

Abdullah MS (2002) Environmentally friendly aquaculture and fisheries. In: Proceedings of a workshop, vol 15, p 17

Abeysinghe KC, Weerarathne AGPT, Colonne DKC, Bellanthudawa BKA (2025) Sustainable development goals (SDGs), climate change, and the development of aquaculture and fisheries industries. Environ Rev 33:1

Agiotis B (2024) Examining the mechanisms for and factors affecting benefit sharing in two South African aquaculture projects

Agoubi B (2021) A review: saltwater intrusion in North Africa's coastal areas current state and future challenges. Environ Sci Pollut Res 28(14):17029–17043

Araya TM (2012) Performance, potential and prospects of fisheries sector in Eritrea

Billi P (2022) Climate variability in the horn of Africa eastern countries: Eritrea, Djibouti, Somalia. In: Landscapes and landforms of the horn of Africa: Eritrea, Djibouti, Somalia, pp 1–39

Blanchard JL, Watson RA, Fulton EA, Cottrell RS, Nash KL, Bryndum-Buchholz A, Jennings S (2017) Linked sustainability challenges and trade-offs among fisheries, aquaculture and agriculture. Nat Ecol Evol 1(9):1240–1249

Brown AR, Lilley M, Shutler J, Lowe C, Artioli Y, Torres R, Tyler CR (2020) Assessing risks and mitigating impacts of harmful algal blooms on mariculture and marine fisheries. Rev Aquac 12(3):1663–1688

Brown-Webb B (2023) Creating a model to develop self-sustainable aquaculture agribusiness enterprises in South Africa. Doctoral dissertation, University of the Free State

Bruckner A (2011) A paradigm shifts for fisheries management to enhance recovery, resilience, and sustainability of coral reef ecosystems in the Red Sea

Bunting SW (2024) Principles of sustainable aquaculture: promoting social, economic and environmental resilience. Routledge

Cataudella S, Crosetti D, Massa F (2015) Mediterranean coastal lagoons: sustainable management and interactions among aquaculture, capture fisheries and the environment. General Fisheries Commission for the Mediterranean. Studies and Reviews, 95

Chithambaran S (2016) Desert aquaculture & environmental sustainability. Indian J Geo-Mar Sci 45:1733–1741

Clavelle T, Lester SE, Gentry R, Froehlich HE (2019) Interactions and management for the future of marine aquaculture and capture fisheries. Fish Fish 20(2):368–388

Corner R, Fersoy H, Crespi V (2020) Integrated agri-aquaculture in desert and arid lands learning from case studies from Algeria, Egypt and Oman. Food & Agriculture Organization (FAO)

Crespi V, Lovatelli A (2011) Global desert aquaculture at a glance. In: Aquaculture in desert and arid lands. FAO, p 25

Cziesielski MJ, Duarte CM, Aalismail N, Al-Hafedh Y, Anton A, Baalkhuyur F, & Aranda M (2021) Investing in blue natural capital to secure a future for the Red Sea ecosystems. Front Mar Sci 7:603722.

Chirdon MA, Mertens KN, Bilien G, Derrien A, Nguyen-Ngoc L, Doan-Nhu H, & Ibrahim NH (2025) Re-isolation of Gonyaulax hyalina from the type locality in the Gulf of Aden: Integrated analysis of morphological, molecular, and toxicological characteristics. J Phycol 61(5): 1486–1502

Darias MJ, Mirera DO (2023) Towards a more sustainable aquatic food system: interdisciplinary research on sustainable marine aquaculture in Africa

De Greef K (2014) Booming illegal abalone fishery in Hangberg: tough lessons for small-scale fisheries governance in South Africa

Dickson M (2022) Regional review on status and trends in aquaculture development in the near East and North Africa–2020. Food & Agriculture Organization (FAO)

Diop S, Scheren PA (2016) Sustainable oceans and coasts: lessons learnt from Eastern and Western Africa. Estuar Coast Shelf Sci 183:327–339

Diop S, Fabres J, Pravettoni R, Barusseau JP, Descamps C, Ducrotoy JP (2014) The western and central Africa land–sea interface: a vulnerable, threatened, and important coastal zone within a changing environment. Springer International Publishing, pp 1–8

Eckardt FD, Maggs-Kölling G, Marais E, De Jager PC (2022) A brief introduction to hot desert environments: climate, geomorphology, habitats, and soils. In: Microbiology of hot deserts, pp 1–36

El Mahrad B, Abalansa S, Newton A, Icely JD, Snoussi M, Kacimi I (2020) Social-environmental analysis for the management of coastal lagoons in North Africa. Front Environ Sci 8:37

Elhetawy AI, Mansour AT, Lotfy AM, Mansour AI, Shahin SA, Zayed MM, Abdel-Rahim MM (2025) Integrated aquaculture of whiteleg shrimp *Litopenaeus vannamei* and European seabass *Dicentrarchus labrax*: impacts on performance, welfare, blood physiological response, carcass traits, productivity, and farm profitability. Aquac Int 33(4):252

El-Sayed AFM (2017) Regional review on status and trends in aquaculture development in the near East and North Africa-2015. FAO Fisheries and Aquaculture Circular C1135

Ennabili A, Libiad M, El Haissoufi M, Khabbach A, Bennas N (2025) Biodiversity, change, and use of Moroccan wetlands. In: Wetlands of Tropical and Subtropical Asia and Africa: biodiversity, livelihoods and conservation. Wiley, pp 265–296

Erasmus VN, Hamutenya SS, Numwa O, Petrus E (2023) The Namibian Mariculture: productivity, challenges and opportunities. In: Emerging sustainable aquaculture innovations in Africa, pp 441–458

Eyayu A, Getahun A, Keyombe JL (2023) A review of the production status, constraints, and opportunities in East African freshwater capture and culture fisheries. Aquac Int 31(4):2057–2078

El Mountassir O, Bahir M, Hakimi S, Faraj TK, & Carreira PM (2024) Assessment of groundwater quality in the Semi-Arid environment: implications of climate change. Limnol Rev 24(4): 520–542

Falcon WP, Naylor RL, Shankar ND (2022) Rethinking global food demand for 2050. Popul Dev Rev 48(4):921–957

Falconer L, Cutajar K, Krupandan A, Capuzzo E, Corner RA, Ellis T, Telfer TC (2023) Planning and licensing for marine aquaculture. Rev Aquac 15(4):1374–1404

Faour-Klingbeil D, Todd E (2020) Prevention and control of foodborne diseases in Middle-East North African countries: review of national control systems. Int J Environ Res Public Health 17(1):70

Farouk ASR, Schuchert P, Farnsworth KD (2024) Pressures on Egyptian Red Sea Fisheries from the artisan fishers' perspective. Ocean Costal Manag 258:107406

Fernandes-Salvador JA, Davidson K, Sourisseau M, Revilla M, Schmidt W, Clarke D, Silke J (2021) Current status of forecasting toxic harmful algae for the north-east Atlantic shellfish aquaculture industry. Front Mar Sci 8:666583

Finke G, Gee K, Kreiner A, Amunyela M, Braby R (2020) Namibia's way to Marine spatial planning–using existing practices or instigating its own approach? Mar Policy 121:104–107
Fisher J, Angel D, Callier M, Cheney D, Filgueira R, Hudson B, Byron CJ (2023) Ecological carrying capacity in mariculture: consideration and application in geographic strategies and policy. Mar Policy 150:105516
Freeman S, Angel D, Freeman S (2019) Integrated aquaculture (INTAQ) as a tool for an ecosystem approach to the marine farming sector in the Mediterranean Sea. Integrated Mariculture 529:133–183
Habte-Tsion HM (2016) Sustainable aquaculture development and its role in the food security and economic growth in Eritrea: trends and prospects a review paper for International Conference on Eritrean Studies (ICES 2016)
Hambrey J (2017) The 2030 agenda and the sustainable development goals: the challenge for aquaculture development and management. FAO fisheries and aquaculture circular
Hutchings L, Van der Lingen CD, Shannon LJ, Crawford RJM, Verheye HMS, Bartholomae CH, Monteiro PMS (2009) The Benguela current: an ecosystem of four components. Prog Oceanogr 83(1–4):15–32
Hzami A, Heggy E, Amrouni O, Mahé G, Maanan M, Abdeljaouad S (2021) Alarming coastal vulnerability of the deltaic and sandy beaches of North Africa. Sci Rep 11(1):2320
Harsh LN (2018) Technological Interventions to Improve the Livelihood of VulnerableDjibouti Communities. Sci and Environment 34:40–55
Idhalla M, Nhhala H, Kassila J, Ait Chattou EM, Orbi A, Moukrim A (2017) Comparative production of two mussel species (*Perna perna* and *Mytilus galloprovincialis*) reared on an offshore submerged longline system in Agadir, Morocco. Int J Sci Eng Res 8(6):1203–1213
Iitembu JA (2005) Analysis of marine aquaculture developments in Namibia. Environmental, economic and legislative considerations. Master's thesis, Universitetet i Tromsø
Iitembu JA, Fitzgerald D, Altintzoglou T, Boudry P, Britz P, Byron CJ, Strand Å (2023) Comparative description and analysis of oyster aquaculture in selected Atlantic regions: production, market dynamics, and consumption patterns. Fishes 8(12):584
Jalludin M (2014) Groundwater and the challenges for the future water supply of the Republic of Djibouti in an arid climate (Horn of Africa). In: Aquifer systems management: Darcy's legacy in a world of impending water shortage, pp 127–135
Kaba AJ (2024a) A survey of the geographic area, altitude, coastline, and climate of African countries and regions: implications for Africa's development. J Sustain Dev 17(3):57–80
Kaba AJ (2024b) The educational attainment of black Americans and gender, 2012 and 2022: resilience, factors, implications and recommendations. Higher Educ Stud 14(4):115–132
Khalil A, Pastor Reyes I, Pinarbaşi K, Halim F, Iglesias Campos A, Cervera Nuñez C, Marsit F (2021) Technical report: current conditions and compatibility of maritime uses in the western Mediterranean
Kolding J, van Zwieten PA, Marttin F, Poulain F (2016) Fisheries in the drylands of Sub-Saharan Africa "Fish come with the rains". Building resilience for fisheries-dependent livelihoods to enhance food security and nutrition in the drylands. Food and Agriculture Organization of the United Nations (FAO)
Lounas R, Kasmi H, Chernai S, Amarni N, Ghebriout L, Meslem-Haoui N, Hamdi B (2020) Towards sustainable mariculture: some global trends. Thalassa's. Int J Mar Sci 36(2):447–456
Mandour M, Sebaq M, Hassan RA (2025) Economic evaluation of bivalves in Egypt: a potential for aquaculture production. Suez Canal Vet Med J 30(1):119–129
Mapfumo B (2011) An overview of desert aquaculture in southern Africa. In: Aquaculture in desert and Arid lands: development constraints and opportunities, pp 119–140
Mapfumo B (2022) Regional review on status and trends in aquaculture development in sub-Saharan Africa–2020
Megahed ME, Aly SM (2009) Challenges facing marine aquaculture and requirements for development in Egypt. In: Proceedings of the 2nd Global Fisheries and Aquaculture Research Conference, Cairo International Convention Center, pp 24–26

Mirera DO (2011) Experimental polyculture of milkfish (*Chanos chanos*) and mullet (*Mugil cephalus*) using earthen ponds in Kenya. West Indian Ocean J Mar Sci 10(1):59–71
Mirera DO, Okemwa D (2023) Salinity tolerance of Nile tilapia (*Oreochromis niloticus*) to seawater and growth responses to different feeds and culture systems. West Indian Ocean J Marine Sci 22(2):75–85
Molotoks A, Smith P, Dawson TP (2021) Impacts of land use, population, and climate change on global food security. Food Energy Secur 10(1):e261
Mouton SF (2019) An assessment of the blue economy in Namibia: a case study of Swakopmund and Walvis Bay
Mupambwa HA, Hausiku MK, Nciizah AD, Dube E (2019) The unique Namib desert-coastal region and its opportunities for climate smart agriculture: a review. Cogent Food Agric 5(1):1645258
Mwaniki A (2006) Achieving food security in Africa: challenges and issues. UN Office of the Special Advisor on Africa (OSAA)
Mwanja WW, Nyandat B (2013) Challenges and issues facing small-scale aquaculture producers: perspectives from Eastern Africa. In: Enhancing the contribution of small-scale aquaculture to food security, poverty alleviation and socio-economic development, p 143
Odunlade OR, Isikwei B (2025) Sustaining Africa's blue and digital economy transition: emerging skills and job roles. In: Securing sustainable futures through blue and green economies. IGI Global, pp 173–194
Olivier D, Heinecken L, Jackson S (2013) Mussel and oyster culture in Saldanha Bay, South Africa: potential for sustainable growth, development and employment creation. Food Secur 5(2):251–267
Ortega A, Cano-Pérez J, Nhhala H, Halla MI, Kara MH, De la Gándara F, Blanco E (2021) Aquaculture in the Alboran Sea. In: Alboran sea-ecosystems and marine resources. Springer, Cham, pp 659–706
Paruğ Ş, Jabbr A, Lazrag FA (2024) Sustainable development of fisheries and aquaculture in North Africa: current status and future prospects. Memba Kastamonu Üniversitesi Su Ürünleri Fakültesi Dergisi 10(1):90–105
Pay MA (2023) Complex ocean ethics: examining human-ocean relationships through South Africa's abalone crisis. Doctoral dissertation, University of East Anglia
Pitcher GC, Louw DC (2021) Harmful algal blooms of the Benguela eastern boundary upwelling system. Harmful Algae 102:101898
Potgieter T (2021) South Africa: the blue economy experience. In: The Blue Economy in Sub-Saharan Africa, pp 115–140
Poynton SL, Lovatelli A (2006) Regional review on aquaculture development near East and North Africa-2005. Food and Agriculture Organization (FAO) of the United Nations
Pulfrich A, Kemper J (2022) Environmental impact assessment for the proposed Meob Bay tourism development, Namibia
Röckmann C, Fernández TV, Pipitone C (2022) Regulation and planning in the Mediterranean Sea. In: Building Industries at Sea-'Blue Growth' and the new maritime economy, pp 365–401
Rothman MD, Anderson RJ, Kandjengo L, Bolton JJ (2020) Trends in seaweed resource use and aquaculture in South Africa and Namibia over the last 30 years. Bot Mar 63(4):315–325
Sadek S (2010) Aquaculture site selection and carrying capacity estimates for inland and coastal aquaculture in the Arab Republic of Egypt. In: Site selection and carrying capacities for inland and coastal aquaculture, pp 6–8
Sasson A (2012) Food security for Africa: an urgent global challenge. Agric Food Secur 1(1):2
Shahin M (2007) Climate of the Arab region. In: Water resources and hydrometeorology of the Arab region, pp 77–134
Sippel SR (2015) All you need is export? Moroccan farmers juggling global and local markets. In: Handbook on the globalisation of agriculture, pp 328–349
Steyl G, Dennis I (2010) Review of coastal-area aquifers in Africa. Hydrogeol J 18(1):217–225
Tesfamichael D, Mohamud S (2016) Eritrea. In: The Red Sea ecosystem and fisheries, pp 49–63

Tolley KA, Da Silva JM, Jansen van Vuuren B, Bishop B, Dalton J, Du Plessis M, Suleman E (2019) South African National Biodiversity Assessment 2018: Technical Report Volume 7: Genetic Diversity

Tran TM (2025) The status of the marine economy in South Africa forward to sustainable development and implications for Vietnam's maritime strategy. J Res Environ Sci Educ 2(1):48–56

Valavanidis A (2023) Feeding a growing world population. In: Critical rethinking of lessons from the "Green Revolution" in the 1960s as a warning for feeding a hungry planet, vol 1, pp 1–47

Van der Knaap M, Ismael FS (2020) Clam and oyster culture in Djibouti. FAO Aquaculture Newsletter 61:35–36

Van der Lingen CD, Hutchings L, Lamont T, Pitcher GC (2016) Climate change, dinoflagellate blooms and sardine in the southern Benguela current large marine ecosystem. Environ Dev 17:230–243

Weldetsion GG (2023) Eritrea's self-reliance policy and the road to sustainable food and water security. Doctoral dissertation, University of London

Chapter 3
The Role of Organic Amendments and Marine-Biomass–Derived Resources in Enhancing the Physicochemical Properties of Horticultural Growing Media

Elia N. M. Ruben, Simeon S. Hamukoshi, and Hupenyu Allan Mupambwa

Abstract Horticultural growing media such as cocopeat are widely used in soilless cultivation systems to support plant growth. While effective in providing physical support and aeration, these substrates are typically nutrient-deficient and chemically inert, necessitating regular supplementation through fertigation. This reliance on synthetic nutrient solutions presents challenges including nutrient leaching, limited water retention, and unstable pH, which may compromise the long-term sustainability of drip-irrigated systems. This book chapter explores the use of organic amendments specifically vermicompost and biochar and inorganic co-amendments, including rock phosphate and fly ash, to improve the physicochemical properties of horticultural media. In addition, it emphasizes the synergistic effects of coapplying fly ash, rock phosphate, and biochar with vermicompost. Biochar enhances pH buffering, cation exchange capacity, and nutrient retention, while fly ash supplies essential minerals, and rock phosphate promotes phosphorus availability, within the vermicompost matrix. The chapter also examines the potential agronomic application of marine-derived resources, including biochar produced from seaweed species, such as *Laminaria pallida* and *Gracilariopsis funicularis*. Marine resource-derived biochar offers promising nutrient composition (N, P, K, Ca, Mg, S) and structural benefits. In contrast, it may pose challenges such as high salinity,

---

E. N. M. Ruben (✉)
Sam Nujoma Marine and Coastal Resources Research Centre, Sam Nujoma Campus, University of Namibia, Henties Bay, Namibia

Department of Crop Production and Agricultural Technologies, Ogongo Campus, University of Namibia, Oshakati, Namibia

S. S. Hamukoshi
Namibia Green Hydrogen Research Institute, University of Namibia, Windhoek, Namibia

National Commission on Research, Science and Technology, Olympia, Windhoek, Namibia

H. A. Mupambwa
Sam Nujoma Marine and Coastal Resources Research Centre, Sam Nujoma Campus, University of Namibia, Henties Bay, Namibia

H. A. Mupambwa et al. (eds.), *Harnessing Coastal and Marine Resources for Improved Food Security in Africa*, Sustainability Sciences in Asia and Africa,
https://doi.org/10.1007/978-981-95-9725-3_3

elevated pH, and the potential presence of heavy metals (e.g., Cd, Cr, Zn, Cu, and Fe). Coapplication strategies with other organic amendments, such as vermicompost, have been evaluated as potential approaches to mitigate these issues. In addition, innovations in green fertilizers synthesized from seawater using hydrogen-based clean energy technologies are discussed as emerging climate-smart alternatives to fossil fuel-based inputs. Drawing on recent empirical studies, this chapter synthesizes the current knowledge on amendment technologies and marine-based resources, identifies key research gaps, and proposes future directions for enhancing the sustainability and efficiency of horticultural growing media in modern soilless agriculture.

Keywords Horticultural growing media · Biochar · Vermicompost · Marine biomass · Physico-chemical properties · Green fertilizers

3.1 Introduction

The progressive loss of arable land due to industrial expansion, population growth, and urbanization has intensified the need for sustainable soil-independent plant production systems (Kraamwinkel et al. 2021; Rajendran et al. 2024). Therefore, horticultural growing media, such as coco peat, perlite, vermiculite, and rockwool, have gained attention in protected horticulture and hydroponic systems, as these substrates effectively substitute soil by offering physical support, moisture retention, and adequate aeration (Gbollie et al. 2021; Messiga et al. 2021). Despite their favorable structural properties, these substrates lack essential plant nutrients, particularly key macro- and micronutrients such as N, P, K, Ca, Mg, S, Fe, Zn, Cu, and Mn, which limits their capacity to optimally support plant growth without additional nutrient supplementation (Badreshiya et al. 2024). Consequently, this limitation has led to increased reliance on nutrient-enriched fertigation solutions, particularly in drip irrigation-based systems, to ensure the consistent and efficient delivery of essential nutrients required for optimal plant growth and productivity (Abedin et al. 2021). In contrast, the high cost and environmental impact associated with the continuous use of synthetic fertilizers in fertigation systems raise significant concerns regarding the long-term sustainability of such practices (Chen et al. 2018).

In recent years, studies have focused on improving the physicochemical properties of horticultural media by adding organic amendments such as compost, vermicompost, biochar, coir, and bark, as well as inorganic amendments such as fly ash, rock phosphate, perlite, and vermiculite (Ajibade et al. 2022; Martins et al. 2023; Mupambwa et al. 2024). Various studies report that amendments such as vermicompost, fly ash, rock phosphate, and biochar can significantly enhance the functionality of low-nutrient substrates such as cocopeat, perlite, and coir by improving their nutrient retention, pH, electrical conductivity (EC), cation exchange capacity (CEC), and microbial activity (Mupambwa et al. 2024; Mupondi et al. 2018). For instance, Messiga et al. (2020) demonstrated that partial substitution of coir and peat with vermicompost and biochar significantly improved nitrogen dynamics and fresh yield of leafy vegetables under greenhouse conditions. Specifically, the addition of vermicompost enhanced nutrient supply, narrowing the yield gap at 50% nitrogen application, while biochar reduced ammonium loss in leachates, likely due to its adsorption capacity. In another study, Messiga et al. (2021) reported that amending peat-based substrates with compost, coir, and biochar, increased microbial biomass and soil respiration, and effectively mitigated nitrous oxide (N_2O) emissions without compromising crop yield. These findings highlight the agronomic and environmental benefits of integrating bio-based and mineral amendments into horticultural growing media, particularly in minimizing N-leaching and greenhouse gas emissions associated with intensive horticulture.

Among the most promising innovations are amendments derived from marine biomass, particularly biochar produced from seaweed species, such as *Laminaria pallida, Gracilariopsis funicularis* (Katakula et al. 2020). Biochar derived from seaweed biomass typically contains high levels of ash and minerals, particularly potassium, calcium, and magnesium, and exhibits a lower carbon to nitrogen (C: N) ratio compared to biochar produced from terrestrial feedstocks (Katakula et al. 2020; Roberts and De Nys 2016). Moreover, biochar derived from seaweed has shown potential to enhance pH buffering capacity, nutrient availability, and water retention when combined with other organic amendments such as vermicompost (Katakula et al. 2022). Complementing this approach is the emerging development of green fertilizers synthesized from seawater using hydrogen-based clean energy technologies, offering a novel pathway to reduce fossil fuel dependency while ensuring a steady and sustainable supply of nutrients to crops (Adeli et al. 2023; Gao et al. 2022). Despite these advancements, research has largely focused on crop yield, with limited attention to how amendments affect pH stability, nutrient retention, microbial activity, and structural integrity in growing media. In addition, there are limited studies that have provided standardized comparative evaluations of different organic and marine-derived amendments under controlled conditions, particularly in drip--irrigated or container-based horticultures.

In response to these gaps, this book chapter examines the major types, properties, and limitations of horticultural growing media, with particular emphasis on those used in drip-irrigated and soilless horticultural systems. It further seeks to evaluate how organic and inorganic amendments, especially those sourced from marine biomass, influence the key physicochemical characteristics of these substrates, such as nutrient retention, pH buffering, water-holding capacity, and microbial activity. Finally, this chapter explores the comparative advantages and limitations of marine biochar and green fertilizers compared to conventional amendments, with the goal of guiding the development of more sustainable and effective media management practices in modern horticulture.

3.1.1 Horticultural Growing Media

Horticultural growing media are materials used in controlled environments (e.g., greenhouses and hydroponic systems) to support plant growth by serving as alternatives to soil, providing root stability, water retention, aeration, and nutrient exchange (Badreshiya et al. 2024). They have gained attention and agronomic importance due to the decline in arable land, which has created a need for sustainable plant production systems, and they comprise organic materials such as peat moss, coir, and composted bark, along with inorganic materials such as perlite, vermiculite, sand, and rockwool (Badreshiya et al. 2024). Peat moss and coir are widely used in modern farming as standard media, with peat moss recognized for its excellent moisture retention though limited by sustainability concerns, while coir is valued for its renewability, sustainability, and structural resilience (Messiga et al. 2020). In contrast, inorganic substrates such as vermiculite and perlite are commonly used as supplements to standard media, with vermiculite enhancing water retention and cation exchange capacity (CEC), whereas perlite mainly improves aeration and prevents compaction (Badreshiya et al. 2024).

As shown in Fig. 3.1, despite these benefits, most horticultural media, including coir, peat moss, and cocopeat, lack essential nutrients and generally exhibit low CEC, with peat moss further characterized by low pH and EC (Hernández-Rodríguez et al. 2017; Nemati et al. 2015). For instance, Nemati et al. (2015) showed that unamended peat has low pH (4.6–4.89), low EC (decreasing from 1.66 to 0.18 dS

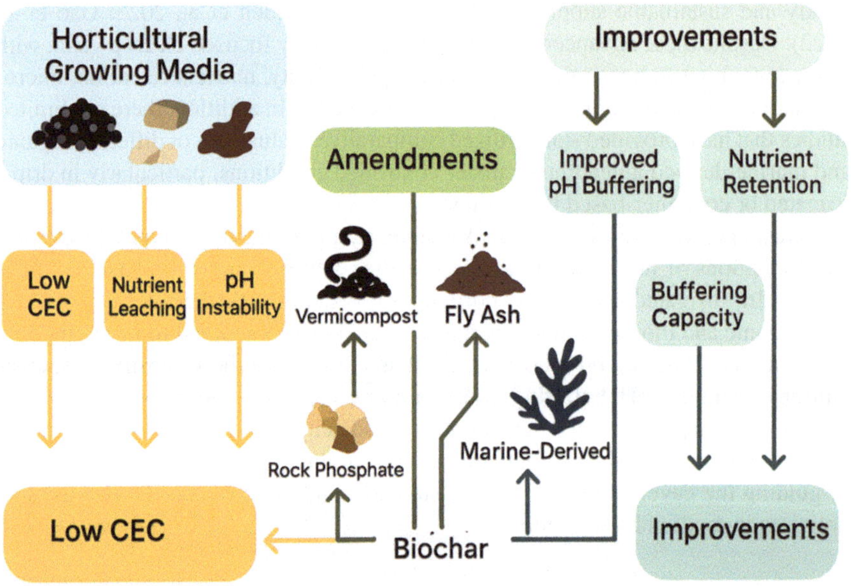

Fig. 3.1 Conceptual overview of horticultural growing media challenges and physicochemical improvements achieved through organic amendments

m^{-1} after leaching), and minimal nitrate and potassium content, while its CEC (100–125 cmolc kg^{-1}) remains low relative to mineral soils. Consistent with these physicochemical limitations, Hernández-Rodríguez et al. (2017) demonstrated that seedlings in peat-only media exhibited significantly lower germination rates and seedling development compared with media amended with vermicompost or semi-compost. Consequently, supplementary fertilization and the incorporation of organic or inorganic amendments have become essential strategies in soilless cultivation, ensuring adequate nutrient supply and compensating for the inherently low CEC of these media.

Organic amendments, including compost, vermicompost, and manure, enhance microbial activity and nutrient retention, whereas inorganic inputs, such as rock phosphate and fly ash, provide targeted nutrient supplementation in horticultural growing media (Table 3.1). For example, Messiga et al. (2020) reported that supplementing coir with vermicompost narrowed the lettuce yield gap from 49% to 28%, while also decreasing nitrate losses in leachates, highlighting vermicompost's ability to improve coir properties. Similarly, Mupambwa et al. (2024) observed that replacing 50% of cocopeat with fly ash-amended vermicompost improved tomato performance compared with the control, highlighting the potential of fly ash to contribute minerals and micronutrients in horticultural media. In a related study, Mupondi et al. (2018) showed that amending cow dung with 2–8% rock phosphate during vermicomposting increased phosphorus availability and enhanced biodegradation, without compromising the agronomic quality of the resulting vermicompost.

Biochar, a product of biomass pyrolysis, is another amendment with potential to enhance the functionality of horticultural media by improving nutrient dynamics and physical properties (Nemati et al. 2015). It is a multifunctional amendment for horticultural growing media, improving porosity, aeration, pH buffering, cation exchange capacity, water retention, and microbial colonization, while also supplying some nutrients depending on the feedstock used (Antileo-Mellado et al. 2024). Although biochar is widely applied in horticultural media modification, its effectiveness is strongly influenced by the feedstock type, application rate, and the co-amendments with which it is combined (Messiga et al. 2020; Roberts and De Nys 2016). For example, Kafle et al. (2024) reported that replacing peat with a full softwood biochar-compost blend raised cucumber shoot weight by 83% and leaf area by 84%, while also improving root density and buffering medium pH and EC. Similarly, Martins et al. (2023) observed that incorporating 12% compost with 10% biochar into coir-based media sustained 90–99% lettuce germination and vigorous seedling growth, demonstrating that peat could be reduced to only 5–10% without loss of performance. Alvarez et al. (2019) found that adding 4–12% biochar with vermicompost to peat substrates reduces nitrogen leaching by about 37% and phosphorus by 30% in petunia, although higher rates doubled potassium leaching, highlighting nutrient trade-offs. On the other hand, Zulfiqar et al. (2021) reported that substituting 30% peat with rice-straw biochar reduced Alpinia shoot and root biomass by more than 40% and 22%, respectively.

Table 3.1 Summary of experiments on horticultural growing media amended with organic and inorganic materials

Media and amendment	Experimental setup	Key results/notes	Reference
Peat (control); control + CaP, Struvite, SSA, Mephrec or rock-P	Tagetes grown in pots in a greenhouse under two pH levels (4.5–5.5 and 5.5–6.0) for 1 month.	Struvite had the highest P uptake (~0.37 mg/plant) at pH 4.5–5.5; Mephrec had the lowest (~0.13 mg/plant).	Hauck et al. (2021)
Peat/perlite (70:30, control); peat/perlite/BC (40:30:30); peat/perlite/BC/compost (30:30:35:5) (v/v).	Alpinia zerumbet grown in pots in a controlled environment for 5 months.	BC alone reduced dry biomass by 42%. BC + compost restored plant growth and was recommended optimal.	Zulfiqar et al. (2021)
Peat: perlite: vermiculite (50:25:25, control); peat partially (12.5% BC + 12.5% compost) or fully (25% BC + 25% compost, v/v) replaced.	Cucumber seedlings grown in a greenhouse for 1 month. Different BC feedstocks were used.	Fully replacing peat with softwood BC–compost was optimal, increasing shoot weight, leaf area, and root length by 83%, 84%, and 40%, respectively.	Kafle et al. (2024)
100% cocopeat (control); 0%, 25%, 50% fly ash-amended VC: cocopeat (w/w).	Tomato and Swiss chard grown in pots in a in a greenhouse, supplemented with either 50% or 100% hydroponics fertilizer.	50% VC + 100% fertilizer increased tomato yield by 10.3%; VC reduced Swiss chard yield; Ni and Cr exceeded permissible limits.	Mupambwa et al. (2024)
Soil-based medium (control); 2.5% or 5% BC, VC, perlite, or ABA (BC + VC + perlite): soil (w/w).	Wheat was grown in pots in a greenhouse for 5 months under well-watered and water-stressed conditions.	At 5% amendment, AWP increased fourfold, and NAR improved 24-fold under drought. Effective for stress resilience	Jahan et al. (2023)
Coir (100%, control); coir +10% BC or VC; peat (100%, control); peat +10% BC or VC (v/v).	Chinese cabbage & lettuce grown in pots in a greenhouse for 2 months under 0%, 50%, and 100% N fertilization levels.	BC + peat and VC + coir sustained yields at 50% N, reducing yield gaps by up to 18–32% & reduce NO_3^--N & NH_4^+-N leachate to 60–130 mg/L and 1–17 mg/L, respectively.	Messiga et al. (2020)
Coir (100%, control); coir +10% BC or VC; peat (100%, control); peat +10% BC or VC (v/v).	Incubation study done in controlled chamber for 2 months to assess gas emissions and N cycling by media.	Peat + BC emitted 2070 μg CO_2-C/g DW and 62.78 μg N_2O-N/g DW. Coir + VC cut N_2O to 47.53 from 111.82 μg/g DW.	Messiga et al. (2021)

(continued)

Table 3.1 (continued)

Media and amendment	Experimental setup	Key results/notes	Reference
Coir (100%, control); coir:compost:bark:peat (73:12:5:10); coir:compost:peat:BC (73:12:5:10) (v/v).	Substrates were characterized prior to planting lettuce seedlings in a greenhouse for 22 days.	Substrate characteristics were within the normal range, leading to overall 90–99% emergence and vigorous growth.	Martins et al. (2023)
Soil (control); soil +RP or FYM at 50% or 100% (w/w)	Mentha arvensis was grown in pots in a screenhouse with or without AMF.	AMF + RP or FYM increased oil content both levels, with spore density reaching 190/100 g soil.	Burni et al. (2023)
GB (100%, control); GBVC +2% ESBC, PMBC, or PLCM (w/w).	5-month vermicomposting in plastic containers after 7-week precomposting, using *E. fetida* worms.	GBVC + ESBC increased VC yield by 25%; respiration decreased by 18% and 20% with ESBC and PMBC, respectively.	Antileo-Mellado et al. (2024)
Commercial mix (control); 7BC:3VC, 8BC:2VC, 7BC:3CM, or 8BC:2CM (v/v).	Ornamentals and vegetables grown in trays (greenhouse) for emergence test.	VC mixes at 70% or 80% BC maintained equal or higher seed emergence and growth than control, for most tested species.	Huang et al. (2019)
Soil (control); soil + RP, SSP, DAP, or PM, and combinations (½RP + ½ PM, ½RP + ½SSP, ½RP + ½DAP).	Chilli was grown in greenhouse pots using substrates incubated for 60 days, with or without PSB.	PM + PSB + RP stimulated P mineralization, releasing up to 25 mg P/kg of soil.	Abbasi et al. (2015)
Peat moss (PS) (control); PS + perlite (7:3 v/v); PS + BC (7:3 v/v)	Potted, additive-enriched substrates characterized in a greenhouse after 8 leaching cycles	BC raised pH, CEC, and reduced nutrient leaching by 11%.	Nemati et al. (2015)
Peat (control); peat +10% or 20%VC; peat +4% or 12% BC; peat +10% VC + 4% BC or 20% VC + 12% BC.	Petunia and Pelargonium grown in pots (greenhouse) for 68 days under standard irrigation in formulated substrates.	N leaching fell by 37% on average; P leaching in Petunia dropped 30% with 10% VC + 4% BC; K leaching rose by 100% with 20% VC + 12% BC.	Alvarez et al. (2019)
Peat moss (PS) (control); PS + VC (1:2, 2:1) or semicompost (1:2, 2:1).	Lettuce and tomato grown in trays (greenhouse) for 21 days in formulated substrates.	In lettuce, 2VC:1PS increased germination by 12.5%. In tomato, 1semicompost:2PS raised germination by 16.8%, performing > control.	Hernández-Rodríguez et al. (2017)

(continued)

Table 3.1 (continued)

Media and amendment	Experimental setup	Key results/notes	Reference
Coco peat; spent mushroom compost; perlite; volcanic tuff; oak sawdust.	Eggplant grown in plastics (greenhouse) to assess its nutrient content under different substrates.	Mushroom compost had the highest leaf K (3.24%) content. Sawdust had the highest leaf P content (0.52%). Volcanic tuff and perlite showed P and K values < organic media.	Erdal et al. (2014)
Soil (control); soil + RP, RP + cow manure VC, or RP + pig manure VC	Swiss chard cultivated in pots in the greenhouse for 8 weeks. Substrates supplemented with P (0, 25, or 50 mg P/kg) and PSB.	VC + PSB + RP at 50 mg P kg^{-1} had the highest fresh weight (39.78 g), leaf area (240.41 cm^2), and total P in tissues (326.91 mg kg^{-1}).	Ajibade et al. (2022)

Abbreviations: *BC* biochar; *VC* vermicompost, *CaP* calcium phosphate, *SSA* single superphosphate, *RP* rock phosphate, *FYM* farmyard/field yard manure, *PSB* phosphate-solubilizing bacteria, *AMF* arbuscular mycorrhizal fungi, *AWP* Apparent Water Productivity, *NAR* Net Assimilation Rate, *ESBC* eucalyptus sawdust biochar, *PMBC* pig manure biochar, *PLCM* poultry-litter carbonaceous material, *GBVC* grape bagasse vermicompost, *CM* chicken manure compost, *PS* peat moss. Maximum permissible limits for heavy metals in vegetables: $Zn \leq 60$ mg/kg, $Cr \leq 2.3$ mg/kg, $Ni = 10$ mg/kg–66.9 mg/kg

These studies demonstrated that biochar can enhance horticultural media functionality by improving physicochemical properties and reducing nutrient losses (Fig. 3.1), but the benefits are context-specific and depend strongly on biochar source, application rate, and coamendments. Importantly, the discussion on biochar also intersects with broader sustainability concerns, particularly the urgent need to find viable alternatives to peat in horticultural media. Literature indicates that most studies have emphasized crop yield over substrate performance, whereas key parameters such as nutrient retention, microbial colonization, and physical integrity under irrigation are often underreported. Research is also fragmented, with varied amendment rates and inconsistent media types, which complicate comparisons across studies. Standardized protocols and long-term evaluations are therefore necessary. At the same time, environmental concerns surrounding peat extraction have intensified the search for sustainable alternatives. Waste-derived media, such as composted grape bagasse and marine biomass biochar, offer promising eco-friendly substitutes (Antileo-Mellado et al. 2024). In conclusion, horticultural growing media are central to sustainable plant production but require targeted amendment to enhance their functionality. The subsequent section examines specific types of growing media compatible with drip irrigation systems, linking media properties to irrigation and fertigation efficiency.

3.1.2 Growing Media for Drip Irrigation Systems

Many conventional soilless substrates used in horticulture, such as peat and coir, exhibit a low water-holding capacity, limited buffering ability, and inadequate nutrient retention (Huang et al. 2019; Nemati et al. 2015). These physical limitations often result in water stress under infrequent irrigation and nutrient leaching under excessive irrigation (Incrocci et al. 2017). In contrast, precision technologies such as drip irrigation mitigate these effects by supplying water and nutrients directly to the rhizosphere (Incrocci et al. 2017). Drip irrigation enhances resource-use efficiency by reducing leaching and surface evaporation, while the concurrent application of fertigation ensures uniform nutrient distribution and minimizes nutrient losses (Incrocci et al. 2017). This approach is particularly important in soilless systems, where growing media lack long-term nutrient reserves; therefore, media with stable pH, moderate EC, and high nutrient retention are essential in drip fertigation systems to balance nutrient supply and minimize leaching (Kafle et al. 2024). The effectiveness of drip irrigation depends on its compatibility with the growing medium, which must maintain structural integrity, retain adequate moisture, and ensure uniform wetting (Incrocci et al. 2017; Kafle et al. 2024). In addition, growing media compatible with drip irrigation must provide consistent nutrient availability and support microbial processes that sustain plant growth (Kafle et al. 2024). Drip--compatible growing media encompasses a broad range of organic, inorganic, and blended substrates, each selected to optimize water retention, nutrient availability, and physical structure for soilless horticultural systems (Zulfiqar et al. 2021).

Organic substrates, such as peat moss, are common because of their fibrous texture and favorable water retention, while inorganic materials, such as perlite and vermiculite, are often added to improve drainage and aeration in dense organic substrates (Kafle et al. 2024; Zulfiqar et al. 2021). In addition, engineered mixtures that include compost, sand, and biochar combine the benefits of individual materials into balanced media suitable for precision fertigation (Huang et al. 2019). Among these substrates, cocopeat is widely used because its high porosity, moderate water--holding capacity, and low bulk density make it particularly suitable for drip irrigation systems (Gbollie et al. 2021). However, Peat moss is also widely used despite sustainability concerns, particularly when modified with amendments such as biochar to enhance its pH and nutrient availability, thereby further optimizing its performance in soilless systems (Nemati et al. 2015; Prasad et al. 2019). Studies have shown that supplementing cocopeat or coir with vermicompost can enhance crop performance under drip fertigation; for example, Messiga et al. (2020) reported improved leafy vegetable yields when coir was partially substituted with vermicompost. Other studies have further explored the use of vermicompost amended with co-amendments, such as biochar, fly ash, and rock phosphate, to enhance the physicochemical properties of growing media, thereby optimizing their performance under drip irrigation systems (Antileo-Mellado et al. 2024; Mupambwa et al. 2024; Mupondi et al. 2018).

Mixed substrates are increasingly favored for drip irrigation systems because they improve cation exchange capacity (CEC) and foster beneficial microbial activity, both of which are essential for optimizing nutrient availability in fertigated crops (Mupondi et al. 2018). Other emerging alternatives involve incorporating marine biomass-derived amendments, such as seaweed biochar, into growing media to enhance their physicochemical properties and nutrient availability (Mupambwa et al. 2024). Seaweed-derived biochar contains high ash content and supplies essential cations, enhancing nutrient retention and stabilizing pH in otherwise inert growing media (Katakula et al. 2020). Despite these developments, challenges such as substrate compaction, emitter clogging, and inconsistent water movement in porous materials affect the system performance (Schindler and Eulenstein 2017). Therefore, the selection of appropriate substrates for drip irrigation involves balancing multiple factors, such as water retention, aeration, nutrient availability, and resistance to compaction or emitter clogging. Lastly, blends enriched with coir, peat, biochar, compost, fly ash, rock phosphate, perlite, vermiculite, or marine-based amendments offer specific advantages; however, ongoing research is needed to evaluate their performance across diverse crops, climates, and irrigation regimes. Among these organic options, vermicompost has attracted growing interest in drip irrigation systems because its fine texture and high microbial activity can improve nutrient retention and structural stability while minimizing leaching losses (Fig. 3.1). Therefore, the next section focuses on vermicompost as a promising organic amendment, both alone and in combination with its coamendments in improving the physical and chemical properties of horticultural growing media.

3.1.3 Vermicompost as an Amendment to Horticultural Media

Vermicompost, the product of organic waste decomposition by earthworms, is recognized for its ability to enhance the microbial activity, nutrient availability, and physical structure of horticultural growing media (Ajibade et al. 2022). Its fine texture and rich organic content improve the cation exchange capacity (CEC), porosity, and moisture retention, which are essential for supporting root health and reducing nutrient loss under irrigation (Ajibade et al. 2022; Messiga et al. 2020). However, vermicompost is prone to nutrient leaching, particularly under frequent irrigation or fertigation regimes, due to its soluble nutrient fractions and limited buffering capacity, which can reduce nutrient use efficiency and potentially contribute to environmental contamination (Messiga et al. 2020).

In addition, vermicompost also faces limitations due to its high moisture content, variable nutrient composition, and potential salinity issues (Kakitis et al. 2017; Mupambwa et al. 2024). These drawbacks necessitate blending vermicompost with coamendments such as biochar, fly ash, or rock phosphate to balance its properties, enhance nutrient retention, and improve the overall media stability (Fig. 3.1).

Several studies have demonstrated that incorporating these coamendments into vermicompost improves nutrient availability and media structure and reduces leaching risks (Antileo-Mellado et al. 2024; Mupambwa et al. 2024; Mupondi et al. 2018).

3.1.3.1 Fly Ash Amended Vermicompost

Fly ash, a coal combustion byproduct, has been increasingly explored as a coamendment for vermicompost because of its abundance in essential minerals, such as calcium, magnesium, potassium, and phosphorus (Mupambwa et al. 2024). Fly ash can help neutralize the slightly acidic pH of vermicompost and improve substrate structure and aeration, thereby enhancing the nutrient profile and buffering capacity of vermicompost-based growing media, making them suitable for high-efficiency fertigation systems (Mupambwa et al. 2024). This integration is beneficial for growing media with low cation exchange capacity or unstable pH, such as cocopeat, perlite, or coir-based substrates, commonly used in drip irrigation systems (Martins et al. 2023; Zulfiqar et al. 2021).

In a study by Mupambwa et al. (2024), integrating fly ash with cow manure-based vermicompost increased Olsen-P to 239.13 mg/kg and raised Zn content by 51.5% in a cocopeat-based growing medium. These improvements in nutrient composition, along with enhanced substrate structure and aeration, supported seedling development and suggest potential for more sustainable outcomes in fertigation systems. Despite the benefits of fly ash, safety measures prior to application are necessary in agronomic systems due to its potential contributions of heavy metals and salts, underscoring the importance of standardizing fly ash application rates (Mupambwa et al. 2024).

3.1.3.2 Rock Phosphate Amended Vermicompost

Rock phosphate (RP) is a naturally occurring mineral that serves as a slow-release phosphorus source, especially in acidic media, where P fixation is common (Ajibade et al. 2022). Although rock phosphate is cost-effective and environmentally friendly, its low solubility, particularly in neutral to alkaline and calcareous media conditions, limits its immediate bioavailability to plants (Burni et al. 2023). Researchers have reported that these challenges can be mitigated through coapplication with biofertilizers or organic amendments, as well as by integrating rock phosphate into vermicomposting systems (Abbasi et al. 2015; Mihreteab et al. 2016). Coapplication of rock phosphate with organic amendments, particularly via vermicomposting, enhances microbial-driven phosphorus solubilization through the production of organic acids and phosphatase enzymes (Abbasi et al. 2015; Mupondi et al. 2018). This interaction significantly enhances phosphorus availability and contributes to sustained nutrient release for plant uptake, ensuring optimal nutrient balance, improved seedling growth, and stability of horticultural media (Ajibade et al. 2022).

Researchers have explored the application of rock phosphate in integration with vermicompost. For example, Ajibade et al. (2022) reported that enriching cow and pig manure vermicomposts with rock phosphate and phosphorus-solubilizing bacteria improved Swiss chard growth and tissue phosphorus content. In another study, Mihreteab et al. (2016) reported that rock phosphate-enriched compost effectively supplied phosphorus to tomato seedlings, comparable to rock phosphate-amended substrates, highlighting the potential of integrating rock phosphate into compost-based growing media. In addition, incorporating rock phosphate into compost media improves structural integrity and phosphorus availability, making it a viable alternative to synthetic fertilizers in sustainable horticulture and a climate-smart strategy for enhancing organic vegetable production (Ajibade et al. 2022; Mihreteab et al. 2016). Existing literature on rock phosphate application in agriculture primarily focuses on its coapplication with organic amendments, such as manure or direct soil incorporation, rather than its direct integration with vermicompost (Abbasi et al. 2015; Burni et al. 2023; Mupondi et al. 2018). This highlights a gap regarding standardized application rates, effects on diverse crops, and potential risks; future studies could address these gaps to optimize vermicompost-based horticultural systems and support sustainable crop production

3.1.3.3 Biochar-Amended Vermicompost

Biochar is a carbon-rich material derived from pyrolyzed biomass that has a porous structure, high surface area, and strong cation exchange capacity (Prasad et al. 2019). These properties enable biochar to improve water retention, enhance nutrient adsorption, and provide habitats for beneficial microbes in the growing media (Kafle et al. 2024). Researchers have demonstrated that biochar can improve structural stability and enhance nutrient retention, particularly under drip fertigation systems, when incorporated into vermicomposting-based growing media (Antileo-Mellado et al. 2024; Kafle et al. 2024). Antileo-Mellado et al. (2024) reported that incorporating biochar into vermicompost improved organic matter stabilization, increased carbon content, and modulated microbial and enzymatic activity, enhancing the overall quality and nutrient-holding capacity of the compost. Similarly, Kafle et al. (2024) demonstrated that a full softwood biochar-compost blend in cucumber growing media increased fresh shoot weight, expanded leaf area, and elevated substrate pH, highlighting its potential to enhance both seedling performance and media properties. Additionally, biochar reduces nitrate leaching and stabilizes moisture dynamics, making it particularly suitable for containerized crops and drip-irrigated horticulture (Messiga et al. 2020). Overall, integrating biochar into vermicompost not only improves water and nutrient retention but also supports long-term substrate performance and ecological resilience, promoting sustainable horticultural practices (Kafle et al. 2024; Prasad et al. 2019).

3.1.4 Marine-Derived Resources in Crop Production

The marine environment is not only a reservoir for aquatic biodiversity but also a source of valuable biological and mineral inputs directly relevant to agriculture (Ammar et al. 2022; Emadodin et al. 2020; Katakula et al. 2020). Various marine-derived materials, such as seaweed, macroalgae, fish waste, crustacean shells, marine sediments, and corals, can serve as amendments, fertilizers, or biostimulants in agriculture (Emadodin et al. 2020; Katakula et al. 2020). These resources hold significant agronomic potential as they are enriched with essential plant macronutrients (N, P, K, Ca, Mg, and S) and micronutrients (Zn, Fe, Mn, and Cu), while also serving as natural sources of plant growth regulators such as auxins, cytokinins, gibberellins, and abscisic acid (Ammar et al. 2022). Together, these properties contribute to improved soil fertility, nutrient cycling, and enhanced plant physiological functions such as chlorophyll synthesis, enzyme activation, and tolerance to abiotic stress, thereby supporting sustainable crop productivity (Emadodin et al. 2020; Hashem et al. 2019). In addition, marine resources are abundantly available, with seaweeds exhibiting rapid growth rates that are further stimulated by rising atmospheric CO_2 dissolving in the oceans, while substantial quantities of waste materials such as fish bones and shells are continuously generated by processing industries (Antu et al. 2024; Katakula et al. 2020). This underscores the potential of these resources to serve as sustainable alternatives to synthetic fertilizers, while simultaneously advancing circular bioeconomy strategies in agriculture.

Several studies have investigated the agronomic potential of diverse marine-derived resources, including macroalgae, microalgae, crustacean shell residues, seagrass wrack, and jellyfish biomass as biostimulants, biofertilizers, and organic amendments to improve soil quality and crop performance (Ammar et al. 2022; Katakula et al. 2020; Xavier James et al. 2024). Xavier James et al. (2024) demonstrated that macroalgae such as *Sargassum wightii*, *S. longifolium*, *Laminaria digitata*, and *Gelidiella acerosa* can function as biofertilizers and biostimulants when applied in powdered form, improving tomato seedling growth and soil microbial populations. Similarly, Katakula et al. (2020) produced biochar from seaweeds (*Laminaria pallida* and *Gracilariopsis funicularis*) and found it to be rich in essential nutrients and stable carbon, underscoring its potential as an organic amendment for improving soil quality. Hashem et al. (2019) further showed that soil amendments with *Ulva lactuca*, *Cystoseira* spp., and *Gelidium crinale* enhanced canola growth and salt stress tolerance, effects linked to improved photosynthetic pigments, osmoprotectants, and plant hormones. Ananthavali et al. (2019) demonstrated that predecomposed seaweeds cocomposted with cow dung produced nutrient-rich vermicompost with higher microbial populations and increased availability of nitrogen (N), phosphorus (P), and potassium (K), confirming the potential of seaweeds as multifunctional organic fertilizers.

Beyond algae, studies highlight the underexplored value of other marine biomasses, for example, Emadodin et al. (2020) reported that seagrass wrack improves soil water-holding capacity and nutrient availability, while jellyfish biomass enhances soil organic matter and seedling survival. Complementary to these findings, Hossain and Shahidi (2024) characterized crab and shrimp shell residues as rich in proteins, chitin, minerals, and carotenoids, underscoring their potential as soil amendments, although direct agronomic applications remain limited. Ammar et al. (2022) reported that microalgae and cyanobacteria, including *Chlorella vulgaris*, *Spirulina platensis*, *Scenedesmus dimorphus*, *Anabaena*, and *Nostoc*, contribute to soil fertility through nitrogen fixation, carbon enrichment, and phosphate solubilization. In addition, they also improve soil texture, aeration, and organic matter content, while reducing heavy metal accumulation and erosion.

Despite the benefits, marine-derived resources may pose challenges such as heavy metal contamination and high salinity, emphasizing the need for careful pretreatment such as vermicomposting or desalination prior to soil application to ensure safety for plants, soil, and humans (Katakula et al. 2020). Many marine materials remain insufficiently characterized; with limited information on their macronutrient and trace element profiles (Table 3.2). Similarly, while seawater and seawater-derived products are being explored as nutrient sources and feedstocks for green fertilizers or saline agriculture, their high salinity limits direct use, and detailed assessments of agriculturally relevant cations extraction or purification remain scarce. Taken together, the literature indicates that marine resources, including algae, seagrass wrack, jellyfish, crustacean shells, sediments, and seawater derivatives, can enhance soil fertility and plant productivity via nutrient supply, organic matter enrichment, hormone modulation, and stress mitigation. Future research should focus on optimizing application rates, mitigating salinity and contamination risks, and thoroughly characterizing different species and material types to support their safe and sustainable integration into crop production systems.

3.1.5 *Biochar from Seaweeds*

The conversion of seaweed biomass into biochar offers a promising pathway for carbon sequestration and the reduction of greenhouse gas emissions, owing to its potential for stable carbon retention and the sustainable use of underutilized marine biomass (Chen et al. 2022). Seaweeds such as *L. pallida*, *G. funicularis*, and *Macrocystis* spp. are promising feedstocks due to their abundance, rapid growth, high nutrient-accumulation capacity, and broad distribution in coastal ecosystems, offering advantages over terrestrial biomass (Dang et al. 2023). In addition, seaweeds are rich in macroelements, particularly N and organic C, as well as minerals such as K, Ca, Mg, and Cl, imparting seaweed-derived biochar with unique characteristics, compared to terrestrial biomass-derived biochar (Katakula et al. 2020). This highlights that seaweeds are eco-friendly and sustainable sources of biochar, with potential applications in wastewater management and soil fertility enhancement, while simultaneously contributing to carbon sequestration (Katakula et al. 2022; Tan et al. 2023).

Table 3.2 Marine resources resources and their potential nutrient composition for agricultural use

Marine resource	Form	Total C (%)	Total N (g/kg)	P (g/kg)	K (g/kg)	Na (g/kg)	Ca (g/kg)	Mg (g/kg)	S (g/kg)	Ash (%)	Protein (%)	Micronutrients (mg/kg)	Reference
Sargassum linearifolium	Dried biomass	–	–	–	68.0 ± 1.9	70.0 ± 1.3	3.0 ± 0.2	9.0 ± 0.75	–	24 ± 2	7 ± 1	Fe: 3 ± 0.01 Zn: 25 ± 0.11, Cd: 1 ± 0.01, Ni: 6.0 ± 0.1	Choudhary et al. (2023)
Ulva lactuca	Dried biomass	–	–	–	30.0 ± 2.2	20.0 ± 1	0.62 ± 0.2	46.0 ± 1.3	–	16 ± 2	6 ± 1	Zn: 40 ± 0.1	Choudhary et al. (2023)
Ulva faciata	Dried biomass	–	–	–	88.0 ± 3	22.0 ± 1	1.60 ± 0.3	48.0 ± 1.3	–	27 ± 1	36 ± 1	Zn: 30 ± 0.1	Choudhary et al. (2023)
Palmaria palmata	Dried biomass	–	2.5	2.35	70.0–90.0	16.0–25.0	5.60–12.0	1.70–6.10	–	12–37	8–35	Fe: 500, Zn: 28.6, Cu: 3.73	Pereira (2011)
Laminaria digitata	Dried biomass	–	–	–	115.79	38.18	10.05	6.59	–	38	8–15	Fe: 3.29, Cu: <5.0, Zn: 17.7	Pereira (2011)
Shrimp Shell powder	Powder	54.54	–	12.6	14	15.3	75.8	3.7	–	36.58	37.78	Cr: 2.2, Fe: 0.04, Cu: <10–100	Antu et al. (2024)
Pangas bone powder	Powder	25.94	–	129	0.1	10.4	222.8	4.3	–	~62.80	32.29	Fe: 0.2, Cu: < 10–100	Antu et al. (2024)
Tuna bone powder	Powder	22.19	–	138.1	0.2	19	245.2	6.4	–	~62.80	22.19	Fe: 0.2, Cu: < 10–100, Cr: 2.0	Antu et al. (2024)
Pink Crab (*Chionoecetes opilio*)	Powder	–	–	–	–	–	>600	–	–	152.3–201.7	76.3–155.0	Cu: 1–13, Fe: 26–141	Hossain et al. (2024)
Northen Shrimp (*Pandalus borealis*)	Powder	–	–	–	–	–	>600	–	–	78.3	107.3	Cu: 1–13, Fe: 26–141	Hossain et al. (2024)

(continued)

Table 3.2 (continued)

Marine resource	Form	Total C (%)	Total N (g/kg)	P (g/kg)	K (g/kg)	Na (g/kg)	Ca (g/kg)	Mg (g/kg)	S (g/kg)	Ash (%)	Protein (%)	Micronutrients (mg/kg)	Reference
Gracilaria gracilis	Dried biomass	–	17.4	0.57	13.8	2.91	4.29	–	–	6.78	10.86	Fe: 152	Rasyid et al. (2019)
Tamban (Sardinella fimbriata)	Bone meal	–	28.1	105.59	1.29	–	125.05	249	21	–	–	Zn: 1260.0, Cu: 1.0, Mn: 680.0	Khairul et al. (2025)
Parang *(Chirocentrus dorab)*	Bone meal	–	26.8	100.48	2.99	–	122.71	2.31	11	–	–	Zn:850.0, Cu: 3.0, Mn: 65	Khairul et al. (2025)
Sealayang *(Decapterus tabl)*	Bone meal	–	30.2	99.94	1.9	–	110.67	2.26	17	–	–	Zn: 710, Cu: 2.0, Mn: 10.0	Khairul et al. (2025)
Macrocystis pyrifera	Dry biomass	–	–	2.97–4.66	102–156	44.3–57.9	17.3	8.69–10.82	–	22–36	7.1	Ni: 0.32, Pb: 0.26, Al: 43.3, Zn: 19.0, Cr: 0.32, Mn: 3.15, Cd: 1.16, As: 121	Pérez-Alva et al. (2024), Biancacci et al. (2022)
Pelagic clay sediment	Sediment	0.45	0.6	1.58	25	28	10	21	2	–	–	Cd:0.42, Cr: 90, Cu: 250, Fe: 65000, Mn: 6700, Ni: 230, Pb: 80	Li and Schoonmaker (2003)
Basal sediment	Sediment	–	–	–	11.5	25.6	14.7	20.8	–	–	–	Cd: 0.4, Cr: 15, Cu: 790, Fe: 200000, Mn: 61000, Ni: 460, Pb:100	Li and Schoonmaker (2003)

Ridge sediment	Sediment	2.1	0.1	138	4.2	4.5	314	1.8	7.2	–	–	Cd:18, Cr: 125, Cu: 5, Fe: 7700, Mn:1200, Ni: 53, Pb: 50	Li and Schoonmaker (2003)
Laminaria pallida	Biochar	26.1	21.5	0.14	185	38.7	23.3	9.3	–	–	–	Cd: 0.5, As: 48.0, Ni: 5.2, Cr: 4.6, Pb: >3.9, Zn: 167.9	Katakula et al. (2020)
Eelgrass (*Zostera marina L.*)	Dried biomass	36.0	25.0	1.7	7.0	10.0	24.8	11.5	6.0	–	–	Fe:4500 mg/kg, Mn:800 mg/kg, Cu:30 mg/kg, Zn:100 mg/kg	Emadodin et al. (2020)
Jellyfish (*Aurelia aurita*)	Dried biomass	27.3	78.0	10.0	4.4	92.0	7.5	10.0	–	–	–	Fe:400 mg/kg, Mn:<100 mg/kg, Cu:<100 mg/kg, Zn:400 mg/kg	Emadodin et al. (2020)
Gracilariopsis funicularis	Biochar	38.3	43.7	1.2	174.9	16.7	31.2	12.4	–	–	–	Cd: 7.5, As: 5.0, Ni: 2.0, Cr: 2.5, Pb: 7.4, Zn: 28.1	Katakula et al. (2020)
Seawater	Seawater	30	420		0.3953	10.679	0.4096	1.278	898	–	–	Fe: 123, Cu: 108, Ni: 54, Cr: 42.5, Al: 500, Cd: 76, Mn: 72, Pb: 2.7	Mohd Nani et al. (2016), Kumar and Kumar Singh (2023)

Note: All macronutrient values are reported in g/kg, micronutrients in mg/kg, and ash/protein as %. Total carbon is presented where available. Micronutrient content includes all trace metals reported by each source (Cd, Pb, Ni, Cr, As, Cu, Al, Mn, Fe). Dashes (–) indicate unavailable data. Values were extracted or converted directly from the original articles. Where multiple treatments were presented, the most agronomically relevant or mid-range value was used

According to Chen et al. (2022), biochar from seaweeds can be produced through several thermochemical pathways, such as conventional pyrolysis, hydrothermal carbonization, microwave pyrolysis, and copyrolysis, each yielding products with distinct properties that suit different applications. Conventional pyrolysis at low temperatures enhances functional groups favorable for soil fertility, while high-temperature processes improve porosity and stability but reduce yield (Chen et al. 2022). Hydrothermal carbonization, which utilizes wet feedstocks, is cost-effective and environmentally friendly but limited by long reaction times; nevertheless, the resulting hydrochar is suitable for agricultural applications due to its nutrient content (Chen et al. 2022). On the other hand, microwave pyrolysis, an alternative method for producing seaweed biochar, offers faster processing and greater energy efficiency; however, the resulting biochar has lower structural quality, which may limit its agronomic applications (Chen et al. 2022).

Generally, biochar from seaweeds, characterized by high N-content, high ash content, alkaline pH, and high EC (Katakula et al. 2020). In contrast, these properties may vary depending on pyrolysis temperature, production pathway, and feedstock used (Katakula et al. 2020). Elevated nitrogen contents in N-rich seaweed-derived biochar make it a promising amendment, with improved electron transfer and strong adhesion capacity, thereby enhancing its catalytic potential in nitrogen cycling processes, which is significant for improving soil fertility (Tan et al. 2023). Despite feedstock type, pyrolysis conditions such as duration and temperature are critical factors that determine the quality and quantity of the resulting seaweed biochar (Katakula et al. 2020; Roberts and De Nys 2016). For instance, within a pyrolysis temperature of 300–450 °C, biochar derived from *Ulva*, *Laminaria pallida*, and *Gracilariopsis funicularis* seaweeds shows better nutrient retention, particularly of N and P, and enhances plant performance (Katakula et al. 2020; Roberts and De Nys 2016). Meanwhile, at higher pyrolysis temperatures (800 °C), biochar from *L. pallida* and *G. funicularis* exhibited an approximate 50% yield reduction, elevated pH and EC, and Cd levels above permissible limits (3.9 mS/cm), indicating agronomic constraints (Katakula et al. 2020).

The alkaline nature of seaweed-derived biochar makes it valuable to use as a liming material to adjust pH in acidic soils. In contrast, its high ash content and electrical conductivity are associated with increased salinity, which may pose risks that hinder agricultural productivity (Katakula et al. 2020). A further concern regarding seaweed-derived biochar is the possible accumulation of heavy metals (Cd, Zn, Cr, Pb, and Cu), which may pose risks to soil, plants, and ultimately human health (Katakula et al. 2020; Roberts and De Nys 2016). To address these concerns, pretreatments such as rinsing, drying, and pelletizing are recommended to lower salt content and enhance performance, while post-production activation improves adsorption capacity and mitigates the mobility of heavy metals (Chen et al. 2022; Roberts and De Nys 2016). Further research has focused on integrating seaweed biochar with other organic amendments to buffer salinity and optimize nutrient dynamics. Katakula et al. (2022) evaluated the incorporation of *G. funicularis*-derived biochar into goat manure and vegetable food waste-derived vermicompost at different rates (0–8%), observing that higher rates reduced the salinity and pH, while improving nutrient availability, particularly N, K, and Mg.

Further studies are needed to explore a wider range of seaweed species beyond *L. pallida* and *G. Funicularis and Ulva* as feedstock biochemical compositions; hence, the physicochemical characteristics of the resulting biochar can vary considerably. In addition, given the possibility of overharvesting, environmental impact assessments are necessary to evaluate the long-term ecological implications of large-scale seaweed biomass extraction, alongside economic assessments of large-scale production. Finally, optimized seaweed biochar-amended fertilizers should be tested on a wide range of vegetables in both soil and soilless systems to ensure the generalizability and scalability of their benefits, with more experimental work on salinity mitigation and regulatory aspects. Collectively, these findings illustrate how marine biomass valorization extends beyond solid forms such as seaweed-derived biochar to include advanced applications like the production of green hydrogen and hydrogen-based fertilizers from seawater. The next section, "Green Fertilizers from Marine Resources," further explores these emerging innovations, revealing complementary pathways for developing climate-resilient and circular agricultural systems.

3.1.6 Green Fertilizers from Marine Resources

Green fertilizers from marine resources have attracted the attention of researchers and farmers as sustainable and ecologically friendly substitutes for synthetic chemical fertilizers, which are unfriendly to both the environment and human health (Alagu et al. 2024; Robinson and Stapelmann 2024). Novel green fertilizers from marine resources encompass a wide spectrum, including the use of renewable marine biomass inputs to produce clean, environmentally friendly, and sustainable fertilizers, such as biofertilizers and organic amendments, as discussed in Sect. 3.1.4 (Emadodin et al. 2020). In contrast, emerging technologies use renewable-energy-driven water electrolysis to produce green hydrogen, which can be converted into green hydrogen-based fertilizers, representing a sustainable pathway for green fertilizer production in agriculture (Adeli et al. 2023; Alagu et al. 2024). The production of green fertilizers involves reacting green hydrogen with atmospheric nitrogen through the Haber-Bosch process to yield green ammonia (NH_3) (Lubello et al. 2025; Szymańska et al. 2019). This ammonia can be applied directly as a nitrogen source or converted into more stable fertilizers, such as ammonium sulfate, thereby providing an environmentally friendly alternative to fossil-fuel-derived fertilizers (Lubello et al. 2025; Szymańska et al. 2019). These transformations occur without fossil fuel inputs, drastically lowering the carbon footprint and environmental degradation compared with conventional fertilizer manufacturing (Adeli et al. 2023). Conventional hydrogen production from fossil fuels for ammonia synthesis releases about 900 Mt. of CO_2 annually, accounting for nearly 1.8% of global emissions, whereas green ammonia production is carbon-free, demonstrating the environmental advantages of green pathways (Adeli et al. 2023; Zhang et al. 2024).

A notable example of green hydrogen-based fertilizer production is the Daures Green Hydrogen Project in Namibia's Erongo Region, which seeks to integrate green hydrogen and green ammonia initiatives and is projected to produce ammonia and ammonium sulfate fertilizers by 2026 (Alagu et al. 2024). In contrast, in the long run, the production of green fertilizers using water-electrolyzed hydrogen may be challenging in Namibia and other arid regions due to the scarcity of freshwater, with approximately 65% of Namibia's freshwater sourced from underground reserves (Alagu et al. 2024). This has prompted the exploration of alternative sustainable water sources, such as the use of seawater for hydrogen production, given that seawater is abundant, constituting about 96.5% of global water and particularly accessible in coastal regions, including Namibia (Zhang et al. 2024). Although seawater offers a promising source for hydrogen, its use is challenged by salts, mineral deposits, microorganisms, and the need for durable catalysts to ensure reliable green fertilizer production (Gao et al. 2022; Zhang et al. 2024). Methods such as pretreatment to remove sediments and microorganisms, desalination, pH stabilization, development of corrosion-resistant and selective catalysts, and advanced electrolyzer designs are being proposed to make seawater-based hydrogen production feasible and sustainable (Gao et al. 2022; Zhang et al. 2024). Frisch et al. (2023) demonstrated that seawater can be used for hydrogen production by employing PGM-free catalysts and an asymmetric-feed electrolyzer, achieving current densities up to 1.0 A cm^{-2} at voltages below 2.0 V, highlighting the feasibility of sustainable seawater electrolysis.

While Frisch et al. (2023) demonstrated the viability of using seawater as a feedstock for green hydrogen production, significant gaps remain. Firstly, there is a lack of country-specific studies, such as those in Namibia that assess localized seawater pretreatment and electrolysis for green hydrogen production as an initial step toward its conversion into green fertilizers. In addition, actual experimental trials focused on producing green ammonium fertilizers such as ammonium sulfate and ammonium nitrate from seawater-derived hydrogen are limited, leaving uncertainties around process efficiency, product quality, and scalability. Another unexplored area concerns the actual agronomic dimension applications of these green fertilizers across a wide range of crops under diverse soil and climatic conditions. Finally, there is limited evidence on how such fertilizers can best be managed through optimized rates, timing, and application methods to maximize yield while minimizing environmental losses. Addressing these gaps will be essential to translate laboratory-scale feasibility into practical, regionally relevant solutions for sustainable agriculture.

3.2 Conclusions and Recommendations for Future Studies

In conclusion, this book chapter evaluated the types, limitations, and amendment strategies of horticultural growing media, focusing on drip-irrigated and horticultural systems. The literature demonstrates the potential of organic and inorganic amendments, including vermicompost, fly ash, rock phosphate, and biochar, to enhance physicochemical properties, such as nutrient retention, pH buffering, water-holding capacity, and microbial colonization. Their combined application, particularly under drip fertigation, has demonstrated superior performance in improving nutrient efficiency and microbial activity. This chapter also explores marine biomass-based resources with agronomic potential, especially seaweed, organic wastes, and seawater-derived green fertilizers, through hydrogen production pathways, which present promising sustainable alternatives despite challenges such as salinity and processing costs. Although the literature offers valuable insights into the potential advantages and limitations of using marine resources in growing media, there is a significant gap in research focused on their application in biochar production for agricultural use. This includes the desalination of marine biomass to enhance its suitability for agriculture and the incorporation of marine-derived biochar into vermicomposting systems to study nutrient dynamics and fertilizer use efficiency in crops. It is crucial to understand the interactions among amendment chemistry, irrigation frequency, and microbial communities under different irrigation regimes. Environmental and economic assessments of marine biomass harvesting and biochar production are required to validate sustainability. In addition, inconsistencies in amendment rates and experimental protocols limit the comparability of findings, and long-term data on structural integrity, microbial dynamics, and nutrient leaching under field conditions are scarce. Therefore, future research should prioritize standardized, long-duration trials, with a particular focus on marine biomass-based amendments, to validate their agronomic and environmental sustainability. Integrating marine-derived fertilizers with organic amendments can advance circular horticultural systems, reduce dependence on synthetic inputs, and enhance resource efficiency.

Declaration of Generative AI and AI-Assisted Technologies in the Writing Process During the preparation of this work, the author(s) used Open Paper Pal in order to improve the English readability of the manuscript. After using this tool/service, the author(s) reviewed and edited the content as needed and take(s) full responsibility for the content of the publication.

Funding The study was funded by the Daures Green Hydrogen Village Project under the University of Namibia (R404), and also partly funded by the New Zealand Government to support the objectives of the Global Research Alliance on Agricultural Greenhouse Gases under the CarouNd project under the University of Namibia 2023-EJPSOILS-C-AROUND-UN (R411).

References

Abbasi MK, Musa N, Manzoor M (2015) Mineralization of soluble P fertilizers and insoluble rock phosphate in response to phosphate-solubilizing bacteria and poultry manure and their effect on the growth and P utilization efficiency of chilli (*Capsicum annuum* L.). Biogeosciences 12(15):4607–4619. https://doi.org/10.5194/bg-12-4607-2015

Abedin T, Yamamoto A, Hayashi T, Hosokawa M (2021) Drip fertigation enhances the growth of hydroponic lettuce (Lactuca sativa) using polyester fiber substrate. Sci Hortic 276:109604. https://doi.org/10.1016/j.scienta.2020.109604

Adeli K, Nachtane M, Faik A, Saifaoui D, Boulezhar A (2023) How green hydrogen and ammonia are revolutionizing the future of energy production: a comprehensive review of the latest developments and future prospects. Appl Sci 13(15):8711. https://doi.org/10.3390/app13158711

Ajibade S, Mupambwa HA, Manyevere A, Mnkeni PNS (2022) Vermicompost amended with rock phosphate as a climate smart technology for production of organic Swiss Chard (Beta vulgaris subsp. Vulgaris). Front Sustain Food Syst 6:757792. https://doi.org/10.3389/fsufs.2022.757792

Alagu AR, Boumrifak C, González LFL (2024) GreeN-H_2 Namibia: green hydrogen production in Namibia [Report]. https://dechema.de/green_hydrogen_namibia_report/_/H2-Report_GreeN-H2_digital.pdf

Alvarez JM, Pasian C, Lal R, López R, Fernández M (2019) Vermicompost and biochar substrates can reduce nutrients leachates on containerized ornamental plant production. Hortic Bras 37(1):47–53. https://doi.org/10.1590/s0102-053620190107

Ammar EE, Aioub AAA, Elesawy AE, Karkour AM, Mouhamed MS, Amer AA, EL-Shershaby NA (2022) Algae as bio-fertilizers: between current situation and future prospective. Saudi J Biol Sci 29(5):3083–3096. https://doi.org/10.1016/j.sjbs.2022.03.020

Ananthavalli R, Ramadas V, John Paul JA, Karunai Selvi B, Karmegam N (2019) Seaweeds as bioresources for vermicompost production using the earthworm, Perionyx excavatus (Perrier). Bioresour Technol 275:394–401. https://doi.org/10.1016/j.biortech.2018.12.091

Antileo-Mellado S, Muñoz C, Sanchez-Hernandez JC, Ginebra M, Sandoval M (2024) Effect of biochar on vermicompost production: chemical, biochemical, and biological properties. Agronomy 14(3):615. https://doi.org/10.3390/agronomy14030615

Antu MAR, Ali MS, Ferdous MJ, Ahmed MT, Ali MR, Suraiya S, Pangestuti R, Haq M (2024) Recovery and characterization of calcium-rich mineral powders obtained from fish and shrimp waste: a smart valorization of waste to treasure. Sustainability 16(14):6045. https://doi.org/10.3390/su16146045

Badreshiya D, Patel GD, Bhatt ST (2024) Different types of nursery growing media. Just Agric 4(8)

Biancacci C, Sanderson JC, Evans B, Callahan DL, Francis DS, Skrzypczyk VM, Cumming EE, & Bellgrove A (2022) Variation in biochemical composition of wild-harvested *Macrocystis pyrifera* (Ochrophyta) from sites proximal and distal to salmon farms in Tasmania, Australia. *Algal Research*, 65:102745. https://doi.org/10.1016/j.algal.2022.102745

Burni T, Hussain F, Bibi S, Ullah R, Lalay G (2023) The salutary impacts of AMF species, rock phosphates (RP), and organic matter (FYM) fertilizers on the development and chemical behavior of Mentha arvensis L. Acta Ecol Sin 43(5):835–841. https://doi.org/10.1016/j.chnaes.2022.12.003

Chen J, Lü S, Zhang Z, Zhao X, Li X, Ning P, Liu M (2018) Environmentally friendly fertilizers: a review of materials used and their effects on the environment. Sci Total Environ 613–614:829–839. https://doi.org/10.1016/j.scitotenv.2017.09.186

Chen B, Gu Z, Wu M, Ma Z, Lim HR, Khoo KS, Show PL (2022) Advancement pathway of biochar resources from macroalgae biomass: a review. Biomass Bioenergy 167:106650. https://doi.org/10.1016/j.biombioe.2022.106650

Choudhary B, Khandwal D, Gupta NK, Patel J, Mishra A (2023) Nutrient composition, physicobiochemical analyses, oxidative stability and antinutritional assessment of abundant tropical seaweeds from the Arabian Sea. Plants 12(12):2302. https://doi.org/10.3390/plants12122302

Dang B-T, Ramaraj R, Huynh K-P-H, Le M-V, Tomoaki I, Pham T-T, Hoang Luan V, Thi Le Na P, Tran DPH (2023) Current application of seaweed waste for composting and biochar: a review. Bioresour Technol 375:128830. https://doi.org/10.1016/j.biortech.2023.128830

Emadodin I, Reinsch T, Rotter A, Orlando-Bonaca M, Taube F, Javidpour J (2020) A perspective on the potential of using marine organic fertilizers for the sustainable management of coastal ecosystem services. Environ Sustain 3(1):105–115. https://doi.org/10.1007/s42398-020-00097-y

Erdal I, Aktas H, Kucukyumuk Z, Daler S, Özen Ö, Gencer K (2014) Effects of different growing mediums on nutrient concentration of eggplant in soilless culture. In: Technological advancement for vibrant agriculture. Athens Institute for Education and Research

Frisch ML, Thanh TN, Arinchtein A, Hager L, Schmidt J, Brückner S, Kerres J, Strasser P (2023) Seawater electrolysis using all-PGM-free catalysts and cell components in an asymmetric feed. ACS Energy Lett 8(5):2387–2394. https://doi.org/10.1021/acsenergylett.3c00492
Gao F-Y, Yu P-C, Gao M-R (2022) Seawater electrolysis technologies for green hydrogen production: challenges and opportunities. Curr Opin Chem Eng 36:100827. https://doi.org/10.1016/j.coche.2022.100827
Gbollie SN, Mwonga SM, Kibe AM (2021) Effects of calcium nitrate levels and soaking durations on cocopeat nutrient content. J Agric Chem Environ 10(03):372–388. https://doi.org/10.4236/jacen.2021.103024
Hashem HA, Mansour HA, El-Khawas SA, Hassanein RA (2019) The potentiality of marine macro-algae as bio-fertilizers to improve the productivity and salt stress tolerance of canola (Brassica napus L.) plants. Agronomy 9(3):146. https://doi.org/10.3390/agronomy9030146
Hauck D, Lohr D, Meinken E, Schmidhalter U (2021) Plant availability of secondary phosphates depending on pH in a peat-based growing medium. Acta Hortic 1305:437–442. https://doi.org/10.17660/ActaHortic.2021.1305.57
Hernández-Rodríguez A, Robles-Hernández L, Ojeda-Barrios D, Prieto-Luévano J, González-Franco AC, Guerrero-Prieto V (2017) Semicompost and vermicompost mixed with peat moss enhance seed germination and development of lettuce and tomato seedlings. Interciencia 42:774
Hossain A, Shahidi F (2024) Upcycling shellfish waste: distribution of amino acids, minerals, and carotenoids in body parts of North Atlantic crab and shrimp. Foods 13(17):2700. https://doi.org/10.3390/foods13172700
Huang L, Yu P, Gu M (2019) Evaluation of biochar and compost mixes as substitutes to a commercial propagation mix. Appl Sci 9(20):4394. https://doi.org/10.3390/app9204394
Incrocci L, Massa D, Pardossi A (2017) New trends in the fertigation management of irrigated vegetable crops. Horticulturae 3(2):37. https://doi.org/10.3390/horticulturae3020037
Jahan S, Ahmad F, Rasul F, Amir R, Shahzad S (2023) Physicochemical analysis of vermicompost-perlite based activated biochar and its influence on wheat (Triticum aestivum L.) growth under water stress. J Soil Sci Plant Nutr 23(3):3034–3050. https://doi.org/10.1007/s42729-023-01258-8
Kafle A, Singh S, Singh M, Venkataramani S, Saini R, Deb S (2024) Effect of biochar-compost amendment on soilless media properties and cucumber seedling establishment. Technol Hortic 4(1). https://doi.org/10.48130/tihort-0023-0029
Kakitis A, Vronskis O, Nulle I (2017, May 24) Experimental study of vermicompost drying process. In: 16th International Scientific Conference Engineering for Rural Development. https://doi.org/10.22616/ERDev2017.16.N229
Katakula AAN, Gawanab W, Itanna F, Mupambwa HA (2020) The potential fertilizer value of Namibian beach-cast seaweed (Laminaria pallida and Gracilariopsis funicularis) biochar as a nutrient source in organic agriculture. Sci Afr 10:e00592. https://doi.org/10.1016/j.sciaf.2020.e00592
Katakula AAN, Gawanab W, Handura B, Itanna F, Mupambwa HA (2022) Seaweed (Gracilariopsis funicularis) biochar incorporation into a goat manure–food waste vermicompost for optimized vermidegradation and nutrient release. Front Sustain Food Syst 6:1005740. https://doi.org/10.3389/fsufs.2022.1005740
Khairul UT, Idris NIM, Shah RM, Nawi IHM, Che Soh N (2025) Evaluation of minerals composition in fish bone meal as organic fertilizer development for sustainable environment. Curr World Environ 19(3):1260–1268. https://doi.org/10.12944/CWE.19.3.17
Kraamwinkel CT, Beaulieu A, Dias T, Howison RA (2021) Planetary limits to soil degradation. Commun Earth Environ 2(1):249. https://doi.org/10.1038/s43247-021-00323-3
Kumar A, Kumar Singh R (2023) Water quality assessment in terms of major and minor elements in surface, ground and sea water and correlating the presence with associated problems. In: Joseph D (ed) Trace metals in the environment. IntechOpen. https://doi.org/10.5772/intechopen.1001129

Li Y-H, Schoonmaker JE (2003) Chemical composition and mineralogy of marine sediments. In: Treatise on geochemistry. Elsevier, pp 1–35. https://doi.org/10.1016/B0-08-043751-6/07088-2

Lubello P, Oduor J, Nganga A, Mutembei M, Njoka F, Akute M, Mungai K, Pye S (2025) Green hydrogen futures in LMICs: opportunities for fertilizer and steel production in Kenya. iScience 28(4):112298. https://doi.org/10.1016/j.isci.2025.112298

Martins T, Machado R, Alves-Pereira I, Ferreira R, Gruda N (2023) Coir-based growing media with municipal compost and biochar and their impacts on growth and some quality parameters in lettuce seedlings. Horticulturae 9(1):105. https://doi.org/10.3390/horticulturae9010105

Messiga AJ, Hao X, Dorais M, Bineng CS, Ziadi N (2020) Supplement of biochar and vermicompost amendments in coir and peat growing media improves N management and yields of leafy vegetables. Can J Soil Sci 102(1):39–52. https://doi.org/10.1139/cjss-2020-0059

Messiga AJ, Hao X, Ziadi N, Dorais M (2021) Reducing peat in growing media: impact on nitrogen content, microbial activity, and CO_2 and N_2 O emissions. Can J Soil Sci 102(1):77–87. https://doi.org/10.1139/cjss-2020-0147

Mihreteab HT, Ceglie FG, Aly A, Tittarelli F (2016) Rock phosphate enriched compost as a growth media component for organic tomato (*Solanum lycopersicum* L.) seedlings production. Biol Agric Hortic 32(1):7–20. https://doi.org/10.1080/01448765.2015.1016114

Mohd Nani SZ, Majid FAA, Jaafar AB, Mahdzir A, Musa MN (2016) Potential health benefits of deep sea water: a review. Evid Based Complement Alternat Med 2016(1):6520475. https://doi.org/10.1155/2016/6520475

Mupambwa HA, Gawanab W, Handura B (2024) Fly ash amended vermicompost as a growing medium for tomato (Solanum lycopersicum) and Swiss chard (Beta vulgaris): implications on growth and heavy metal uptake. Discover Agric 2(1):96. https://doi.org/10.1007/s44279-024-00119-x

Mupondi LT, Mnkeni PNS, Muchaonyerwa P, Mupambwa HA (2018) Vermicomposting manure-paper mixture with igneous rock phosphate enhances biodegradation, phosphorus bioavailability and reduces heavy metal concentrations. Heliyon 4(8):e00749. https://doi.org/10.1016/j.heliyon.2018.e00749

Nemati MR, Simard F, Fortin J-P, Beaudoin J (2015) Potential use of biochar in growing media. Vadose Zone J 14(6):1–8. https://doi.org/10.2136/vzj2014.06.0074

Pereira L (2011) A review of the nutrient composition of selected edible seaweeds. In: Seaweed: ecology, nutrient composition and medicinal uses. Nova Science Publishers

Pérez-Alva A, Ramírez-Rodrigues MA, Baigts-Allende DK, Martínez-Jiménez LA, Ramírez-Rodrigues MM (2024) *Macrocystis pyrifera*: a potential source of bioactive compounds. Curr Nutr Food Sci 20(4):505–512. https://doi.org/10.2174/1573401319666230601110804

Prasad M, Chrysargyris A, McDaniel N, Kavanagh A, Gruda NS, Tzortzakis N (2019) Plant nutrient availability and pH of biochars and their fractions, with the possible use as a component in a growing media. Agronomy 10(1):10. https://doi.org/10.3390/agronomy10010010

Rajendran S, Domalachenpa T, Arora H, Li P, Sharma A, Rajauria G (2024) Hydroponics: exploring innovative sustainable technologies and applications across crop production, with emphasis on potato mini-tuber cultivation. Heliyon 10(5):e26823. https://doi.org/10.1016/j.heliyon.2024.e26823

Rasyid A, Ardiansyah A, Pangestuti R (2019) Nutrient composition of dried seaweed Gracilaria gracilis. ILMU KELAUTAN: Indones J Mar Sci 24(1):1. https://doi.org/10.14710/ik.ijms.24.1.1-6

Roberts DA, De Nys R (2016) The effects of feedstock pre-treatment and pyrolysis temperature on the production of biochar from the green seaweed Ulva. J Environ Manag 169:253–260. https://doi.org/10.1016/j.jenvman.2015.12.023

Robinson C, Stapelmann K (2024) Plasma treating water for nitrate based nitrogen fertilizer—A review of recent device designs. Curr Opin Green Sustain Chem 50:100978. https://doi.org/10.1016/j.cogsc.2024.100978

Schindler UG, Eulenstein F (2017) Hydraulic properties of horticultural substrates. Open Data J Agric Res 3. https://doi.org/10.18174/odjar.v3i1.15765

Szymańska M, Sosulski T, Szara E, Wąs A, Sulewski P, Van Pruissen GWP, Cornelissen RL (2019) Ammonium Sulphate from a bio-refinery system as a fertilizer—agronomic and economic effectiveness on the farm scale. Energies 12(24):4721. https://doi.org/10.3390/en12244721

Tan X, Zhang F, Wang H, Ho S-H (2023) The magic of algae-based biochar: advantages, preparation, and applications. Bioengineered 14(1):2252157. https://doi.org/10.1080/21655979.2023.2252157

Xavier James VC, Pushpa Thiraviam AG, Al-Dosary MA, Hatamleh AA, Bukhari NA, Arokiyaraj S, Kalaiyarasi M (2024) Evaluation of nutrient composition and biostimulant properties of seaweeds for improving soil microbial population and tomato plant growth. Bioresources 20(1):1431–1451. https://doi.org/10.15376/biores.20.1.1431-1451

Zhang W, Wei Y, Li J, Xiao H (2024) Harvesting energy from marine: seawater electrolysis for hydrogen production. Fuel 377:132782. https://doi.org/10.1016/j.fuel.2024.132782

Zulfiqar F, Wei X, Shaukat N, Chen J, Raza A, Younis A, Nafees M, Abideen Z, Zaid A, Latif N, Naveed M, Siddique KHM (2021) Effects of biochar and biochar–compost mix on growth, performance and physiological responses of potted Alpinia zerumbet. Sustainability 13(20):11226. https://doi.org/10.3390/su132011226

Chapter 4
Impacts of Aquatic Pollution on Marine Food Resources

Emem Okon Mbong, Inibehe Ime Okon, and Peter Chukwudi

Abstract Marine habitats across Africa are increasingly threatened by pollution, posing severe risks to biodiversity, ecosystem health, and food security for millions of coastal inhabitants. This chapter examines the major drivers and consequences of marine pollution across the continent, emphasizing how chemical contaminants such as heavy metals, pesticides, hydrocarbons, and untreated industrial effluents degrade aquatic ecosystems and compromise marine food resources. In African waters, these pollutants bioaccumulate and biomagnify along the food chain, resulting in toxic contamination of fish and shellfish consumed by humans. Oil spills in the Niger Delta, industrial discharges along the Gulf of Guinea, and nutrient runoff in East African estuaries have collectively reduced fish stocks and disrupted breeding habitats. Eutrophication from agricultural waste promotes algal blooms and oxygen depletion, while the spread of microplastics from the Mediterranean coast to the Indian Ocean islands introduces new physiological and ecological stress on marine organisms. The loss of vital habitats such as coral reefs and mangroves further undermines the productivity and resilience of fisheries, threatening the livelihoods and nutritional security of millions of Africans dependent on seafood. Addressing these challenges requires stronger pollution control measures, effective enforcement of environmental laws, and sustainable management of marine resources. Regional frameworks such as the Abidjan and Nairobi Conventions provide a basis for collaborative governance but need to be supported by national commitment, public awareness, and technological innovation. Integrating ecosystem-based management, improving waste disposal systems, and fostering international partnerships are critical for restoring marine health and ensuring long-term food security. Ultimately, protecting Africa's marine ecosystems demands a holistic approach that combines scientific monitoring, policy reform, and community

E. O. Mbong · P. Chukwudi
Department of Biology and Forensic Sciences, Admiralty University of Nigeria, Ibusa, Delta State, Nigeria

I. I. Okon (✉)
Department of Biochemistry, Heritage Polytechnic, Eket, Nigeria

H. A. Mupambwa et al. (eds.), *Harnessing Coastal and Marine Resources for Improved Food Security in Africa*, Sustainability Sciences in Asia and Africa, https://doi.org/10.1007/978-981-95-9725-3_4

participation to build resilient coastal systems capable of sustaining both ecological integrity and human well-being in the face of growing pollution pressures.

4.1 Introduction

Coastal and marine environments are a lifeline for millions across Africa, offering not only food but employment, cultural identity, and ecological protection (FAO 2022a, b, c, d). Yet, beneath their shimmering surfaces, these waters face an escalating threat, aquatic pollution. The continent's reliance on marine food resources is becoming increasingly precarious as pollution intensifies, undermining both the safety of what we consume and the sustainability of what we harvest (UNEP 2021a, b, c).

Aquatic pollution, in its many forms, has long been underestimated in public discourse, often overshadowed by more visible crises like drought or conflict. However, pollutants, whether chemical, plastic, biological, or oil-based, have quietly crept into rivers, estuaries, and oceans, where they interact with and disrupt food webs (Khan et al. 2020). The result? A growing crisis that not only endangers marine biodiversity but also jeopardizes the food security of coastal populations who depend on fisheries as their primary protein source (Béné et al. 2016).

Africa's vulnerability to aquatic pollution is worsened by several factors, including poor waste management infrastructure, rapid urbanization, oil exploration, and weak enforcement of environmental regulations (Nwilo and Badejo 2019). From the Niger Delta's oil-soaked creeks to the plastic-strewn shores of East Africa, the problem is both vast and complex. Many communities, unaware of the invisible toxins seeping into their waters, continue to consume contaminated seafood, exposing themselves to long-term health risks such as mercury poisoning, reproductive disorders, and gastrointestinal diseases (Olisah et al. 2021).

Moreover, the economic implications of pollution on marine food systems are profound. When pollution diminishes fish stocks or leads to fishing bans, it is not just ecosystems that suffer. Livelihoods collapse, food prices surge, and tensions mount over dwindling resources. For many artisanal fishers and coastal households, the line between sustainability and survival grows thinner by the day (OECD 2020).

Despite these threats, the connection between aquatic pollution and food security remains underexplored in mainstream policy and academic discussions, especially within the African context. While global attention has turned toward climate change and overfishing, pollution often slips through the cracks—a silent saboteur that amplifies existing vulnerabilities (FAO 2020). Understanding how different pollutants affect marine food chains is not just an environmental concern; it is a social and economic imperative.

This chapter aims to unpack the multifaceted impacts of aquatic pollution on marine food resources, with a particular focus on Africa's coastal regions. It explores the types and sources of pollution, the pathways through which contaminants affect marine organisms, and the cascading effects on human health and food systems. Drawing on African case studies, the chapter will also examine existing policy responses and highlight opportunities for more sustainable, inclusive solutions.

4.2 Types and Sources of Aquatic Pollution

Aquatic pollution is not a monolith; it is a web of contaminants, each with distinct sources, behaviours, and consequences. To truly understand how marine food resources are being undermined, we must first break down the types of pollution affecting aquatic environments across Africa and trace their origins. The continent's coastal and inland waters are being fed a steady stream of pollutants from industries, agriculture, households, and urban runoff, creating a toxic mix that jeopardizes both aquatic ecosystems and the food systems tied to them (UNEP 2021a, b, c; FAO 2022a, b, c, d).

4.2.1 Chemical Pollution

Chemical pollutants are among the most damaging to marine environments, especially because they often go unnoticed until their effects become widespread. Pesticides, herbicides, industrial effluents, heavy metals, and oil residues fall under this category. Coastal agriculture, often practised with minimal environmental safeguards, is a major contributor. Farmers frequently use synthetic fertilisers and pesticides, which eventually wash into nearby rivers and estuaries during rainfall, making their way into the ocean (Nkedi-Kizza 2020).

Another significant contributor is oil exploration. Countries like Nigeria and Angola, rich in crude oil, face the recurring nightmare of oil spills. In the Niger Delta, oil pollution has been a persistent problem for decades, contaminating waterways and depleting fish populations (Ogri 2021a, b). Crude oil and its derivatives affect marine life in several ways—they suffocate fish, destroy breeding grounds, and introduce carcinogenic compounds into the food chain (Nriagu 2019a, b). What makes chemical pollution particularly dangerous is that many of these substances bioaccumulate, meaning they concentrate as they move up the food chain, ending up in seafood consumed by humans (Clark et al. 2022a, b).

Heavy metals like mercury, lead, and cadmium are also increasingly found in African waters, largely due to mining and industrial discharges. These metals are notorious for their toxicity and long-term effects. Mercury, for instance, is neurotoxic and can cause severe developmental defects, especially in children and pregnant women who consume contaminated seafood (Wepener et al. 2020a, b).

4.2.2 Plastic Pollution

Plastics have become a defining feature of modern pollution, and nowhere is their presence more visually apparent than in coastal Africa. From plastic bags and bottles to fishing gear and microplastics, the continent's waterways are becoming dumping grounds for nonbiodegradable waste. Urban centres lacking formal waste management systems often see plastic waste directly entering rivers and drains, eventually ending up in the sea (Jambeck et al. 2015a, b; Biney et al. 2023).

Once in the marine environment, plastics pose multiple threats. Large debris can entangle marine animals, while smaller fragments are ingested by fish and shellfish. These microplastics—often invisible to the naked eye have been detected in the stomachs and tissues of fish sold in local markets (Akindele et al. 2021). The danger lies not only in the plastic itself but also in the toxic chemicals it carries, such as phthalates and bisphenol A (BPA), which mimic hormones and can disrupt biological processes in both marine organisms and humans (Rochman et al. 2013a, b).

The fishing industry is also complicit in plastic pollution. Abandoned, lost, or discarded fishing gear, often referred to as "ghost gear," continues to trap marine life long after it's been lost at sea. The result is a silent, ongoing loss of biodiversity and a reduction in the availability of fish for human consumption (Macfadyen et al. 2009).

4.2.3 Organic and Nutrient Pollution

Not all pollution is industrial. In many African coastal towns and cities, untreated domestic sewage is a primary pollutant. Raw or partially treated waste is often discharged directly into rivers and oceans, introducing high levels of nutrients such as nitrogen and phosphorus. These nutrients fuel algal blooms, which in turn deplete oxygen in the water, a process known as eutrophication (Anderson et al. 2017).

Eutrophication creates dead zones where marine life struggles to survive. Fish either migrate or die off, shellfish beds become unproductive, and biodiversity plummets. In extreme cases, toxins released by harmful algal blooms can contaminate fish and shellfish, posing serious health risks to humans (Heisler et al. 2008).

Animal waste from coastal livestock farming also contributes significantly to nutrient pollution, especially when waste is allowed to flow untreated into nearby water bodies. The cumulative effect of all these organic loads is a subtle, chronic degradation of water quality and marine life productivity (Scheren et al. 2021).

4.2.4 Microbial and Pathogen Pollution

Marine pollution is not always chemical or physical; biological contamination is equally dangerous. In many parts of Africa, faecal contamination of water bodies is a serious issue, particularly where sanitation infrastructure is weak or nonexistent. Pathogens such as *E. coli*, Salmonella, and Vibrio species thrive in polluted waters, contaminating fish and shellfish, especially filter feeders like oysters and mussels (WHO 2022a, b).

The risk here is twofold: not only do these pathogens cause disease outbreaks ranging from cholera to gastrointestinal infections, but they also damage the reputation of local fisheries. Coastal communities that rely on shellfish harvesting may face trade restrictions or loss of market access due to contamination concerns (FAO and WHO 2020a, b).

Fish farming operations, which are expanding rapidly in Africa, are also susceptible to microbial pollution. Poorly managed aquaculture facilities can become hotspots for disease, especially when water exchange is limited and waste accumulates. Infected fish often do not show symptoms until after they've been harvested and consumed (Bondad-Reantaso et al. 2021a, b).

4.2.5 Thermal and Sediment Pollution

While less discussed, thermal and sediment pollution are particularly relevant in regions experiencing rapid industrialization or infrastructure development. Thermal pollution occurs when industries, particularly power plants, discharge heated effluents into natural water bodies. Elevated temperatures reduce the water's oxygen-holding capacity and can disturb the natural balance of aquatic life (Dudgeon et al. 2006).

Sediment pollution, on the other hand, results from land clearing, construction, mining, and poor farming practices. Increased sediment loads cloud the water, reducing sunlight penetration and disrupting photosynthesis in aquatic plants. It also smothers coral reefs and clogs fish gills, further weakening aquatic biodiversity (Thrush et al. 2004; FAO 2023).

4.3 Pathways of Pollution into Marine Food Chains

Pollutants in aquatic environments do not remain isolated in water or sediment—they move. They enter living organisms, travel through food webs, and eventually land on our plates. Understanding the mechanisms through which pollution infiltrates marine food chains is crucial, not just for environmental protection, but for public health and sustainable food security across Africa.

4.3.1 Bioaccumulation and Biomagnification

At the heart of the pollution-to-food process are two key concepts: bioaccumulation and biomagnification. Bioaccumulation refers to the gradual build-up of pollutants in an organism over time, while biomagnification is the increase in concentration of those pollutants as they move up the food chain (Clark et al. 2022a, b). This means a small fish that has ingested contaminated plankton becomes more toxic, and when a bigger fish eats that smaller fish, it accumulates even more toxins. By the time it reaches humans, especially in the form of top predators like tuna or barracuda, the contamination levels can be dangerously high (Wepener et al. 2020a, b).

Heavy metals like mercury and cadmium are classic examples. These metals are introduced into aquatic systems through mining, industrial runoff, and fossil fuel combustion. Once in the water, they attach to particles, are ingested by plankton, and slowly climb the food web. Mercury, for instance, is converted by bacteria into methylmercury, a potent neurotoxin, which is easily absorbed by fish tissues (FAO 2022a, b, c, d).

The danger is compounded by the fact that these pollutants are not easily broken down or excreted. They lodge in fatty tissues and organs, accumulating with every meal. This is particularly concerning in regions where communities depend heavily on fish for daily protein needs, often consuming species at higher trophic levels (Ogri 2021a, b).

4.3.2 Ingestion of Microplastics

Microplastics fragments less than 5 mm in size are increasingly infiltrating marine food webs. Unlike other pollutants, they do not chemically dissolve or degrade in the environment. Instead, they are physically ingested by marine organisms, often mistaken for plankton or other natural food (Akindele et al. 2021). Microplastics have been found in the stomachs of fish, sea cucumbers, molluscs, and even zooplankton, which are at the very base of the food chain (Jambeck et al. 2015a, b).

Once ingested, these microplastics can cause internal injuries, block digestive tracts, or leach toxic chemicals like BPA and phthalates into the organisms' systems. These chemicals mimic hormones, leading to developmental and reproductive issues in marine life and, eventually, in humans (Rochman et al. 2013a, b).

More alarmingly, microplastics can act as carriers for other contaminants. Studies have shown that persistent organic pollutants (POPs) such as polychlorinated biphenyls (PCBs) and dioxins adhere to the surfaces of microplastics. When these particles are consumed by marine animals, the attached toxins are transferred along with the plastic, thereby amplifying the chemical load in the food chain (Clark et al. 2022a, b).

4.3.3 Filter Feeding and Pathogen Uptake

Some marine species, especially bivalves like oysters, clams, and mussels, feed by filtering large volumes of water through their bodies. While this makes them efficient at cleaning up their environment, it also makes them vulnerable to contamination. These filter feeders tend to concentrate pathogens, heavy metals, and toxins within their tissues, making them both a bioindicator and a high-risk food item (WHO 2022a, b).

Coastal waters contaminated with sewage or animal waste often contain bacteria such as Vibrio cholerae and *E. coli*, as well as viruses and protozoa. Bivalves harvested from such areas, if consumed raw or undercooked, can cause severe foodborne illnesses (FAO and WHO 2020a, b). In some cases, entire fisheries have been shut down due to pathogen detection, disrupting livelihoods and regional economies.

Aquaculture, especially when poorly regulated, also contributes to this dynamic. Overcrowded conditions, combined with limited water circulation and antibiotic overuse, can promote the spread of disease and antimicrobial resistance, both of which can pass up the food chain to humans (Bondad-Reantaso et al. 2021a, b).

4.3.4 Trophic Transfer of Pollutants

Beyond the well-known pathways of plastic ingestion and heavy metal build-up, there is a more subtle but equally important route: trophic transfer. This occurs when pollutants are transferred from prey to predator without direct environmental exposure. For example, a zooplankton species may ingest pesticide-laden algae. That zooplankton is eaten by a small fish, which is then consumed by a larger predator. Even if that predator never swims near a contaminated site, it still accumulates toxins by consuming contaminated prey (Nriagu 2019a, b).

This is how pollution from inland sources like agricultural runoff hundreds of kilometres away can eventually affect open ocean species. In African marine ecosystems, migratory fish species such as mackerel and tuna can carry these pollutants across borders, complicating traceability and policy enforcement.

4.4 Impacts on Marine Biodiversity and Food Availability

Marine biodiversity forms the backbone of coastal and oceanic food systems. It determines not only the variety of species available for harvest but also the resilience of ecosystems to environmental pressures. When pollution infiltrates marine environments, it disrupts this delicate balance, weakening ecosystems, killing off species, and reducing overall productivity. This has direct implications for food security, especially in African coastal communities where seafood plays a major nutritional and economic role.

4.4.1 Habitat Destruction and Species Decline

Pollution-driven habitat degradation is one of the most immediate threats to marine biodiversity. Sensitive ecosystems like coral reefs, mangroves, and seagrass beds known as nurseries for juvenile fish, are particularly vulnerable. Nutrient-rich runoff from agricultural lands promotes algal blooms that suffocate corals and deplete oxygen levels in the water (UNEP 2021a, b, c). These eutrophic zones, or "dead zones," become uninhabitable for many marine species, forcing fish to migrate or perish.

Heavy metal contamination also alters sediment chemistry and water quality, leading to stunted growth, reproductive failure, and increased mortality in fish and invertebrates. For instance, cadmium has been shown to damage gill structures and interfere with ion regulation in marine organisms (Adebayo et al. 2020). Over time, this causes population collapses, particularly among slow-growing or late-breeding species, which cannot recover quickly enough to withstand constant pressure.

The loss of biodiversity leads to a simplified food web, a dangerous condition that makes the entire ecosystem more susceptible to collapse. If one or two key species disappear, the consequences ripple throughout the trophic levels, reducing the overall availability of marine food resources (Pauly et al. 2002a, b).

4.4.2 Disruption of Reproductive Cycles and Recruitment

Pollutants such as endocrine-disrupting chemicals (EDCs), found in pesticides, plastics, and industrial effluents, interfere with hormone functions in marine organisms. These substances can cause hermaphroditism, infertility, and abnormal development in fish and shellfish (Jobling et al. 2006). In extreme cases, they lead to complete reproductive failure in local populations.

Fish exposed to EDCs may develop smaller gonads, exhibit erratic spawning behaviour, or produce eggs with low survival potential. When recruitment, the process of young fish reaching maturity, declines, fish stocks shrink, leaving fewer adult fish for harvest and sale. This is already visible in parts of the Gulf of Guinea, where persistent pollution has coincided with declining fish catches over the past decade (FAO 2022a, b, c, d).

Additionally, larval stages of marine organisms are particularly sensitive to toxins. Exposure to pollutants at this vulnerable stage can cause deformities or mortality before the young organisms even reach the juvenile phase, further compounding the issue.

4.4.3 *Invasive Species and Ecological Imbalance*

Pollution can also facilitate the invasion of non-native species, especially in estuarine environments where water quality is compromised. For example, species that tolerate low-oxygen or high-nutrient conditions can outcompete native species that are less adaptable. This results in ecological imbalance, often reducing the diversity and productivity of local fish stocks (Molnar et al. 2008).

Such invasions disrupt traditional fishing patterns and diets. In some cases, invasive species may be unpalatable, carry parasites, or lack nutritional value, thus failing to compensate for the loss of native fish. Moreover, their presence complicates fisheries management, as regulations are often designed around known, native species and their habitats.

4.4.4 *Impact on Fisheries Yields and Economic Livelihoods*

For millions of Africans living along the coasts, fishing is more than a food source; it is a livelihood. The cumulative effects of pollution include species decline, loss of biodiversity, and habitat degradation which ultimately result in reduced fish yields. Fisherfolk find themselves travelling farther and working longer for smaller catches. This not only raises costs but also increases risk, especially for small-scale or artisanal fishers who lack the equipment and capital to adapt (African Development Bank 2023).

Oil spills and industrial pollution in the Niger Delta have contaminated water, soil, and air, resulting in toxic levels of petroleum hydrocarbons and heavy metals in fish and crops. Some fish species now contain lead and cadmium far above WHO safety limits, posing risks of neurotoxicity, organ damage, and cancer to consumers. The contamination has rendered many traditional food sources unsafe, with 91% of local fishers reporting that their catch smells and tastes of oil, reducing both edibility and market value (Anyanwu et al. 2023; Amatorio et al. 2025).

The ripple effects stretch beyond the coast. As marine food becomes scarcer or more expensive, inland communities that depend on dried or smoked fish from coastal areas also suffer. Thus, aquatic pollution has a direct line to household food insecurity, malnutrition, and economic instability across vast geographic regions.

4.5 Human Health Risks from Polluted Marine Foods

The health of marine food resources is intrinsically linked to the health of the people who consume them. Across coastal Africa, seafood is not only a major protein source but also a cultural staple and economic necessity. However, when marine organisms become contaminated due to pollution, the danger extends from sea to table. The ingestion of polluted seafood poses serious health risks, some immediate, others long-term, and these risks are often magnified in populations with limited access to healthcare or alternative protein sources (FAO 2021).

4.5.1 Bioaccumulation and Biomagnification

One of the most insidious dangers of aquatic pollution lies in bioaccumulation, the gradual buildup of toxic substances, such as mercury, cadmium, lead, and polychlorinated biphenyls (PCBs), in the tissues of marine organisms. As these contaminants move up the food chain, from plankton to small fish to larger predators consumed by humans, they become more concentrated, a process known as biomagnification (Selin 2011).

Mercury, particularly in its methylated form, is a well-documented neurotoxin. It can impair brain development in children, cause memory loss, tremors, and vision problems in adults, and lead to complications in pregnancy (WHO 2017). Shellfish and top predator fish like tuna or barracuda are particularly prone to high mercury levels, making them dangerous in polluted environments (Grandjean et al. 2010).

PCBs and dioxins, commonly released from industrial waste, are linked to hormone disruption, immune suppression, and cancers. Because these substances are fat-soluble, they persist in marine animals and resist breakdown, ensuring their presence in seafood long after the pollution source is removed (UNEP 2019).

4.5.2 Bacterial and Viral Contamination

Sewage discharge and agricultural runoff introduce a variety of pathogenic microorganisms into marine waters. These include bacteria like Vibrio cholerae, Salmonella spp., and *Escherichia coli*, as well as viruses such as norovirus and hepatitis A. Filter feeders like oysters and mussels are especially vulnerable, since they accumulate microbes from the surrounding water as they feed (Lipp et al. 2001; Pinto et al. 2009).

Consumption of raw or undercooked contaminated seafood can lead to outbreaks of foodborne illnesses. These illnesses are often severe in areas with poor sanitation and limited medical infrastructure, where diarrhoeal diseases already place a heavy burden on public health systems (WHO 2020).

4.5.3 Chemical Residues and Allergic Reactions

Pesticides, industrial solvents, detergents, and pharmaceuticals all find their way into the sea through runoffs and discharge. These chemicals not only alter the physiology of marine species but can also remain in edible tissues. Some chemicals can act as allergens or irritants, triggering reactions in sensitive individuals, while others have been linked to endocrine disruption, reproductive toxicity, and carcinogenicity (FAO/WHO 2020; Ng and Obbard 2006).

Even in low doses, chronic exposure to these chemicals through regular seafood consumption can lead to cumulative health effects. This is particularly concerning in fishing communities with a high reliance on seafood, or among pregnant women and children, whose bodies are more susceptible to chemical damage (EFSA 2012).

4.5.4 Lack of Monitoring and Regulatory Oversight

In many African countries, regular monitoring of marine pollutants and seafood safety is limited or entirely absent. Informal fish markets dominate the seafood trade, and food safety standards are often poorly enforced (Obirikorang et al. 2015). Consumers rarely have access to information about the origins or quality of the fish they purchase, leaving them vulnerable to health risks without knowing it.

Where regulatory frameworks exist, they are frequently underfunded, poorly coordinated, or politically undermined. This institutional weakness allows contaminated seafood to enter and remain in local markets, creating a cycle of exposure and illness that remains largely invisible (Mensah et al. 2017).

The lack of effective oversight also means that data on seafood-related health impacts are sparse, making it difficult to quantify or respond to the crisis. Without robust surveillance systems and enforcement mechanisms, public health protection remains inadequate. Further entrenching the risks posed by polluted marine foods (FAO 2021).

4.6 Impact on Marine Biodiversity and Ecosystem Services

The ecological integrity of marine ecosystems is vital for sustaining not only biodiversity but also the ecosystem services that underpin food security, livelihoods, and climate resilience. Unfortunately, aquatic pollution is a major disruptor of these natural systems. The introduction of pollutants ranging from plastic debris and oil spills to heavy metals, agrochemicals, and untreated sewage alters the delicate balance of marine life, often with irreversible consequences (Halpern et al. 2008; UNEP 2016).

4.6.1 Disruption of Marine Food Webs

Pollution affects marine organisms at every trophic level, from phytoplankton and zooplankton to fish and apex predators. The death or reduced productivity of primary producers like phytoplankton, often due to oil spills, nutrient overload, or chemical poisoning disrupts the base of the marine food web (Paerl et al. 2014). When primary consumers like zooplankton feed on toxic phytoplankton or absorb pollutants directly, the impacts cascade upward through the food chain, compromising fish stocks and the predators that depend on them (Boersma et al. 2015).

Heavy metals such as cadmium and lead have been shown to reduce reproductive success and cause morphological deformities in fish, molluscs, and crustaceans (Fitzgerald et al. 2007; Wang et al. 2020). These disruptions reduce species populations, alter predator-prey relationships, and ultimately destabilize the entire marine ecosystem.

4.6.2 Coral Reefs and Seagrass Beds Under Threat

Coral reefs and seagrass beds: critical habitats for fish breeding and nursery grounds are particularly vulnerable to aquatic pollution. Sedimentation from coastal development, chemical runoff from agriculture, and increased water turbidity from sewage discharges reduce the sunlight reaching these habitats, leading to bleaching and death of coral polyps and seagrass (Fabricius 2005; Orth et al. 2006).

Nutrient enrichment from fertilizers leads to algal blooms that smother coral reefs and outcompete seagrass for light. Additionally, toxic substances such as herbicides and hydrocarbons can impair coral reproduction and larval settlement, ultimately reducing reef resilience and biodiversity (Negri et al. 2011; Hughes et al. 2017). As coral reef structures deteriorate, fish populations that rely on them for shelter and food decline, further exacerbating food insecurity for coastal communities.

4.6.3 Plastic Pollution and Microplastic Ingestion

The widespread presence of plastic pollution, especially microplastics, poses a significant and growing threat to marine biodiversity. Marine species ingest plastic particles either directly or indirectly through contaminated prey. This ingestion can lead to internal injuries, starvation, and death (Wright et al. 2013; Lusher et al. 2017). Furthermore, microplastics can act as vectors for toxic chemicals such as PCBs and persistent organic pollutants (POPs), which adhere to their surfaces and enter the bodies of marine organisms (Teuten et al. 2009).

Studies have found microplastics in the stomachs of commercial fish and shellfish, raising alarms about the potential bioaccumulation of these particles in humans through seafood consumption (Barboza et al. 2018). The ecological implications are equally dire: impaired feeding, altered reproductive cycles, and behavioural changes are being documented across multiple marine species (Galloway et al. 2017).

4.6.4 *Loss of Ecosystem Services*

Marine ecosystems offer a range of critical services—including fishery production, carbon sequestration, water purification, and shoreline protection. When these systems are compromised by pollution, the services they provide are significantly diminished (MEA 2005; Barbier et al. 2011). For instance, the degradation of mangrove forests due to oil contamination or solid waste dumping reduces their ability to act as carbon sinks and nurseries for juvenile fish.

Similarly, polluted wetlands lose their capacity to filter water, leading to further downstream contamination. The economic cost of losing these services is enormous, particularly for low-income African nations where alternative infrastructure is either unaffordable or unavailable (Costanza et al. 2014). The ripple effects extend to food production, tourism, and disaster risk reduction, further undermining the resilience of coastal societies as can be seen in Fig. 4.1.

4.7 Health Implications for Human Populations

The impact of aquatic pollution extends beyond environmental degradation to pose significant threats to human health, especially among populations that rely heavily on marine resources for food. Consumption of contaminated seafood, exposure to polluted waters, and dependence on degraded coastal ecosystems increase the vulnerability of millions of Africans to both chronic and acute health conditions (UNEP 2016; WHO 2019).

4.7.1 *Consumption of Contaminated Seafood*

One of the most direct health risks associated with aquatic pollution is the consumption of seafood tainted with toxic substances. Heavy metals such as mercury, cadmium, arsenic, and lead are bioaccumulative and magnify through trophic levels, meaning that predatory fish often contain dangerously high concentrations of these toxins (Burger and Gochfeld 2005; Clarkson and Magos 2006). In regions with

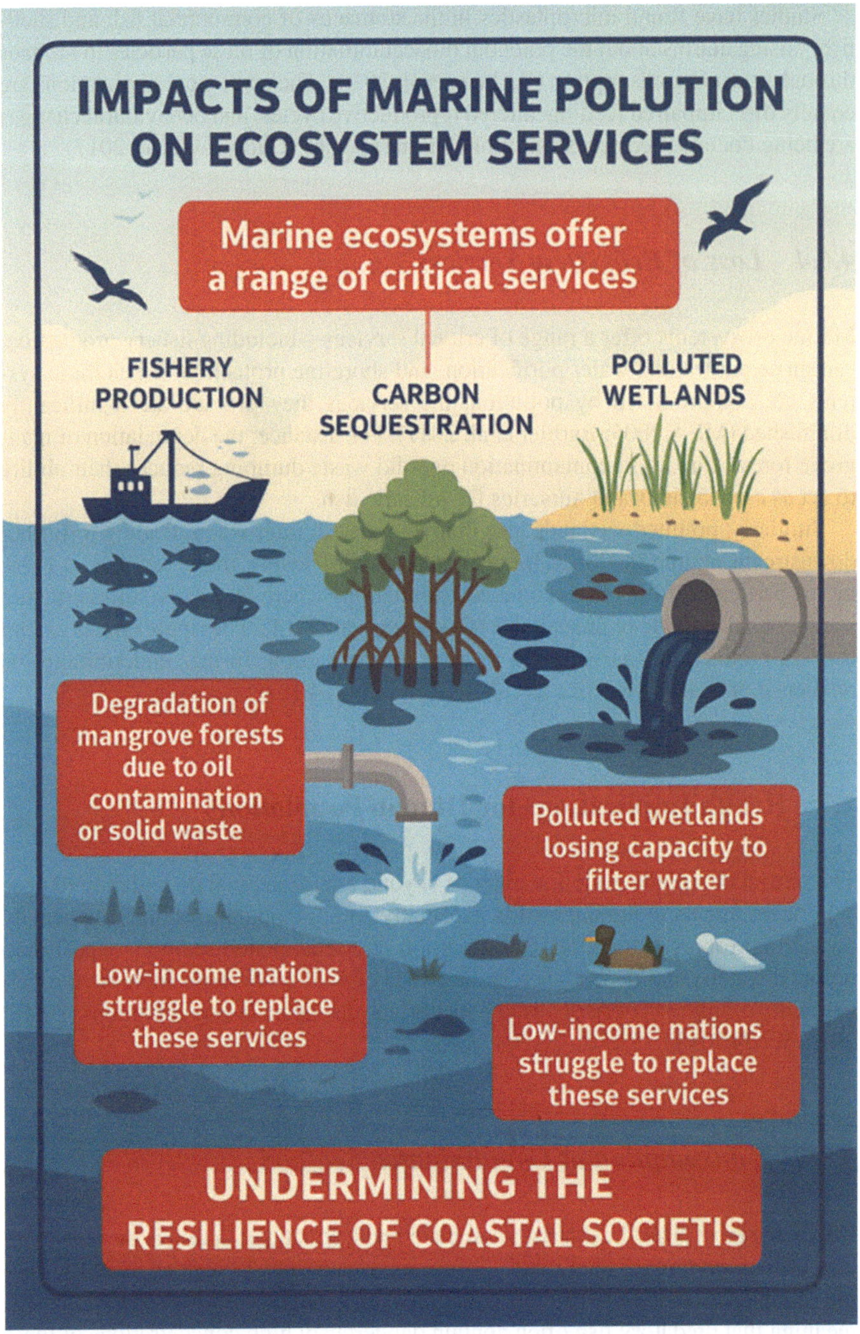

Fig. 4.1 Some key ecosystem services that marine ecosystems provide

limited health infrastructure, communities often lack the capacity to test or regulate seafood quality, making exposure almost inevitable.

Prolonged ingestion of these metals has been linked to a range of health complications, including kidney damage, neurological disorders, impaired cognitive development in children, and certain cancers (Karagas et al. 2012; Grandjean and Landrigan 2014). Mercury is known for its neurotoxic effects and is especially hazardous to pregnant women and young children due to its interference with brain development.

4.7.2 Microplastics and Chemical Contaminants

Another emerging concern is the ingestion of microplastics and the chemical contaminants they carry. Marine organisms consumed by humans have been found to contain microplastics, which can translocate from the digestive tract into human tissues (Smith et al. 2018). While the long-term health implications are still being studied, there is growing evidence that microplastic particles can trigger inflammation, disrupt endocrine function, and act as carriers of toxicants such as polycyclic aromatic hydrocarbons, bisphenol A (BPA), and phthalates (Wright and Kelly 2017; Schwabl et al. 2019).

The endocrine-disrupting chemicals often associated with plastics interfere with hormonal systems and have been linked to reproductive abnormalities, metabolic disorders, and certain cancers (Diamanti-Kandarakis et al. 2009). This presents a serious health burden for communities dependent on shellfish and filter-feeding species, which are particularly prone to plastic accumulation.

4.7.3 Waterborne Diseases and Direct Exposure

Polluted coastal waters also increase the incidence of waterborne diseases. Untreated sewage, agricultural runoff, and industrial effluents introduce pathogens such as *Vibrio cholerae*, *Salmonella*, *Shigella*, and *Escherichia coli* into marine environments (Farooqui et al. 2009; Ahmed et al. 2021). Direct contact with or ingestion of these contaminated waters can lead to gastrointestinal infections, skin diseases, and respiratory complications.

In informal fishing communities where sanitation infrastructure is poor, children and women are often the most affected. Skin infections and parasitic diseases such as schistosomiasis are common in areas where waste is dumped directly into the water and residents bathe or fish nearby (Steinmann et al. 2006).

4.7.4 *Undermining of Nutritional Security*

Beyond toxicity, pollution also reduces the availability and quality of seafood, which is a vital protein source in many African diets. Declining fish stocks due to habitat degradation or bioaccumulation of pollutants results in nutritional deficits, particularly among children in coastal communities (Golden et al. 2016). This is further exacerbated by food safety concerns that lead to reduced fish consumption, even when fish is available.

The ripple effect includes increased reliance on alternative protein sources, often less nutritious and more expensive, thereby worsening food insecurity and economic vulnerability (FAO 2020). In countries with high rates of malnutrition and undernutrition, the degradation of marine Food resources due to pollution represents a silent public health crisis.

4.8 Economic Consequences of Polluted Marine Resources

Aquatic pollution not only degrades ecosystems and threatens public health, but it also carries profound economic consequences for coastal and marine-based economies across Africa. As pollution disrupts fisheries, damages biodiversity, and compromises the quality of marine products, it erodes the livelihood base of millions who depend directly or indirectly on marine ecosystems (Barbier 2017; UNEP 2018).

4.8.1 *Collapse of Small-Scale Fisheries*

Small-scale fisheries, which form the economic backbone of many coastal communities, are particularly vulnerable to aquatic pollution. Polluted waters lead to fish kills, migration of species to cleaner habitats, and a decrease in reproductive success among marine species (Pauly et al. 2002a, b; Boesch 2002). These disruptions reduce daily catches and income for fishers who lack the resources to travel further offshore in search of cleaner waters.

In regions like West Africa where artisanal fishing contributes significantly to GDP and household nutrition, the decline in fishery productivity has ripple effects, increasing poverty, unemployment, and rural-urban migration (Belhabib et al. 2015). The lack of effective marine spatial planning or regulatory enforcement further compounds these challenges, leaving coastal livelihoods increasingly fragile.

4.8.2 *Trade Barriers and Loss of Export Revenue*

Pollution also threatens international trade in marine products. Seafood exports are subject to stringent safety and quality standards, particularly in markets such as the European Union and North America. High levels of contaminants—like heavy metals, pesticide residues, or microbial contamination—often result in rejections at borders, recalls, or outright bans (FAO 2020; OECD 2019).

Such trade restrictions diminish export revenue for African countries that rely heavily on marine exports such as tuna, shrimp, and shellfish. Ghana, Senegal, and Namibia, for instance, have faced scrutiny and restrictions due to lapses in marine hygiene standards, many of which stem from pollution (Asiedu and Nunoo 2013). The economic costs of losing access to premium markets can be devastating, especially for small producers and processors who lack the capacity to upgrade safety systems.

4.8.3 *Decline in Coastal Tourism and Recreation*

Coastal and marine pollution also jeopardizes the tourism industry, which is a major revenue generator in countries like Kenya, South Africa, and Mauritius. Polluted beaches, oil spills, floating plastics, and algal blooms degrade the aesthetic and recreational value of marine environments, discouraging tourism (Dodds and Holmes 2019).

Tourism-related jobs, from hospitality to recreational fishing and water sports, suffer from falling tourist numbers. In areas where tourism infrastructure overlaps with fishing or aquaculture, conflicts over space and pollution sources often emerge, compounding the economic loss (Said and MacMillan 2004). This dual blow to fisheries and tourism underscores how aquatic pollution stifles multiple economic streams.

4.8.4 *Cost of Environmental Remediation*

The cost of cleaning up polluted marine environments is often prohibitive. Oil spill clean-ups, dredging of polluted sediments, and restoration of degraded coastal habitats require significant investment, which many African governments cannot afford (UNEP 2016). Where polluters are not held accountable or enforcement is weak, these costs are borne by the state or left unaddressed, allowing further degradation.

Moreover, pollution undermines the return on investments in marine conservation and blue economy initiatives. Programmes aimed at promoting aquaculture, eco-tourism, or sustainable fisheries struggle to meet goals in environments already compromised by pollution (World Bank 2017). The result is a stagnation or reversal of development gains, particularly in coastal regions that already face socioeconomic marginalization.

4.9 Legal and Institutional Gaps in Addressing Aquatic Pollution

Despite the growing awareness of aquatic pollution and its devastating effects on marine food resources, significant legal and institutional shortcomings persist across many African nations. These gaps hinder the effectiveness of environmental governance, weaken regulatory enforcement, and compromise the continent's ability to sustainably manage coastal and marine ecosystems (Mensah 2019; UNEP 2021a, b, c).

4.9.1 Weak Regulatory Frameworks

A major challenge lies in the inadequacy or outdated nature of environmental laws governing marine pollution. In some countries, environmental legislation does not fully account for emerging pollutants such as microplastics, pharmaceutical residues, and nutrient runoff from intensified agriculture (UNEP 2021a, b, c). This leaves regulatory agencies ill equipped to manage the complex and evolving nature of pollution threats.

Even where appropriate laws exist, there is often limited political will or administrative capacity to implement them effectively. Poor interagency coordination, bureaucratic delays, and fragmented responsibilities across different government ministries lead to ineffective policy execution (Kokutse 2016). The absence of clear marine spatial plans further complicates decision-making on pollution control and resource use.

4.9.2 Inadequate Monitoring and Enforcement Mechanisms

Another critical gap is the lack of robust monitoring systems to detect, report, and act upon pollution incidents. Many African countries suffer from insufficient technical infrastructure, limited environmental laboratories, and a lack of trained personnel for regular marine quality assessments (Adeleke et al. 2020). This undermines early detection, timely responses, and the prosecution of polluters.

In regions with high levels of industrial or port activity, enforcement is often undermined by corruption, lack of transparency, or political interference. As a result, industries that discharge untreated waste into rivers or the sea often escape penalties, setting a dangerous precedent and weakening deterrent measures (Akpalu et al. 2011).

4.9.3 Gaps in Transboundary Governance

Aquatic pollution does not respect national borders. Rivers, ocean currents, and air flows carry pollutants across countries, necessitating regional cooperation. However, existing frameworks for transboundary governance—such as the Abidjan Convention or regional fisheries management organizations—often lack sufficient funding, enforcement powers, or political support (van der Elst and Everett 2015).

This leads to gaps in collective monitoring, inconsistent pollution standards, and difficulty in holding upstream polluters accountable. Coastal countries downstream in river basins, for example, often bear the brunt of pollution caused by upstream mining or agriculture in neighbouring countries (FAO 2018). The failure to harmonize pollution control standards regionally hampers efforts to protect shared marine resources.

4.9.4 Limited Community and Stakeholder Involvement

Top-down legal and institutional responses often overlook the role of local communities, civil society, and traditional governance structures. Many coastal populations possess indigenous knowledge and customary laws that could enhance environmental stewardship if recognized and integrated into formal systems (Njoh 2006). Unfortunately, most legal frameworks do not provide adequate platforms for inclusive participation.

This exclusion breeds mistrust, reduces compliance, and results in conflict over resource access and pollution blame. Strengthening legal mechanisms to ensure public participation, accountability, and access to environmental justice is crucial for meaningful progress in pollution governance.

4.10 Climate Change and Its Amplifying Effects on Aquatic Pollution

Climate change is not only a standalone environmental crisis, but it is also a potent amplifier of aquatic pollution. The interplay between changing climate patterns and pollution dynamics intensifies the degradation of marine ecosystems and further threatens food security in Africa (IPCC 2022; UNEP 2021a, b, c).

4.10.1 Altered Rainfall and Increased Runoff

One of the most direct effects of climate change is the alteration of rainfall patterns, often leading to more intense and erratic precipitation. These changes increase surface runoff, washing more pollutants—such as agricultural fertilizers, pesticides, sewage, and oil residues into rivers and coastal waters (Nyenje et al. 2010). The heavier the rainfall, the more sediment and pollutants are flushed from urban and rural landscapes into aquatic environments, causing eutrophication and the proliferation of harmful algal blooms (FAO 2021).

Low-lying coastal cities with poor drainage systems, such as Lagos or Dar es Salaam, are particularly vulnerable. The increased influx of pollutants overwhelms natural filtration capacities, contributing to oxygen-depleted zones that severely impact marine life and fisheries (Rabalais et al. 2014).

4.10.2 Rising Sea Temperatures and Chemical Interactions

Higher sea surface temperatures caused by climate change affect the chemical behaviour of pollutants in marine environments. For instance, warmer waters can increase the solubility and toxicity of certain pollutants, such as heavy metals and organic compounds (Bindoff et al. 2019). This means that even previously tolerable concentrations of contaminants may become more harmful to marine organisms, disrupting food chains and affecting species consumed by humans.

Additionally, thermal stress can weaken the immune systems of fish and shellfish, making them more susceptible to disease and the toxic effects of pollutants. This leads to declining fish stocks and quality, with direct consequences for nutrition and food supply (Nye et al. 2009).

4.10.3 Ocean Acidification and Heavy Metal Bioavailability

As oceans absorb more carbon dioxide from the atmosphere, their pH levels decline, a process known as ocean acidification. This change affects the bioavailability and mobility of pollutants like mercury and cadmium, making them more likely to accumulate in marine organisms (Fitzgerald et al. 2007). Shell-forming species such as molluscs and crustaceans are especially vulnerable, and their decline can disrupt marine ecosystems and reduce available protein sources for coastal communities.

4.10.4 Sea Level Rise and Coastal Infrastructure Risks

Rising sea levels inundate and degrade coastal infrastructure, including sewage treatment plants, industrial zones, and landfills. When such facilities are flooded, untreated waste can spill directly into the ocean, introducing pathogens, nutrients, and hazardous chemicals into the marine environment (Nicholls and Cazenave 2010). In countries with limited resources for upgrading coastal defences, such events are becoming more frequent, putting marine food resources at risk.

4.10.5 *Compounded Impacts on Vulnerable Communities*

The combination of aquatic pollution and climate stress disproportionately affects vulnerable populations. Small-scale fishers, who lack the means to adapt or relocate, face dwindling catches and reduced income (FAO 2021). Women, often responsible for fish processing and trade in many African coastal communities, bear additional burdens as climate-amplified pollution diminishes their access to safe, marketable seafood (Porter et al. 2020). These intersecting vulnerabilities underscore the urgent need for integrated approaches that consider climate change and pollution as interconnected challenges.

4.11 Emerging Technologies and Innovations in Pollution Management

As the challenges of aquatic pollution intensify, technological innovation offers promising pathways for more effective monitoring, prevention, and remediation of contaminants in marine environments. Several emerging technologies are being piloted and deployed across Africa, demonstrating potential to enhance marine food resource protection.

4.11.1 *Remote Sensing and GIS for Pollution Monitoring*

Satellite remote sensing and Geographic Information Systems (GIS) have revolutionized environmental monitoring by providing real-time data on water quality, pollutant spread, and ecosystem health (Scheren et al. 2020). These technologies enable early detection of harmful algal blooms, sediment plumes, and oil spills, helping authorities respond more rapidly and efficiently.

For example, initiatives in countries like Kenya and South Africa use remote sensing data combined with in situ water sampling to track nutrient pollution hotspots, informing targeted interventions (Mwangi et al. 2021). These tools also facilitate long-term trend analysis essential for policy formulation.

4.11.2 *Bioremediation and Phytoremediation*

Bioremediation, the use of microorganisms to degrade pollutants, and phytoremediation, which employs plants to absorb or break down contaminants, gaining traction as eco-friendly solutions. Certain species of bacteria and algae have been shown to remove heavy metals and hydrocarbons from water bodies (Abdel-Shafy and Mansour 2016).

In coastal wetlands of Nigeria, mangrove reforestation projects not only restore biodiversity but also naturally filter pollutants, improving water quality for adjacent fisheries (Olayanju et al. 2020). These nature-based solutions offer sustainable alternatives to costly chemical or mechanical cleanup methods.

4.11.3 Smart Wastewater Treatment Systems

Technological advancements have led to the development of smart wastewater treatment plants equipped with sensors and automated controls. These systems optimize pollutant removal processes and reduce operational inefficiencies (Zhao et al. 2019). Implementation in urban coastal areas, though currently limited, is expanding as governments recognize their value in controlling pollution at source.

Pilot projects in Cape Town and Accra illustrate how sensor-enabled treatment facilities can adapt to variable pollutant loads, reducing nutrient and pathogen discharge into marine environments (Aidoo et al. 2022).

4.11.4 Community-Based Digital Platforms

Digital platforms that enable local communities to report pollution incidents and track environmental changes are emerging as powerful participatory tools. Mobile apps and social media channels facilitate crowdsourcing of data increasing transparency and accountability (Mwangi et al. 2022).

These platforms also serve educational purposes, raising awareness about pollution impacts and empowering stakeholders to advocate for cleaner marine environments. In East Africa, community groups use such tools to document illegal dumping and lobby for stronger enforcement (Kiptoo and Wanjiru 2020).

4.11.5 Challenges and Future Prospects

While these innovations are promising, challenges remain in scaling them across Africa due to funding constraints, technical capacity gaps, and uneven access to digital infrastructure (UNEP 2021a, b, c). Successful adoption requires collaboration between governments, research institutions, private sector, and local communities.

Integrating technological advances with traditional knowledge and governance framework will be essential for creating adaptive and resilient systems capable of safeguarding marine food resources amidst growing pollution threats.

4.12 Conclusion

Aquatic pollution presents a formidable challenge to the sustainability of marine food resources across Africa's coastal regions. The multifaceted nature of pollution, from chemical contaminants and plastic debris to nutrient overload and emerging pollutants, interacts with socioeconomic pressures and climate change to threaten the health of marine ecosystems and the communities that depend on them. The consequences of these pressures are far reaching, impacting food security, livelihoods, and public health.

Effective management of aquatic pollution requires an integrated approach that combines scientific innovation, policy enforcement, community engagement, and sustainable development. Emerging technologies such as remote sensing, bioremediation, and smart wastewater treatment offer valuable tools for monitoring and mitigating pollution, yet their success depends on adequate investment and cross-sector collaboration.

Moreover, addressing the causes of pollution, including inadequate waste management infrastructure, unsustainable agricultural practices, and weak regulatory framework, remains critical. Strengthening governance, fostering regional cooperation, and empowering local communities to participate in decision making are essential to building resilience in coastal ecosystems.

Ultimately, safeguarding marine food resources against the growing threats of aquatic pollution is pivotal in advancing food security and sustainable livelihoods in Africa. As this chapter has highlighted, a holistic and adaptive strategy that aligns environmental protection with socioeconomic development offers the best prospect for nurturing the continent's vital coastal and marine assets for present and future generations.

Authors' Contributions Conceptualization: E.O.M; critical thinking: I.I.O; methodology: I.I.O; validation: E.O. M; investigation: I.I.O; data curation: I.I.O; writing—original draft preparation: I.I.O; review and editing: E.O.M, I.I.O, and P.C; supervision: E.O.M. project administration: E.O.M.

References

Abdel-Shafy HI, Mansour MSM (2016) A review on polycyclic aromatic hydrocarbons: source, environmental impact, effect on human health and remediation. Egypt J Pet 25(1):107–123. https://doi.org/10.1016/j.ejpe.2015.03.011

Adebayo A, et al (2020) Study on heavy metal contamination effects on marine organisms

Adeleke A, Aduwo E, Ayoola A (2020) Environmental monitoring challenges in African coastal zones. Afr J Mar Sci 42(3):345–359

African Development Bank (2023) Report on economic impacts of declining fisheries

Ahmed W, Bertsch PM, Bivins A, Bibby K, Farkas K, Gathercole A et al (2021) Comparison of virus concentration methods for the RT-qPCR-based recovery of murine hepatitis virus, a surrogate for SARS-CoV-2 from untreated wastewater. Sci Total Environ 739:139960. https://doi.org/10.1016/j.scitotenv.2020.139960

Aidoo K, Mensah M, Boateng A (2022) Smart wastewater treatment technologies and their potential in Africa: a case study of Ghana. Environ Technol Innovation 27:102545. https://doi.org/10.1016/j.eti.2022.102545

Akindele EO, Ehlers SM, Koop JHE (2021) First empirical study of freshwater microplastics in West Africa using gastropods from Nigeria as bioindicators. Environ Pollut 268:115642. https://doi.org/10.1016/j.envpol.2020.115642

Akpalu W, Dasmani I, Aglobitse P (2011) Demand for environmental quality: applications of the hedonic pricing method in Ghana. Environ Dev Econ 16(3):343–360. https://doi.org/10.1017/S1355770X10000517

Amatorio C, Estrada C, Estorico G (2025) Human health risks of oil spill contaminants in Nigeria: a systematic review of fish contamination and its impact on public health. Int J Innov Sci Res Technol. https://doi.org/10.38124/ijisrt/25apr231

Anderson DM, Burkholder JM, Cochlan WP, Glibert PM, Gobler CJ, Heil CA et al (2017) Harmful algal blooms and eutrophication: examining linkages from selected coastal regions of the United States. Harmful Algae 65:139–157. https://doi.org/10.1016/j.hal.2017.05.009

Anyanwu I, Beggel S, Sikoki F, Okuku E, Unyimadu J, Geist J (2023) Pollution of The Niger Delta with total petroleum hydrocarbons, heavy metals and nutrients in relation to seasonal dynamics. Sci Rep 13. https://doi.org/10.1038/s41598-023-40995-9

Asiedu B, Nunoo FKE (2013) Fish handling and sanitation practices in Ghana: implications for export. Food Control 34(1):151–156. https://doi.org/10.1016/j.foodcont.2013.04.019

Barbier EB (2017) Marine ecosystem services. Curr Biol 27(11):R507–R510. https://doi.org/10.1016/j.cub.2017.03.020

Barbier EB, Hacker SD, Kennedy C, Koch EW, Stier AC, Silliman BR (2011) The value of estuarine and coastal ecosystem services. Ecol Monogr 81(2):169–193. https://doi.org/10.1890/10-1510.1

Barboza LGA, Vieira LR, Branco V, Figueiredo N, Carvalho F, Guilhermino L (2018) Microplastics cause neurotoxicity, oxidative damage and energy-related changes and interact with the bioaccumulation of mercury in the European seabass, Dicentrarchus labrax (Linnaeus, 1758). Aquat Toxicol 195:49–57. https://doi.org/10.1016/j.aquatox.2017.12.008

Belhabib D, Sumaila UR, Pauly D (2015) Feeding the poor: contribution of West African fisheries to employment and food security. Ocean Coastal Manag 111:72–81. https://doi.org/10.1016/j.ocecoaman.2015.04.010

Béné C, Barange M, Subasinghe R, Pinstrup-Andersen P, Merino G, Hemre G-I, Williams M (2016) Feeding 9 billion by 2050—putting fish back on the menu. Food Secur 7(2):261–274. https://doi.org/10.1007/s12571-015-0427-z

Bindoff NL, Cheung WWL, Kairo JG, Arístegui J, Guinder VA, Hallberg R et al (2019) Changing ocean, marine ecosystems, and dependent communities. In: Pörtner H-O et al (eds) IPCC Special Report on the Ocean and Cryosphere in a Changing Climate. Intergovernmental Panel on Climate Change, pp 447–587. https://www.ipcc.ch/srocc/

Biney CA, Ofori D, Amoako J (2023) Assessment of plastic waste management practices in west African coastal cities. Mar Pollut Bull 190:114934. https://doi.org/10.1016/j.marpolbul.2023.114934

Boersma M, Wiltshire KH, Malzahn AM (2015) Toxic cyanobacteria and their effects on aquatic zooplankton and fish. In: Chorus L, Bartram J (eds) Toxic cyanobacteria in water: a guide to their public health consequences, monitoring and management, 2nd edn. CRC Press, pp 191–203

Boesch DF (2002) Challenges and opportunities for science in reducing nutrient over-enrichment of coastal ecosystems. Estuaries 25(4b):886–900. https://doi.org/10.1007/BF02804914

Bondad-Reantaso MG, Arthur JR, Subasinghe RP, Surachetpong W (2021a) Improving biosecurity through prudent and responsible use of veterinary medicines in aquatic food production. FAO Fisheries and Aquaculture Technical Paper No. 670. FAO

Bondad-Reantaso MG, et al (2021b) Study on aquaculture, disease spread, and antimicrobial resistance

Burger J, Gochfeld M (2005) Heavy metals in commercial fish in New Jersey. Environ Res 99(3):403–412. https://doi.org/10.1016/j.envres.2005.02.001
Clark JR, Cole M, Lindeque PK, Fileman ES, Blackford J, Lewis C et al (2022a) Marine microplastic debris: a targeted plan to understand and assess ecological risk. Front Environ Sci 10:809688. https://doi.org/10.3389/fenvs.2022.809688
Clark R, et al (2022b) Research on bioaccumulation, biomagnification, and microplastics as contaminant carriers
Clarkson TW, Magos L (2006) The toxicology of mercury and its chemical compounds. Crit Rev Toxicol 36(8):609–662. https://doi.org/10.1080/10408440600845619
Costanza R, de Groot R, Sutton P, van der Ploeg S, Anderson SJ, Kubiszewski I et al (2014) Changes in the global value of ecosystem services. Glob Environ Chang 26:152–158. https://doi.org/10.1016/j.gloenvcha.2014.04.002
Diamanti-Kandarakis E, Bourguignon JP, Giudice LC, Hauser R, Prins GS, Soto AM et al (2009) Endocrine-disrupting chemicals: an endocrine society scientific statement. Endocr Rev 30(4):293–342. https://doi.org/10.1210/er.2009-0002
Dodds R, Holmes M (2019) Beach tourists; what factors satisfy them and drive them to return. Ocean Coastal Manag 168:158–166. https://doi.org/10.1016/j.ocecoaman.2018.11.005
Dudgeon D, Arthington AH, Gessner MO, Kawabata ZI, Knowler DJ, Lévêque C et al (2006) Freshwater biodiversity: importance, threats, status and conservation challenges. Biol Rev 81(2):163–182. https://doi.org/10.1017/S1464793105006950
EFSA (2012) Report on chronic exposure to chemical contaminants in seafood
van der Elst R, Everett B (2015) Towards a western Indian Ocean regional fisheries strategy. South West Indian Ocean Fisheries Commission. https://www.fao.org
Fabricius KE (2005) Effects of terrestrial runoff on the ecology of corals and coral reefs: review and synthesis. Mar Pollut Bull 50(2):125–146. https://doi.org/10.1016/j.marpolbul.2004.11.028
FAO (2018) Transboundary river basins and the role of regional governance in Africa. Food and Agriculture Organization of the United Nations. https://www.fao.org/publications
FAO (2020) The state of world fisheries and aquaculture 2020: sustainability in action. Food and Agriculture Organization of the United Nations. https://doi.org/10.4060/ca9229en
FAO (2021) Report on seafood safety monitoring and regulatory challenges
FAO (2022a) Report on heavy metals in aquatic systems and methylmercury
FAO (2022b) Report on pollution effects and declining fish catches in the Gulf of Guinea
FAO (2022c) The state of world fisheries and aquaculture 2022: towards blue transformation. Food and Agriculture Organization of the United Nations
FAO (2022d) The state of world fisheries and aquaculture 2022: towards blue transformation. Food and Agriculture Organization of the United Nations. https://doi.org/10.4060/cc0461en
FAO (2023) Water pollution from agriculture: a global review. Food and Agriculture Organization of the United Nations. https://www.fao.org/documents/card/en/c/ca0146en
FAO & WHO (2020a) Guidelines on pathogens in bivalves and food safety
FAO & WHO (2020b) Risk assessment of microbiological hazards in foods. Microbiological Risk Assessment Series No. 36. FAO/WHO
FAO/WHO (2020) Assessment of chemical residues and health impacts in seafood
Farooqui A, Khan A, Kazmi SU, Khan E, Kazmi SZH (2009) Microbial contamination of drinking water in Pakistan—a review. Environ Forensic 10(1):25–35. https://doi.org/10.1080/15275920802635608
Fitzgerald WF, Lamborg CH, Hammerschmidt CR (2007) Marine biogeochemical cycling of mercury. Chem Rev 107(2):641–662. https://doi.org/10.1021/cr050353m
Galloway TS, Cole M, Lewis C (2017) Interactions of microplastic debris throughout the marine ecosystem. Nature Ecol Evol 1(5):0116. https://doi.org/10.1038/s41559-017-0116
Golden CD, Allison EH, Cheung WW, Dey MM, Halpern BS, McCauley DJ et al (2016) Fall in fish catch threatens human health. Nature 534(7607):317–320. https://doi.org/10.1038/534317a
Grandjean P, Landrigan PJ (2014) Neurobehavioural effects of developmental toxicity. Lancet Neurol 13(3):330–338. https://doi.org/10.1016/S1474-4422(13)70278-3

Grandjean P, et al (2010) Study on mercury exposure and health effects
Halpern BS, Walbridge S, Selkoe KA, Kappel CV, Micheli F, D'Agrosa C et al (2008) A global map of human impact on marine ecosystems. Science 319(5865):948–952. https://doi.org/10.1126/science.1149345
Heisler J, Glibert PM, Burkholder JM, Anderson DM, Cochlan W, Dennison WC et al (2008) Eutrophication and harmful algal blooms: a scientific consensus. Harmful Algae 8(1):3–13. https://doi.org/10.1016/j.hal.2008.08.006
Hughes TP, Barnes ML, Bellwood DR, Cinner JE, Cumming GS, Jackson JB et al (2017) Coral reefs in the Anthropocene. Nature 546(7656):82–90. https://doi.org/10.1038/nature22901
IPCC (2022) Climate change 2022: impacts, adaptation and vulnerability. In: Working Group II contribution to the Sixth Assessment Report of the Intergovernmental Panel on Climate Change. Cambridge University Press. https://www.ipcc.ch/report/ar6/wg2/
Jambeck JR, Geyer R, Wilcox C, Siegler TR, Perryman M, Andrady A et al (2015a) Plastic waste inputs from land into the ocean. Science 347(6223):768–771. https://doi.org/10.1126/science.1260352
Jambeck JR, et al (2015b) Study on microplastics presence in marine organisms
Jobling S, et al (2006) Research on endocrine-disrupting chemicals and reproductive effects in marine life
Karagas MR, Choi AL, Oken E, Horvat M, Schoeny R, Kamai E et al (2012) Evidence on the human health effects of low-level methylmercury exposure. Environ Health Perspect 120(6):799–806. https://doi.org/10.1289/ehp.1104494
Khan FR, Syberg K, Shashoua Y, Bury NR (2020) Influence of polyethylene microplastic beads on the uptake and localization of silver in zebrafish (Danio rerio). Environ Pollut 265:114870. https://doi.org/10.1016/j.envpol.2020.114870
Kiptoo M, Wanjiru P (2020) Digital tools and citizen participation in marine pollution control: the case of East Africa. Afr J Mar Sci 42(3):305–312. https://doi.org/10.2989/1814232X.2020.1770342
Kokutse F (2016) Environmental laws exist in Africa, but enforcement remains weak. SciDev.Net. https://www.scidev.net/global/news/environmental-laws-africa-enforcement/
Lipp EK, et al (2001) Research on bacterial contamination in shellfish
Lusher AL, Hollman PC, Mendoza-Hill JJ (2017) Microplastics in fisheries and aquaculture: status of knowledge on their occurrence and implications for aquatic organisms and food safety. FAO Fisheries and Aquaculture Technical Paper No. 615. Food and Agriculture Organization of the United Nations
Macfadyen G, Huntington T, Cappell R (2009) Abandoned, lost or otherwise discarded fishing gear. FAO Fisheries and Aquaculture Technical Paper No. 523. FAO
Mensah J (2019) Environmental law in Ghana: evolving legal responses to marine pollution. Afr J Int Compar Law 27(2):209–229. https://doi.org/10.3366/ajicl.2019.0266
Mensah P, et al (2017) Analysis of institutional weaknesses in seafood safety enforcement
Millennium Ecosystem Assessment (MEA) (2005) Ecosystems and human well-being: synthesis. Island Press
Molnar JL, et al (2008) Study on invasive species and ecological imbalance
Mwangi JW, Otieno DO, Obiero KO (2021) GIS and remote sensing applications in monitoring water pollution in East Africa: current practices and future prospects. J Environ Inf 37(1):12–22. https://doi.org/10.3808/jei.202100040
Mwangi JW, Ochieng A, Akinyi M (2022) Community-driven digital platforms for environmental monitoring in coastal Kenya. J Environ Manag 308:114560. https://doi.org/10.1016/j.jenvman.2022.114560
Negri AP, Flores F, Röthig T, Uthicke S (2011) Herbicides increase the vulnerability of corals to thermal and low salinity stress. Mar Pollut Bull 62(5):1130–1135. https://doi.org/10.1016/j.marpolbul.2011.02.029
Ng KKS, Obbard JP (2006) Study on chemical pollutants and marine species

Nicholls RJ, Cazenave A (2010) Sea-level rise and its impact on coastal zones. Science 328(5985):1517–1520. https://doi.org/10.1126/science.1185782
Njoh AJ (2006) Tradition, culture and development in Africa: historical lessons for modern development planning. Ashgate Publishing
Nkedi-Kizza P (2020) The environmental risks of pesticides and fertilisers on water quality in Africa. Environ Sci Pollut Res 27(3):2364–2375. https://doi.org/10.1007/s11356-019-07123-4
Nriagu JO (2019a) Oil pollution and human health risks in The Niger Delta. Sci Total Environ 671:411–424. https://doi.org/10.1016/j.scitotenv.2019.03.378
Nriagu JO (2019b) Research on trophic transfer of pollutants
Nwilo PC, Badejo OT (2019) Impacts and management of oil spill pollution along the Nigerian coastal areas. Int J Environ Stud 76(5):785–797. https://doi.org/10.1080/00207233.2019.1627285
Nye JA, Link JS, Hare JA, Overholtz WJ (2009) Changing spatial distribution of fish stocks in relation to climate and population size on the Northeast United States continental shelf. Mar Ecol Prog Ser 393:111–129. https://doi.org/10.3354/meps08220
Nyenje PM, Foppen JW, Uhlenbrook S, Kulabako R, Muwanga A (2010) Eutrophication and nutrient release in urban areas of sub-Saharan Africa—a review. Sci Total Environ 408(3):447–455. https://doi.org/10.1016/j.scitotenv.2009.10.020
Obirikorang KA, et al (2015) Research on informal fish markets and food safety
OECD (2019) Fishing for tomorrow: ensuring sustainability in global seafood trade. Organisation for Economic Co-operation and Development. https://doi.org/10.1787/seafood-2019-en
OECD (2020) The ocean economy in 2030. Organisation for Economic Co-operation and Development Publishing. https://doi.org/10.1787/9789264251724-en
Ogri O (2021a) Study on fish consumption and pollutant accumulation in human diets
Ogri OR (2021b) A review of the environmental impacts of oil production and spill incidents in The Niger Delta. Environ Manag Sustain Dev 10(2):37–54. https://doi.org/10.5296/emsd.v10i2.18697
Olayanju A, Olowoporoku A, Okunola O (2020) Mangrove restoration as a tool for pollution mitigation in coastal Nigeria. Afr J Ecol 58(4):456–463. https://doi.org/10.1111/aje.12728
Olisah C, Okoh OO, Okoh AI (2021) Occurrence of organochlorine pesticide residues in biological and environmental matrices in Africa: a two-decade review. Heliyon 7(4):e06691. https://doi.org/10.1016/j.heliyon.2021.e06691
Orth RJ, Carruthers TJ, Dennison WC, Duarte CM, Fourqurean JW, Heck KL Jr et al (2006) A global crisis for seagrass ecosystems. Bioscience 56(12):987–996. https://doi.org/10.1641/0006-3568(2006)56[987:AGCFSE]2.0.CO;2
Paerl HW, Hall NS, Calandrino ES (2014) Controlling harmful cyanobacterial blooms in a world experiencing anthropogenic and climatic-induced change. Sci Total Environ 470–471:243–253. https://doi.org/10.1016/j.scitotenv.2013.09.062
Pauly D, Christensen V, Guénette S, Pitcher TJ, Sumaila UR, Walters CJ, Watson R, Zeller D (2002a) Towards sustainability in world fisheries. Nature 418(6898):689–695. https://doi.org/10.1038/nature01017
Pauly D, et al (2002b) Research on biodiversity loss and simplified food webs
Pinto FR, et al (2009) Study on viral contamination in seafood
Porter JR, Xie L, Challinor AJ, Cochrane K, Howden SM, Iqbal MM et al (2020) Food security and food production systems. In: Climate change 2020: impacts, adaptation, and vulnerability. IPCC Fifth Assessment Report (AR5), Working Group II
Rabalais NN, Turner RE, Díaz RJ, Justić D (2014) Global change and eutrophication of coastal waters. ICES J Mar Sci 66(7):1528–1537. https://doi.org/10.1093/icesjms/fsp047
Rochman CM, Hoh E, Kurobe T, Teh SJ (2013a) Ingested plastic transfers hazardous chemicals to fish and induces hepatic stress. Sci Rep 3:3263. https://doi.org/10.1038/srep03263
Rochman CM et al (2013b) Research on toxic chemicals leached from microplastics and their effects

Said A, MacMillan D (2004) The importance of coastal resources in the livelihood of coastal communities in Zanzibar. Ocean Coastal Manage 47(11–12):531–545. https://doi.org/10.1016/j.ocecoaman.2004.12.002
Scheren P, Savenije H, Lemmens A, Bonzi J (2020) Monitoring pollution in West African coastal waters using remote sensing and GIS. Environ Monit Assess 192(3):152. https://doi.org/10.1007/s10661-020-8112-3
Scheren PAGM, Zanting HA, Lemmens AMC (2021) Estimation of water pollution sources in Africa using population and economic data. Ecol Model 354:34–50. https://doi.org/10.1016/j.ecolmodel.2021.109342
Schwabl P, Köppel S, Königshofer P, Bucsics T, Trauner M, Reiberger T, Liebmann B (2019) Detection of various microplastics in human stool: a prospective case series. Ann Intern Med 171(7):453–457. https://doi.org/10.7326/M19-0618
Selin NE (2011) Review on bioaccumulation and biomagnification of marine toxins
Smith M, Love DC, Rochman CM, Neff RA (2018) Microplastics in seafood and the implications for human health. Curr Environ Health Rep 5:375–386. https://doi.org/10.1007/s40572-018-0206-z
Steinmann P, Keiser J, Bos R, Tanner M, Utzinger J (2006) Schistosomiasis and water resources development: systematic review, meta-analysis, and estimates of people at risk. Lancet Infect Dis 6(7):411–425. https://doi.org/10.1016/S1473-3099(06)70521-7
Teuten EL, Saquing JM, Knappe DR, Barlaz MA, Jonsson S, Björn A et al (2009) Transport and release of chemicals from plastics to the environment and to wildlife. Philos Trans R Soc B: Biol Sci 364(1526):2027–2045. https://doi.org/10.1098/rstb.2008.0284
Thrush SF, Hewitt JE, Cummings VJ, Ellis JI, Hatton C, Lohrer A, Norkko A (2004) Muddy waters: elevating sediment input to coastal and estuarine habitats. Front Ecol Environ 2(6):299–306. https://doi.org/10.1890/1540-9295(2004)002[0299:MWEIST]2.0.CO;2
UNEP (2011) Report on oil spill pollution impacts in Nigeria's Niger Delta
UNEP (2016) Marine litter in the African region: a regional overview. United Nations Environment Programme
UNEP (2018) Africa's blue economy: a policy handbook. United Nations Environment Programme. https://wedocs.unep.org/handle/20.500.11822/25430
UNEP (2019) Report on PCBs, dioxins, and industrial waste impacts
UNEP (2021a) Report on nutrient runoff, eutrophication, and habitat degradation
UNEP (2021b) From pollution to solution: a global assessment of marine litter and plastic pollution. United Nations Environment Programme. https://www.unep.org/resources/pollution-solution
UNEP (2021c) Making peace with nature: a scientific blueprint to tackle the climate, biodiversity and pollution emergencies. United Nations Environment Programme. https://www.unep.org/resources/making-peace-nature
United Nations Environment Programme (UNEP) (2016) Marine plastic debris and microplastics: global lessons and research to inspire action and guide policy change. UNEP
Wang J, Tan Z, Peng J, Qiu Q, Li M (2020) The behaviors of microplastics in the marine environment. Mar Environ Res 113:7–17. https://doi.org/10.1016/j.marenvres.2015.10.014
Wepener V, et al (2020a) Study on contamination levels in top predator fish
Wepener V, Mbhele S, Cyrus D (2020b) Assessment of heavy metals in estuarine environments of South Africa. Afr J Aquat Sci 45(1):43–56. https://doi.org/10.2989/16085914.2020.1712213
WHO (2017) Report on mercury toxicity and health risks
WHO (2020) Report on foodborne illnesses linked to contaminated seafood
WHO (2022a) Report on filter feeders as bioindicators and high-risk foods
WHO (2022b) Guidelines for safe recreational water environments: volume 1—coastal and fresh waters. World Health Organization. https://www.who.int/publications/i/item/9241545801
World Bank (2017) The potential of the blue economy: increasing long-term benefits of the sustainable use of marine resources for small Island developing states and coastal least developed countries. World Bank. https://www.worldbank.org/en/news/feature/2017/06/06/blue-economy

World Health Organization (WHO) (2019) Microplastics in drinking water. https://www.who.int/publications/i/item/9789241516198

Wright SL, Kelly FJ (2017) Plastic and human health: a micro issue? Environ Sci Technol 51(12):6634–6647. https://doi.org/10.1021/acs.est.7b00423

Wright SL, Thompson RC, Galloway TS (2013) The physical impacts of microplastics on marine organisms: a review. Environ Pollut 178:483–492. https://doi.org/10.1016/j.envpol.2013.02.031

Zhao Y, Zhang B, Wang D (2019) Smart wastewater treatment: a review of current technologies and future directions. Water Res 148:60–72. https://doi.org/10.1016/j.watres.2018.10.046

Chapter 5
Unlocking the Potential of Namibia's Small-Scale Coastal Marine Fisheries: Historical Context, Socioeconomic Impact, and Sustainable Management

Tamuka Nhiwatiwa, Martin Tjipute, and Michael N. Mateus

Abstract Small-scale fisheries (SSF) are vital to global food security and livelihoods, contributing to 40% of the world's fishery catch and supporting nearly 500 million people. In Namibia, the small-scale coastal marine fisheries (SSCMF) sustains coastal communities through employment, nutrition, and cultural practices, yet they remain marginalized in policy and resource access. This paper examines Namibia's SSCMF through three lenses: historical exploitation under colonial rule, socioeconomic contributions to local economies, and pathways toward sustainable management. Despite its significance, the SSCMF faces systemic challenges, including competition with industrial fisheries, climate change, and inadequate legal recognition under Namibia's Marine Resources Act (2000). Drawing on case studies from other countries (Mozambique, Tanzania, Senegal, Ghana, and South Africa), this paper highlights the successful models of community-based management and gender-inclusive policies that Namibia could adopt. Key recommendations include enacting dedicated SSF legislation, improving infrastructure, and securing funding for climate change adaptation. By addressing these gaps, Namibia can unlock the sector's potential to enhance food security, livelihoods, and marine conservation. These findings underscore the urgent need for inclusive policies that prioritize small-scale fishers and ensure resilience in the face of global and environmental pressures.

Keywords Small-scale coastal marine fisheries (SSCMF) · Namibia · Coastal livelihoods · Food security · Sustainable fisheries management

---

T. Nhiwatiwa (✉) · M. Tjipute · M. N. Mateus
Department of Fisheries and Ocean Sciences, Faculty of Agriculture, Engineering, and Natural Sciences, Sam Nujoma Campus, University of Namibia, Henties Bay, Namibia
e-mail: tnhiwatiwa@unam.na

H. A. Mupambwa et al. (eds.), *Harnessing Coastal and Marine Resources for Improved Food Security in Africa*, Sustainability Sciences in Asia and Africa,
https://doi.org/10.1007/978-981-95-9725-3_5

5.1 Introduction

Small-scale fisheries (SSF) are a cornerstone of global marine fisheries, contributing approximately 40% of the global catch and supporting the livelihoods of over 492 million people, including 45 million women (FAO 2021). Beyond its economic significance, SSF plays a critical role in food security, providing essential micronutrients to vulnerable coastal and inland communities, particularly in developing economies (DECs) where alternative protein sources are often scarce (Thilsted et al. 2016). In Namibia, SSF, encompassing both marine and freshwater fisheries, sustains over 280,000 people (11% of the population) through employment, nutrition, and cultural practices (MFMR 2021). Despite their importance, the SSF remains systematically marginalized in policy and resource allocation, a pattern seen across much of Africa.

This marginalization is not unique to Namibia. In Mozambique, for example, small-scale coastal marine fisheries (SSCMF) account for 75% of total marine catches, yet their contributions have historically been underreported in official FAO data by a factor of 6.2 due to inadequate monitoring and political neglect (Jacquet et al. 2010). Similarly, Namibia's Marine Resources Act (2000) has long prioritized industrial fisheries, leaving SSCMF in an informal and vulnerable state (MFMR 2021). Reconstructed catch data from Mozambique reveal that SSCMF landings peaked in the mid-1980s before declining—a trend driven by overexploitation and weak governance (Jacquet et al. 2010). Without intervention, Namibia risks following the same trajectory.

The challenges faced by SSCMF are compounded by competition from industrial fleets, habitat degradation, and climate change. In Tanzania, SSCMF employs over 43,000 fishers using traditional methods, such as basket traps and canoes; however, destructive practices, such as dynamite fishing and industrial trawler encroachment, have severely depleted fish stocks (Jiddawi and Öhman 2002). Namibia faces similar pressures, with coastal communities experiencing declining catches due to habitat loss and overfishing (MFMR 2021). These parallels across African nations highlight the urgent need for legal recognition of SSCMF, inclusive governance models, and stronger environmental protection (March and Failler 2014, 2022).

Despite these vulnerabilities, SSCMF remains indispensable for local food systems. Although global fisheries catches have declined since the mid-twentieth century, SSCMF landings have increased to 95% of which are consumed domestically, underscoring their vital role in nutrition security (Pauly and Zeller 2016). However, 64% of SSCMF in DECs are now struggling to meet local dietary needs due to overexploitation, climate stressors, and policy neglect (Canty and Deichmann 2022). Namibia's coastal communities reflect this troubling trend in which diminishing fish stocks threaten livelihoods and food access.

This chapter seeks to address critical gaps in the literature by examining Namibia's SSCMF through historical, socioeconomic, and policy perspectives.

While existing research, such as Nayak et al. (2014), links SSCMF poverty to socio-ecological disconnection, and Sowman et al. (2014) advocate for rights-based reforms in South Africa, a few studies have explored Namibia's colonial legacy and its enduring impact on fisheries governance. Although Namibia has adopted the FAO's SSCMF Guidelines (2015) and launched a National Plan of Action for SSCMF (2022–2026), its implementation remains weak and unmonitored, particularly in gender equity and community participation (Kadhila et al. 2024).

To bridge these gaps, this chapter employs a mixed-methods approach, combining historical analysis, socioeconomic data, and policy reviews to:

1. Examine the historical exploitation and marginalization of the SSCMF in Namibia under colonial and post-independence policies
2. Assess the socioeconomic contributions of SSCMF, including employment, nutrition, and gender roles in coastal communities
3. Evaluate gaps in Namibia's legal frameworks (e.g., the Marine Resources Act) compared to successful models in Senegal, Ghana, and South Africa
4. Analyze climate change impacts on SSCMF resilience and food security
5. Propose actionable recommendations for sustainable management, emphasizing cogovernance, gender inclusion, and climate finance

By addressing these objectives, this chapter aims to support Namibia's transition toward equitable and sustainable SSCMF policies, aligning with global sustainability goals (for example, SDG 14. b) while preserving cultural heritage and ecological resilience.

To help structure the complex and multidimensional analysis presented in this chapter, an organizing framework was developed (Fig. 5.1). This framework maps the progression of Namibia's small-scale coastal marine fisheries (SSCMF) sector from its historical roots and legal marginalization through its socioeconomic and environmental challenges, culminating in forward-looking strategies for sustainable management. It also reflects the logical flow of the chapter, with each thematic block representing a major section. This approach enables a holistic understanding of the interlinked drivers, impacts, and policy pathways necessary to unlock the full potential of Namibia's SSCMF.

Fig. 5.1 Presentation organizing framework for Namibia's Small-scale coastal marine fisheries sector

5.2 Methods

Figure 5.2 shows Namibia's coastline along the South Atlantic Ocean, which stretches for approximately 1500 km from the mouth of the Orange River in the south to the Kunene River on the northern border with Angola. The map highlights key coastal small-scale fisheries (SSCMF) locations, including Lüderitz in the south, Walvis Bay and Swakopmund in the central region, and Henties Bay further north. These towns are focal points for landing, processing, and distributing marine resources, and are strategically positioned along major fishing grounds influenced by the Benguela Current system. The inland capital, Windhoek, while not a fishing

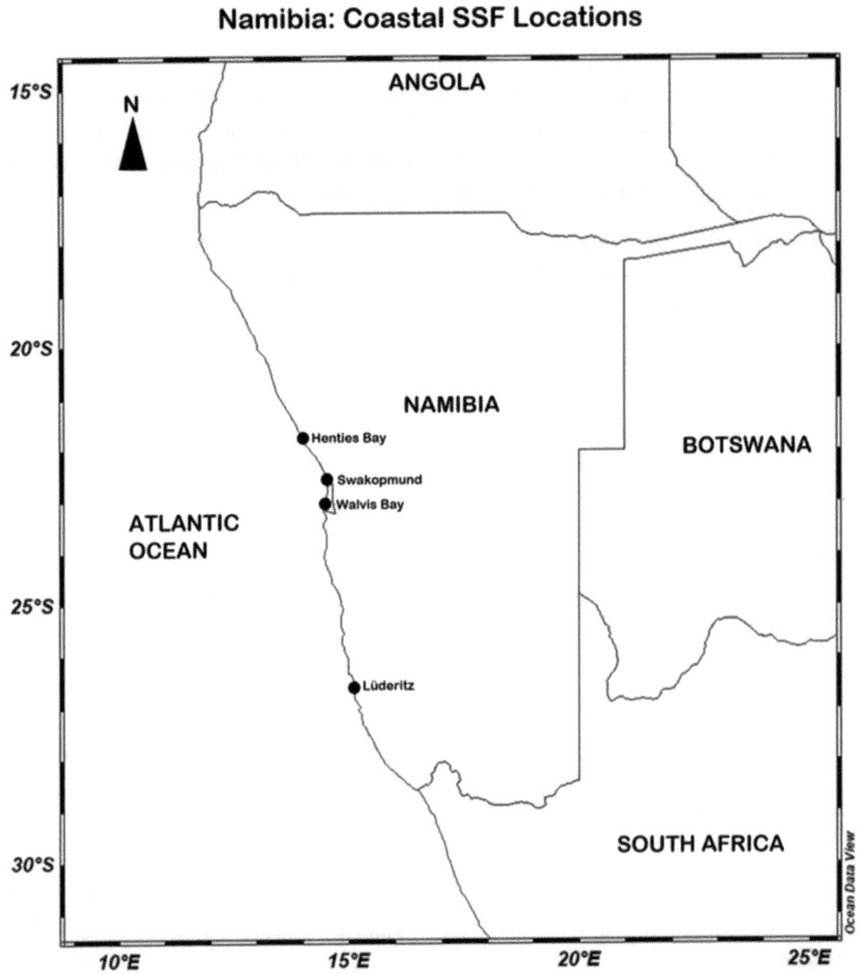

Fig. 5.2 Map of Namibia showing the locations of most coastal SSCMF activities. (Source: Ocean data view)

hub, plays a critical role in policy-making, trade coordination, and infrastructure support for the sector. This geographic context underscores the spatial distribution of Namibia's SSCMF activities and their relationship to both domestic markets and regional fisheries management zones.

The paper adopted a qualitative, interdisciplinary review approach to analyze Namibia's coastal small-scale fisheries (SSCMF) within both regional and global policy contexts. The literature review drew from peer-reviewed publications spanning 1990 to 2025 that addressed SSCMF governance, socioeconomic impacts, gender roles, and climate vulnerability in Namibia and comparable African contexts. It also incorporated key policy and legal documents, including Namibia's Marine Resources Act (2000), Inland Fisheries Resources Act (2003), the National Plan of Action for SSCMF (2022–2026), and the FAO Voluntary Guidelines for Securing Sustainable Small-Scale Fisheries. Regional case studies from Mozambique, Tanzania, South Africa, Senegal, and Ghana were selected if they met at least two inclusion criteria: the SSCMF sector accounted for at least 20% of national marine catch; there was a documented history of policy reform relevant to SSCMF; explicit integration of gender equity or community-based management was evident; and/or data were available in FAO reports or peer-reviewed sources.

A thematic coding approach was employed to organize the literature into five core themes: historical exploitation and marginalization; socioeconomic contributions and gender roles; legal and policy frameworks; environmental and climate challenges; and occupational health and safety. Within each theme, Namibia's context was compared with selected regional case studies to identify converging challenges and divergent policy responses. Comparative lessons were then derived using a policy gap analysis, which mapped Namibia's existing policy instruments and implementation status against successful or promising approaches in comparator countries.

5.3 Historical Context and Marginalization

Namibia's fisheries sector was profoundly shaped by colonial exploitation under German and later South African rule, which systematically prioritized commercial fishing operations at the expense of local communities. Post-independence reforms, including the Marine Resources Act (2000), sought to revive the sector but largely overlooked small-scale fisheries (SSF), relegating them to informal status without legal recognition or resource access (MFMR 2021). The marginalization of the SSF is a recurring theme across Africa, and this pattern of marginalization is not unique to Namibia. In South Africa, decades of industrial prioritization and institutionalized racism have systematically excluded small-scale fishers from resource access (Sowman et al. 2014). For instance, in Tanzania, SSCMF, which employs over 43,000 fishers using traditional methods such as basket traps and canoes, has similarly been neglected by policies favoring industrial fleets (Jiddawi and Öhman 2002). Destructive practices such as dynamite fishing and drag nets have further degraded habitats and reduced yields, compounding the challenges faced by artisanal fishers (Jiddawi and Öhman 2002). Namibia's SSCMF faces parallel struggles, including habitat loss and competition from industrial trawlers, highlighting the urgent need for legal recognition and habitat restoration (MFMR 2021). Despite systemic exclusion, indigenous fishing practices such as seasonal floodplain fishing (efundja) in northern Namibia and women-led fish drying have persisted, demonstrating the resilience of SSF communities (FAO 2021). Namibia's recent National Plan of Action for SSF (NPOA-SSF 2022–2026) marks a potential turning point, formally recognizing SSF as critical for livelihoods and food security, with technical support from the FAO (MFMR 2022). However, meaningful implementation remains a challenge.

5.3.1 Exclusion from Decision-Making and Lessons from Regional Reforms

A key issue across African SSCMF sectors is their exclusion from the decision--making processes. In Kenya and Tanzania, port development projects have displaced fishers and degraded critical nearshore habitats such as coral reefs and mangroves, often without fisher consultation (Thoya et al. 2022). Namibia's Marine Resources Act (2000) similarly overlooks the SSCMF, exacerbating its informal status and limiting access to fishing grounds. Lessons from South Africa's 2012 Small-Scale Fisheries Policy, which adopted a rights-based approach to redress historical inequities, can inform Namibia's reform (Sowman et al. 2014). Community--led initiatives, such as Senegal's Ngaparou village comanagement system, further demonstrate how integrating fishers' committees into governance structures can mitigate conflicts and enhance sustainability (March and Failler 2022).

5.3.2 Poverty as a Process: Institutional Neglect and Ecological Degradation

The challenges facing SSCMF are not merely economic but rooted in intertwined socioecological and institutional factors. Nayak et al. (2014) argue that poverty in small-scale fisheries stems from a vicious cycle of ecological degradation (e.g., habitat loss), economic exclusion, and political disempowerment. Their studies in the Chilika Lagoon (India) and Paraty (Brazil) revealed how caste and class hierarchies exacerbate vulnerability, trapping fishers in cycles of indebtedness and displacement. These findings resonate with Namibia's context, where colonial legacies and industrial competition continue to marginalize local fishers (Nayak et al. 2014).

5.3.3 *Underreporting and Systemic Neglect of SSCMF Catches*

Systemic neglect of SSCMF is further evidenced by chronic underreporting of catches. Jacquet et al. (2010) found that SSCMF landings in Mozambique and Tanzania were underreported in FAO data by factors of 6.2 and 1.7, respectively, due to inadequate monitoring and policy bias toward industrial fisheries. This mirrors Namibia's experience that the Marine Resources Act (2000) prioritizes industrial fleets while leaving SSCMF catches unrecorded and unregulated (Jacquet et al. 2010). Accurate data collection and inclusive policy frameworks are essential to reversing historical oversight and ensuring sustainable SSCMF management.

5.3.4 *Pathways to Equitable and Sustainable SSCMF Management*

Namibia's SSCMF sector is at a crossroads. While the NPOA-SSCMF (2022–2026) represents progress, its success hinges on addressing historical inequities, integrating SSCMF into governance, and adopting lessons from regional models such as South Africa's rights-based approach and Senegal's comanagement systems. Recognizing the SSCMF's contributions to food security, livelihoods, and cultural heritage while tackling challenges such as climate change, habitat loss, and industrial competition will be critical for achieving equitable and sustainable fisheries management in Namibia.

5.4 Marine-Sector Information (Status, Statistics, and Context)

Namibia's coastal small-scale fisheries (SSCMF) are primarily concentrated in the Erongo and!Karas Regions, with key hubs in Walvis Bay, Swakopmund, Henties Bay, and Lüderitz. These fisheries emerged in the mid- to late-twentieth century as an economic alternative for work-seekers who were unable to secure industrial jobs, leading to the development of fishing, shellfish gathering, and processing activities that continue today, involving both men and women.

The sector was divided into several fishery categories, each targeting a different species. Shellfish and seaweed collection, predominantly carried out by women, occurs in the intertidal zones of Walvis Bay, Swakopmund, and Lüderitz. The line-fishery, based on Henties Bay and Lüderitz, catches species such as Kabeljou, Galjoen, Steenbras, black tail, sea catfish, elf/shad, and snoek. Rock-lobster fishing (*Jasus lalandii*) is centered in Lüderitz, while Mullet fisheries and recreational angling further complement the sector (Tables 5.1 and 5.2).

Table 5.1 Primary species and ecosystems utilized by coastal small-scale fishers

Cluster/ regional ecosystem	Description	Primary species
Coastal North and South Erongo and!Karas	The coastal SSCMF predominantly utilizes marine resources (Henties Bay, Swakopmund, Walvis Bay, and Lüderitz).	Shellfish; seaweed; Kabeljou (*Argyrosomus japonicus*); Galjoen (*Dichistius capensis*); Steenbras (*Lithognathus lithognathus*); blacktail (*Diplodus sargus*); sea catfish; rock lobster (*Jasus lalandii*); snoek; mullets; elf/shad (*Pomatomus saltatrix*); limpets & mussels

Source: NPOA SSCMF

Table 5.2 Legal, economic, environmental, social, and occupational challenges constraining Namibia's small-scale marine fisheries sector, with interlinked impacts

Dimension	Challenges
Legal and institutional	No dedicated SSCMF legislation; exclusion from decision-making
Economic	Limited market access; underreporting of catches; low investment
Environmental	Habitat degradation; climate change impacts
Social and gender	Women's roles undervalued; safety risks; cultural disconnection
Occupational health	Unsafe conditions; inadequate protection; long-term impairments

In terms of scale and economic impact, coastal SSCMF directly employs approximately 1500 fishers and supports the livelihoods of approximately 9500 people, including dependents (Table 5.3). Preliminary estimates place the annual economic output at N$ 27.6 million, although the landing data remain incomplete. For example, the Hanganeni Artisanal Fishers Association (HAFA) in Henties Bay lands roughly 300 tonnes annually, valued at N$ 2.6 million, while the Benguela Ski-Boat Fishers Association (BSBFS) in Swakopmund contributes an estimated N$ 25 million per year from line-fish catches, employing 140 crew across 35 vessels. However, under-reporting of catches obscures the sector's true contribution to the GDP.

Table 5.2 summarizes the key interconnected challenges confronting Namibia's small-scale fisheries, spanning legal, economic, environmental, social, and occupational dimensions. These issues are not isolated; weak legal recognition exacerbates economic marginalization, environmental degradation compounds social vulnerability, and inadequate safety measures reflect deeper systemic neglect. Understanding how these dimensions interact provides a foundation for targeted policy reforms that address root causes rather than symptoms.

Organizationally, the sector includes groups such as HAFA (140 members), the Namibia Shore Angling Association (NSAA, 410 members), BSBFS (35 vessels/45 members, 140 jobs), and the Namibian Women's Fisheries Network (NAMFISHNET, 60 members). Despite these associations, governance remains weak with the need for stronger comanagement structures. Legally, Namibia's Marine Resources Act

Table 5.3 Status of small-scale marine fisheries organizations in Namibia

Subsector cluster/ regional ecosystem	Region/locality	Organization/initiative	Membership	Employment
Coastal SSF	Henties Bay, Erongo	Hanganeni Artisanal Fisheries Organisation (HAFA)	140 members	
Coastal SSF	Erongo	Namibia Shore Angling Association (NSAA)	410 members	6–40 part/ full-time jobs
Coastal SSF	Swakopmund, Erongo	Benguela Ski Boat Fisheries Association (BSBFS)	45 members (35 vessels)	140 direct jobs
Coastal SSF	Swakopmund, Erongo	Centre for Marine Environmental Education & Sustainability (CEMEES) Trust	8 members	
Coastal SSF	Lüderitz, /!Karas	Namibian Women's Fisheries Network (NAMFISHNET)	60 members	

Source: NPOA—SSF

(2000) focuses on industrial fisheries and neglects small-scale fisheries, leaving coastal SSCMF in a regulatory gap. The National Plan of Action for Small-Scale Fisheries (NPOA-SSF) recommends a dedicated legal framework to address access conflicts with industrial fleets and unclear tenure rights. Gender dynamics highlight women's critical roles in postharvest processing and inshore gatherings. The plan advocates for gender-transformative support, including training, finance, and market access, alongside a subprogram targeting gender equity and climate resilience in food systems.

Climate change poses significant risks, with rapid vulnerability assessments in the Benguela Current Large Marine Ecosystem identifying socioecological threats requiring localized adaptation strategies. Additionally, data gaps persist in catch reporting, economic valuation, and socioeconomic impact assessments, necessitating systematic monitoring to align marine SSCMF with inland fisheries for policy considerations.

5.5 Sociocultural Dimensions: Identity, Livelihoods, and Gender Roles

Small-scale fisheries in Namibia represent far more than just an economic activity; they form the fabric of cultural identity and community heritage for coastal populations. Recent field research in Swakopmund and Walvis Bay has revealed that SSCMF communities view the ocean as central to their way of life, with fishing practices deeply embedded in cultural traditions and daily rhythms (Kadhila et al.

2024). However, this intrinsic connection faces growing threats from restrictive policies such as Namibia's Marine Resources Act (2000), which imposes stringent regulations, including daily catch limits (10 fish per day) and fishing zone restrictions without meaningful community consultation. This top-down governance approach has not only disrupted traditional fishing practices but also reflects a broader global pattern of institutional neglect that perpetuates cycles of poverty in SSCMF communities (Nayak et al. 2014).

The cultural significance of SSCMF extends beyond mere subsistence in Namibia. Kadhila et al. (2024) demonstrate how communities like the Topnaar people relocated to areas such as Gobabeb and Sesfontein and maintained their identity through deep ecological connections—whether through processing the!Nara plants or preservation songs from natural water sources. This challenges conventional notions of identity tied solely to ethnicity, highlighting how fishing communities construct their identities through their relationships with marine ecosystems. These findings align with the Nayak et al. (2014) framework of socioecological systems, where the loss of access to traditional fishing grounds, often due to competing interests such as tourism development in Swakopmund, simultaneously erodes both economic livelihood and cultural continuity.

Despite progressive policy instruments such as Namibia's National Plan of Action for SSCMF (2022–2026), implementation gaps remain significant. Fisher communities reported minimal meaningful engagement with the Ministry of Fisheries and Marine Resources, and interviews revealed strong demands for participatory governance structures that balance conservation needs with livelihood security (Kadhila et al. 2024). This mirrors South Africa's experience, where SSCMF communities have only achieved legal recognition after decades of persistent activism (Sowman et al. 2014). While emerging fisher associations like Hanganeni Fishing Association (HAFA) in Henties Bay and the Benguela Infinite Fisheries Harvesting Association show promise as platforms for collective bargaining, systemic barriers to inclusion persist throughout Namibia's fisheries governance framework.

The critical yet undervalued role of women in SSCMF systems represents another layer of marginalization. Globally, women's contributions to fisheries, particularly in postharvest activities such as processing and marketing, remain systematically underrepresented in official statistics and policy frameworks (FAO 2013; Bradford and Katikiro 2019). This pattern holds across Africa, with Tanzania serving as a clear example that women dominate fish processing while being excluded from formal recognition (Bradford and Katikiro 2019). Namibia reflects this same dynamic, where women undertake 70–80% of dried catfish processing yet face persistent barriers to market access and policy inclusion (FAO 2021). The Marine Resources Act's industrial focus has particularly disadvantaged women's fishing activities, reinforcing gendered economic disparities (MFMR 2021; Kadhila et al. 2024).

Regional case studies have revealed the extent of this gap in gender data. In Mozambique, the crucial role of women in holothurian (sea cucumber) processing was completely omitted from official FAO reports until retrospective data correction included these activities (Doherty et al. 2015). Similarly, Mozambique's 2009–2010 fisheries data revisions demonstrated how gender-disaggregated reporting can align policies with on-the-ground realities; a lesson Namibia must urgently apply through its NPOA-SSF implementation (MFMR 2022). The multidimensional nature of women's fishing activities becomes particularly evident in northern Mozambique, where Samoilys et al. (2019) documented artisanal fishers using 11 different gear types to harvest 153 taxa. Their study highlighted how women predominantly engage in gleaning and mosquito net fishing methods, yielding modest per-capita catches (2–4 kg/trip), but providing essential subsistence and income. These findings directly parallel Namibia's context, where women dominate intertidal gleaning and postharvest processing, underscoring the need for gender-sensitive approaches to SSCMF policy development (Kadhila et al. 2024).

The Namibian experience ultimately reflects broader regional patterns where SSCMF systems, with their complex cultural, economic, and gender dimensions, continue to be marginalized despite their vital contributions. Addressing these challenges requires moving beyond tokenistic participation to establish genuine comanagement structures that recognize SSCMF communities as equal partners in governance. Learning from regional examples such as Mozambique's data reforms and South Africa's rights-based approach could help Namibia develop more inclusive policies that protect both the marine resources and communities that depend on them. As Namibia implements its NPOA-SSF, centering on community knowledge, addressing gender disparities, and ensuring meaningful participation is critical for achieving sustainable and equitable fisheries management.

5.6 Socioeconomic Contributions and Challenges

By learning from regional experiences and centering on community voices, Namibia can develop SSCMF policies that honor cultural heritage while ensuring ecological and economic sustainability. The ocean's rhythm must continue to beat in harmony with communities that depend on it. Freshwater small-scale fisheries play a crucial role in Namibia's economy and food security, supporting approximately 280,000 people (approximately 11% of the population) through direct employment and nutrition (MFMR 2021). These fisheries are particularly vital along the Zambezi and Kavango Rivers, where they contribute an estimated $ 230 million annually to the local economy, although this significant contribution often goes unrecognized in national accounting systems (MFMR 2022). The national sector exhibits clear gender divisions, with men predominantly engaged in active fishing, while women process 70–80% of the dried catfish sold in markets such as Katima Mulilo (FAO 2021). This gendered labor distribution mirror pattern is seen throughout Africa, such as in Tanzania, where SSCMF contributes 1.4% to mainland GDP and 5.7% in Zanzibar, while supplying 30% of animal protein (Jiddawi and Öhman 2002; Onyango and Yahya 2022).

However, Namibian SSCMF faces multiple threats that jeopardize its sustainability. Overfishing by both industrial and artisanal fleets, combined with climate change impacts such as warming waters and shifting fish stocks, has led to significant resource depletion. Coastal development exacerbates this problem via habitat degradation. Policy barriers compounded these challenges, particularly the Marine Resources Act (2000), which disproportionately favors industrial operators while marginalizing small-scale fishers. Women in the sector face additional hurdles: they lack access to formal markets and their catches are often diverted to export markets at the expense of local food security. These issues are worsened by systemic data gaps that lead to underreporting of SSCMF contributions and exclusion of fishers from the decision-making processes.

The situation in Namibia reflects broader regional trends. Mozambique's experience demonstrates how official catch reports can dramatically underestimate SSCMF contributions; reconstructed data show that actual catches were 6.2 times higher than reported figures (Jacquet et al. 2010). Across developing nations, 64% of SSCMF now struggle to meet nutritional needs because of overexploitation and climate stressors (Canty and Deichmann 2022). Mozambique's shifting catch composition, where demersal fish declined from 35% to 10% of catches between 1950 and 2010 while small pelagics increased, serves as a warning for Namibia regarding potential ecosystem stress (Doherty et al. 2015).

Regional examples offer valuable lessons for Namibia's path. Despite the ongoing challenges of youth unemployment and postharvest losses (Onyango and Yahya 2022), Tanzania's Beach Management Units (BMUs) show promise in balancing conservation needs with livelihood protection. Zanzibar's women-led seaweed farming initiatives demonstrate how alternative livelihoods can enhance community resilience (Msuya and Hurtado 2017), a model that can be adapted to Namibia's aquaculture sector. Nayak et al. (2014) emphasize that the collapse of SSCMF represents more than just economic loss; it erodes the cultural identity and intergenerational knowledge that are fundamental to coastal communities.

Comprehensive policy reforms are required to ensure the success of Namibia's SSCMF. These should include formal recognition of SSCMF in national accounts and legislation, adoption of ecosystem-based management approaches to address trophic cascades, and specific measures to empower women through improved market access and decision making. Developing climate-resilient alternative livelihoods, such as seaweed farming or eco-tourism, could help diversify income sources while reducing pressure on fish stocks. Most crucially, Namibia must strengthen comanagement systems that actively involve fishing communities in resource governance. By learning from both regional successes and failures, Namibia can develop SSCMF systems that are equitable, productive, and resilient enough to withstand mounting environmental and economic challenges while preserving the cultural heritage embedded in traditional fishing practices.

5.7 Legal and Policy Frameworks

Namibia's current fisheries legislation reveals significant gaps in supporting marine small-scale coastal marine fisheries (SSCMF). The Marine Resources Act (2000) primarily serves commercial fishing interests, whereas the Inland Fisheries Resources Act (2003) vaguely addresses subsistence fishing (MFMR 2021). This legislative overview mirrors the challenges faced in Africa. For instance, in Tanzania, the Marine Parks Act (1994) and the Zanzibar Fisheries Act (2010) similarly fail to incorporate community rights or gender equity (de la Torre-Castro et al. 2017), highlighting a regional pattern of SSF marginalization in fishery governance.

The exclusion of SSF from decision-making processes is particularly detrimental. Mozambique's experience demonstrates how top-down governance and fragmented data systems obscured SSF contributions for decades, with reconstructed catch data revealing underreporting by a factor of 6.2 in official statistics (Jacquet et al. 2010). Namibia risks similar outcomes in the absence of policy reforms. The country's National Plan of Action for SSF (NPOA-SSF 2022–2026), while aligned with the FAO's SSF Guidelines (2015), lacks the legal teeth to secure tenure rights or promote gender equity in practice (FAO 2015; MFMR 2021).

Several African models offer valuable lessons for Namibia's policy development. Tanzania's Beach Management Units (BMUs) in Lake Victoria demonstrate how local comanagement can reduce illegal fishing, although challenges such as corruption persist (LVFO 2020). Senegal's Ngaparou village approach combines artificial reefs with fisher committees to restore habitats and secure tenure (March and Failler 2022). Perhaps the most instructive is South Africa's 2012 Small-Scale Fisheries Policy, which adopted a human rights-based approach by allocating fishing rights to collective entities rather than individuals and mandating gender-balanced representation (Isaacs 2006; Sowman et al. 2014).

However, the implementation of these policies faces substantial challenges. Traditional compliance measures often criminalize small-scale fishers without addressing root causes such as inequitable resource access (Hauck 2008). Namibia's experience with the Hanganeni Artisanal Fishing Association (HAFA) in Henties Bay shows promise, demonstrating how organized marketing can secure livelihoods (MFMR 2021). However, broader issues persist, particularly regarding the impact of industrial fishing. In Mozambique, industrial shrimp trawlers generate discards at ratios of 2.3:1 to 5:1 relative to landings, with weak enforcement of bycatch regulations (Banks and Macfayden 2011; Doherty et al. 2015); a pattern Namibia must avoid through stronger policy frameworks.

The integration of robust data collection and participatory approaches is critical to successful SSCMF governance. Mozambique's use of creel surveys, focus group discussions, and Geographic Information Systems (GIS)-based participatory mapping has revealed important disconnections between local fishers' perceptions and formal policies (Samoilys et al. 2019). These methods could help Namibia develop more inclusive legislation that addresses gender gaps, a persistent issue, as seen in

Tanzania's Zanzibar Fisheries Act (2010) and Namibia's Marine Resources Act (de la Torre-Castro et al. 2017).

Namibia's Marine Resources Act (2000) serves as the primary legal framework governing fisheries, yet it overwhelmingly favors industrial and commercial operations, leaving small-scale coastal marine fisheries (SSCMF) systematically marginalized (MFMR 2021). The Act broadly defines harvesting to include catching, taking, or attempting to catch marine resources, along with preparatory activities (Section 1). However, it imposes stringent requirements on commercial fishing rights, exploratory rights, and fisheries agreements (Section 32), which are often financially and bureaucratically prohibitive for small-scale fishers. Crucially, the Act does not explicitly differentiate the SSCMF from industrial fisheries or establish mechanisms to secure tenure rights or quotas for small-scale operators. This oversight reflects a long-standing policy bias toward large-scale export-driven fisheries rooted in colonial and post-independence economic priorities (Sowman et al. 2014).

Gender disparities in SSCMF are another critical gap in legislation. Despite women's central role in postharvest activities, such as processing and trading (FAO 2021), the Act contains no provisions to ensure their representation or equitable access to resources. While it establishes the Marine Resources Advisory Council (Section 24) to advise the fisheries minister, the council's composition lacks mandatory SSCMF representation or gender balanced participation (Section 25). Similarly, the Fisheries Observer Agency (Section 8) focuses primarily on monitoring industrial operations with no explicit mandate to support SSCMF data collection or compliance (Section 9). The licensing regime (Section 40) further disadvantages SSCMF by requiring formal vessel registration and quotas, barriers that many artisanal fishers, who often operate informally, cannot overcome (MFMR 2022).

The Act also fails to incorporate community-based management or cogovernance models that are essential for SSCMF sustainability. For instance, while the minister has the authority to declare marine reserves (Section 51), the Act does not integrate SSCMF access rights or traditional knowledge into reserve management plans. This contrasts sharply with South Africa's 2012 Small-Scale Fisheries Policy, which explicitly allocates fishing rights to collective entities and mandates gender equity–a model that Namibia could adapt (Sowman et al. 2014). Additionally, the Act's punitive measures, including fines of up to N$2 million for unlicensed harvesting (Section 52), risk criminalizing SSCMF livelihoods without addressing the systemic barriers that force many fishers into informality (Hauck 2008).

The financial provisions under the Act further entrench SSCMF marginalization. The Marine Resources Fund (Section 45) was designed to support industrial fishery research and development, with no dedicated funding for SSCMF infrastructure, training, or climate adaptation (MFMR 2022). Although the minister has the discretionary power to exempt certain activities from regulations (Section 62), this provision is rarely applied to the SSCMF, perpetuating their exclusion from formal recognition.

Recent policy developments, such as Namibia's National Plan of Action for SSF (NPOA-SSF 2022–2026), acknowledge these legislative shortcomings and advocate for reforms aligned with the FAO's Voluntary Guidelines for Securing Sustainable Small-Scale Fisheries (FAO 2015). Key recommendations include amending the Act to formally recognize SSCMF as a distinct sector, introducing gender-sensitive quota systems, and establishing community-led comanagement frameworks (MFMR 2022).

Namibia's Marine Resources Act (2000) inadequately addresses the needs of the SSCMF, reinforcing their exclusion through restrictive licensing, a lack of representation, and the absence of targeted financial support. Legislative reforms must prioritize tenure security, gender equity, and inclusive governance to harness the sector's full potential for food security, poverty reduction, and sustainable resource management. Without such changes, the SSCMF will remain vulnerable to overexploitation, climate change, and continued marginalization in Namibia's fishery sector.

5.8 Environmental and Climate Challenges

Climate change is a growing threat to Namibia's SSCMF sector, with rising ocean temperatures, habitat degradation, and shifting fish distributions that have already impacted coastal communities (FAO 2021). These environmental changes compounded the existing pressures from industrial encroachment and overfishing, creating a crisis that demanded urgent policy attention. The National Plan of Action for SSF (NPOA-SSF) recognizes these threats and promotes an Ecosystem Approach to Fisheries (EAF), along with community-led vulnerability assessments to build resilience (FAO 2021). However, its implementation faces significant hurdles, including limited state capacity, market competition, and inadequate climate financing, which mirror those experienced by South Africa's SSCMF sector (Sowman et al. 2014; FAO 2021).

The impact of climate change on SSCMF has also been observed across the region. In Mozambique, fishermen report declining inshore catches because of ecological changes, forcing them to invest in larger boats and undertake longer, more dangerous fishing trips, a pattern emerging in Namibia's fisheries (Blythe et al. 2013). These environmental shifts disrupt traditional fishing patterns and threaten food security, with studies showing that 64% of SSCMF in developing economies now struggle to meet nutritional needs owing to combined environmental and anthropogenic pressures (Canty and Deichmann 2022). Regional disparities are stark, with Southeast Asian fisheries showing greater resilience than the significant declines noted in Central and Southern Africa.

Habitat loss is a critical issue. In Tanzania, for example, mangrove deforestation for aquaculture and tourism has destroyed vital fish nurseries, while rising ocean temperatures are reducing seaweed yields which is a crucial income source for women in coastal communities (Semesi 1991; Msuya and Hurtado 2017). Namibia faces similar threats, with Walvis Bay's shellfish collectors losing traditional harvesting grounds for industrial development (One Ocean Hub 2023). These environmental changes are creating what Nayak et al. (2014) term "human-environment disconnection," whereby the erosion of fishers' physical and cultural ties to their ecosystems through processes such as aquaculture encroachment and exclusionary conservation policies.

The gendered dimensions of these climatic impacts are also of particular concern. Women in SSCMF communities, who typically rely on near-shore resources for processing and trading activities, face disproportionate vulnerability to habitat loss and changes in the fish stock (Fröcklin et al. 2014). In Tanzania, female seaweed farmers are already experiencing declining yields due to rising ocean temperatures (Suckall et al. 2014), a trend likely to emerge in Namibia's coastal fisheries in the near future. These challenges underscore the need for gender-responsive adaptation strategies in Namibia's climate planning.

Regional cooperation through mechanisms such as the Benguela Current Commission (BCC) offers potential solutions, with climate adaptation plans providing a framework for mangrove restoration and community-based resource management (BCC 2017). However, significant gaps remain in translating these plans into concrete actions that will benefit the SSCMF communities. Namibia must prioritize the integration of local ecological knowledge with scientific research, strengthen community-based monitoring systems, and secure dedicated climate financing for SSCMF adaptation measures. Without such interventions, climate change threatens to further marginalize small-scale fishers and undermine their critical role in Namibia's food security and coastal economy.

5.9 Occupational Hazards in Small-Scale Coastal Marine Fisheries: Lessons for Namibia

Small-scale coastal marine fisheries (SSCMF) serve as a crucial source of livelihood, food security, and economic stability for Namibia's coastal communities. Nevertheless, the fishermen who sustain this essential sector are confronted with a hazardous work environment, characterized by health risks, physical injuries, and long-term impairments that jeopardize their well-being and productivity. Global experiences, ranging from the sun-exposed decks of Egyptian Mediterranean boats to the storm-ravaged vessels of Bangladesh, and including the deep-diving divers of Chile and Peru, reveal patterns of hazards that Namibia must not overlook. By integrating these insights, Namibia can develop a future where fishing is not only productive but also safe and sustainable.

The daily grind of small-scale fishing exacts a heavy toll on human health. In Bangladesh, 76% of fishers report fever, and 60% battle skin diseases, contaminated water, and relentless sun exposure turn work into a health crisis (Zakaria et al. 2022). Egypt's fishers fare no better, with musculoskeletal disorders, hearing loss, and respiratory illnesses rampant due to roaring engines, backbreaking labor, and unforgiving weather (Zytoon and Basahel 2017). These are not isolated tragedies but systemic failures that Namibia can preempt. Simple measures, such as mobile clinics for regular check-ups, waterproof gloves to prevent skin lesions, and ear protection to combat noise-induced hearing loss, could improve fisher health. Clean water stations at landing sites and sanitary facilities on boats, modelled after successful interventions in Bangladesh, would further curb disease spread, turning harbors from hazard zones into hubs of safety.

When the sea becomes violent, fishermen face extreme danger. Egyptian studies show how unstable vessels become death traps, with sprains, fractures, and lacerations plaguing crews, especially the young and the elderly (Zytoon and Basahel 2017). In Bangladesh, 78% of fishermen suffer injuries due to falls or gear mishaps (Zakaria et al. 2022). Namibia's waters demand equally rigorous safeguards: life jackets and first-aid kits as non-negotiable gear, weather alert systems to prevent storms, and safety drills to prepare crew for emergencies. However, protection goes beyond gear-training fishers to handle equipment safely, as Chile's peer-to-peer programs have demonstrated, reducing accidents by over half (Garrido et al. 2019). Such practical knowledge, paired with enforced vessel inspections, could transform Namibia's fleets from vulnerable to resilient ones.

The real cost of a lifetime in fishing often becomes apparent only after many years. Aging Egyptian fishers suffer from chronic ailment-failing vision, urinary disorders, and hearing damage from decades of cold, wet labor (Zytoon and Basahel 2017). Chilean shellfish divers face a different spectrum of cognitive decline from repeated decompression sickness, with 68% showing impaired memory and decision-making (Garrido et al. 2020). Namibia has the opportunity to act before these crises take root. Routine health screenings could capture conditions early, whereas ergonomic tools and modified work practices might spare the next generation of fishers from debilitating wear-and-tear. For high-risk roles, such as diving, pre-emptive depth limits, and neurological monitoring learned from Peru's community health programs could safeguard both bodies and minds.

Climate change is a multiplier of these risks. Rising seas and fiercer storms, already exacerbating Egypt's deck hazards (Zytoon and Basahel 2017), will also test Namibia's fleets. Sustainable catch limits and seasonal closures may preserve fish stocks; however, fishers also need alternatives when the ocean becomes barren. Aquaculture or eco-tourism ventures paired with early warning systems for extreme weather could offer buffers against the unpredictability of nature. However, resilience is not just ecological—it is economic. In Egypt, financial desperation drives fishers to overload boats or brave deadly waves (Zytoon and Basahel 2017). Namibia could break this cycle by diversifying incomes, perhaps through training in fish processing or coastal tourism, ensuring that bad seasons do not become life threatening.

None of these solutions were achieved without the fishers themselves being involved. Chile's success in slashing diving injuries came when crews codesigned safety rules (Garrido et al. 2019), which is a model Namibia could replicate through fisher-led safety councils. Similarly, Bangladesh's community health partnerships show how local buy-ins can turn policies into practice (Zakaria et al. 2022). Namibia's reforms must be rooted in the collaboration between mandatory safety gear and culturally tailored training workshops. The goal is clear for a fishing sector where risks are managed, health is protected, and livelihoods thrive not just today but for generations to come.

While health and safety risks underscore the immediate human costs of small-scale fishing, they also expose deeper structural inequalities that may limit Namibia's SSCMF sector from participating in high-value fisheries. Inadequate safety equipment, limited training, and chronic health issues are symptoms of broader exclusionary systems rooted in gaps in infrastructure, financing, and governance. Nowhere is this more evident than in the small pelagic fishery, where small-scale operators face technological, financial, and regulatory barriers that restrict access and profitability. Addressing occupational safety must therefore be coupled with reforms that dismantle these systemic constraints, enabling equitable participation and sustainable growth within Namibia's marine economy.

5.10 Barriers to Small-Scale Fisher Participation in Namibia's Small Pelagic Fishery

Namibia's small-scale fisheries face significant challenges in accessing small pelagic fisheries, owing to limitations in fishing gear, technology, and vessel capacity. The sector primarily targets sardines (*Sardinops sagax*), Cape horse mackerel (*Trachurus capensis*), and anchovies (*Engraulis encrasicolus*), but the decline in sardine biomass has led to a fishing moratorium from 2018 to 2020, severely impacting small-scale fishers (Cochrane et al. 2020). The industrial sector dominates this fishery, leaving small-scale fishers marginalized because of the high costs of purse-seine gear and the need for larger, often outdated vessels. Unlike Namibia, neighboring Angola and South Africa have more diversified small pelagic fisheries that better integrate the artisanal and semi-industrial sectors. For instance, Angola allocates a portion of its total allowable catch to artisanal fishers who use smaller vessels and simpler gear, contributing significantly to food security and livelihoods (Cochrane et al. 2020). Similarly, South Africa's small pelagic fishery supports small-scale operators through adaptable dual-purpose vessels and a regulatory framework that facilitates participation, despite persistent financial constraints.

However, in Namibia, small-scale fishers struggle because of restrictive policies, inadequate infrastructure, and a lack of quota allocation. The government's focus on industrial fishing leaves small-scale operators with limited access to horse mackerel quotas, thereby further excluding them from fishing (Cochrane et al. 2020). Additionally, financial barriers prevent investment in necessary equipment, as small-scale fishers often lack access to credit and licensing costs are prohibitive (Chiripanhura and Teweldemedhin 2016). The regulatory framework further disadvantaged them as quota allocations prioritize large companies that can meet the employment and value-addition criteria that small-scale fishers cannot easily satisfy. Market access is another hurdle, with larger firms dominating export channels and small-scale fishermen being relegated to lower-value local markets. The absence of cold storage and processing facilities further limits their ability to add value to their catches, thereby reducing their profitability.

Environmental pressures such as oil exploration and phosphate mining exacerbate these challenges by disrupting fish stocks and fishing operations (Chiripanhura and Teweldemedhin 2016). Moreover, skill gaps hinder the adoption of modern fishing techniques and compliance with international standards, thereby restricting access to higher-value markets. To enhance small-scale fisher participation, Namibia must implement reforms such as equitable quota allocation, financial support mechanisms, simplified regulations, and investments in infrastructure and training. By learning from the more inclusive approaches of Angola and South Africa, Namibia can empower small-scale fishermen and ensure their sustainable integration into small pelagic fisheries.

5.11 Regional Case Studies: Lessons from Africa

Small-scale fisheries serve as vital economic and nutritional pillars across Africa, supporting millions of livelihoods through traditional, low-mechanized fishing practices rooted in local ecological knowledge. These fisheries demonstrate remarkable diversity in target species and fishing methods while facing the common challenges of overexploitation and climate pressure. Several notable examples highlight their socioeconomic importance and potential lessons for sustainable management.

In Ghana, wooden canoe fisheries dominate the small-scale sector, contributing approximately 70% of the country's marine catch (FAO 2022). These canoe crews of 10–20 fishers primarily harvest small pelagics, including sardines (*Sardinella aurita* and *S. maderensis*), anchovies (*Engraulis encrasicolus*), and mackerels (*Scomber colias*) using purse seines, drift gillnets, and encircling nets. While the adoption of outboard motors has increased efficiency, it has also intensified concerns regarding stock depletion (Atta-Mills et al. 2004).

Senegal's artisanal fisheries showcase impressive diversity, targeting species ranging from sardinella (*Sardinella* spp.) and bonga shad (*Ethmalosa fimbriata*) to high-value demersals like groupers (*Epinephelus aeneus*). Fishers employ motorized pirogues (traditional wooden boats) with various gear types including purse seines, gillnets, and handlines. A specialized octopus (*Octopus vulgaris*) fishery using traps provides crucial export income, although the sector grapples with declining stocks from both climatic and anthropogenic pressures (Thiao et al. 2012).

South Africa's small-scale coastal fisheries employ distinct techniques for species, such as snoek (*Thyrsites atun*), yellowtail (*Seriola lalandi*), and cape bream (*Pachymetopon blochii*). The use of handlines, rod-and-reel, and traditional fyke nets (hoop nets) reflects the sector's artisanal character, whereas small trawls operate in limited areas for shrimp and lobsters. Despite government efforts to implement rights-based management, tensions persist between small-scale and commercial operators (DAFF 2016).

Mozambique's SSCMFs operate across coastal and estuarine environments, harvesting shrimp (*Penaeus indicus*), reef fish (*Lethrinus* spp. and *Lutjanus* spp.), and mollusks using gillnets, beach seines, and ingeniously crafted bamboo/mangrove fish traps. Octopus harvesting via spears and hooks provides additional income; however, industrial trawling and reef degradation pose significant threats (Jacquet et al. 2010).

Community-led initiatives demonstrate innovative adaptations, particularly in women-led enterprises. Tanzania's Tweyambe Fishing Enterprise and Zanzibar's Seaweed Cluster Initiative exemplify how collective action through savings schemes, value-added production, and advocacy can strengthen livelihoods (Onyango and Jentoft 2011; Msuya and Hurtado 2017). Such models offer valuable templates for Namibia, where supporting organizations such as the Hanganeni Artisanal Fishing Association (HAFA) with improved infrastructure (e.g., swakopmund fish markets) and climate finance could significantly reduce postharvest losses (MFMR 2022).

These diverse African SSCMF systems underscore the sector's resilience and vulnerability. Their collective experiences highlight the urgent need for management approaches that balance ecological sustainability with the protection of traditional livelihoods, while providing models for innovation in governance, gender equity, and climate adaptation.

5.12 Critical Success Factors for Sustainable Small-Scale Fisheries in Africa

Small-scale fisheries (SSFs) in Africa serve as vital economic and nutritional lifelines to support millions of coastal and inland communities through food security, poverty alleviation, and employment opportunities. Their long-term success depends on several interconnected factors that balance ecological, economic, social and governance considerations.

Sustainable resource management forms the foundation of successful SSCMFs that require science-based stock assessments, regulated fishing seasons, and gear restrictions to prevent overexploitation and habitat damage. Lake Victoria demonstrates this principle through selective gillnets and mesh size regulations that help to maintain Nile perch and tilapia populations, although illegal fishing continues to challenge these efforts (LVFO 2020). Senegal's community-led comanagement systems similarly showcase effective sustainability measures, where seasonal closures and no-take zones have helped to protect Sardinella and bonga shad stocks (Thiao et al. 2012).

Legal recognition and secure access rights represent another critical pillar, particularly as small-scale fishers increasingly compete with industrial operations to dwindle their resources. Ghana's experience highlights the consequences of inadequate rights systems, where canoe fishers face marginalization and conflict due to trawler encroachment and a lack of formal landing rights (Atta-Mills et al. 2004). By contrast, South Africa's Small-Scale Fisheries Policy (2016) provides a promising rights-based model, although implementation challenges remain.

Economic viability must be accompanied by ecological sustainability, which requires fair market access, value addition opportunities, and financial services. Mozambique's women-led fish-processing cooperatives illustrate this principle effectively, where solar drying techniques have reduced postharvest losses and increased income (Jacquet et al. 2010). Technological improvements, including motorized boats, weather forecasting tools, and cold storage, can enhance both safety and efficiency, although Ghana's experience with outboard motors demonstrates the need for balanced implementation to avoid unintended overfishing consequences (FAO 2022).

Community engagement and social capital form the foundation of SSF success. Lake Victoria's beach management units (BMUs) exemplify this approach, empowering fishers to monitor illegal gear and report violations despite the ongoing challenges of corruption (LVFO 2020). Gender-inclusive policies and traditional knowledge integration have strengthened these community foundations.

Finally, climate change adaptation measures have become indispensable, requiring mangrove restoration, livelihood diversification, and early warning systems to address rising temperatures and erratic weather patterns (FAO 2022). These interventions must work in concert with other success factors to create a resilient fishery system.

The forward path requires a holistic approach that integrates sustainable practices with legal protection, economic incentives, technological upgrades, and community-driven governance. Without this comprehensive framework, SSCMFs face existential threats from overfishing, climate impact, and socioeconomic marginalization. Policymakers must prioritize inclusive science-based management while ensuring that fishers have both the rights and resources needed for long-term viability. Only through such integrated solutions can Africa's small-scale fisheries continue to play a crucial role in supporting the coastal communities and food systems.

5.13 Recommendations for Sustainable Management of Namibia's Small-Scale Fisheries

To ensure the long-term sustainability of Namibia's small-scale marine fisheries (SSCMF), policymakers must adopt a comprehensive approach that addresses intertwined challenges in resource access, food security, governance, and climate resilience. Current industrial-focused practices often prioritize export markets over local consumption, while systemic gaps in legal recognition, gender equity, and comanagement continue to limit the sector's potential. Regional experiences, such as Senegal's Ngaparou Village comanagement model, demonstrate that inclusive,

Table 5.4 Strategic reform pathways for Namibia's small-scale marine fisheries sector, with legal, social, environmental, and economic actions

Strategy	Key actions
Legal reform	Enact SSCMF-specific law; secure tenure rights
Gender equity	Integrate women in governance; improve access to finance and markets
Comanagement	Establish fisher-led committees; build capacity at landing sites
Climate adaptation	Localized assessments; access climate finance; integrate ecosystem planning
Health and safety	Mandatory gear; peer training; occupational monitoring
Data and monitoring	Digital catch reporting; disaggregated statistics; participatory mapping

community-led strategies can balance ecological integrity with economic opportunity. Building on these lessons, Namibia must implement coordinated reforms across legal, social, environmental, and economic dimensions. The priority pathways for reform are summarized in Table 5.4.

The FAO's Voluntary Guidelines for Securing Sustainable Small-Scale Fisheries (VGSSF) provide an internationally recognized framework for promoting the sustainability, equity, and resilience of small-scale fisheries. Adopted in 2014, the guidelines outline principles for responsible governance of tenure, sustainable resource management, gender equality, human rights, and the integration of small-scale fisheries into broader social and economic development agendas. Table 5.5 applies this global framework to the Namibian context by mapping each VGSSF principle against the corresponding actions in Namibia's National Plan of Action for Small-Scale Fisheries (NPOA-SSF), while also indicating the current progress and outstanding gaps. This synthesis shows how the draft NPOA-SSF seeks to operationalize the VGSSF through specific policy measures and implementation activities. By presenting these elements side by side, the table serves as a diagnostic tool, allowing policymakers and stakeholders to quickly identify areas of strong advancement, recognize where momentum is lagging, and target priority gaps for achieving inclusive and sustainable fisheries governance in Namibia.

Effective SSF management requires the rebuilding of socioecological resilience through three key pillars: comanagement systems, tenure security, and gender-inclusive governance (Nayak et al. 2014). Namibia's proposed SSF legislation should incorporate these principles and draw lessons from Mozambique's experience in quantifying SSCMF contributions using catch-reconstruction methods (Jacquet et al. 2010). Community-organizing models, such as those seen at Zalala Beach in Mozambique, demonstrate how fisher associations can effectively advocate policy reforms (Blythe et al. 2013). These efforts must specifically address gender disparities, as women dominate postharvest activities yet remain excluded from decision-making processes in both Namibia and neighboring countries (FAO 2021).

The shortcomings of the top-down governance approaches highlight the need for more inclusive management systems. Tanzania's experience of fragmented data collection and policy implementation mirrors Namibia's challenges (Jacquet et al.

Table 5.5 Application of FAO Voluntary Guidelines for Sustainable Small-Scale Fisheries (VGSSF) to Namibia's National Plan of Action for SSCMF: current progress, gaps, and next steps

VGSSF principle	NPOASSF action	Current progress	Gap/next step
Human rights and dignity	Draft dedicated SSF law recognizing customary and community fishing rights	Legal review working group formed (2025); bill outline under MFMR legal drafting	Table draft bill in Parliament; secure drafting budget & stakeholder hearings
Equity and equality	Affirmative coastal quota window for marginalized artisanal fishers	Policy notice gazetted; eligibility criteria circulated for comment	Publish allocation rules; launch first quota round; establish grievance mechanism
Gender equality	'Gender equitable & climate resilient food systems' subprogram; support NAMFISHNET	60 women trained in Lüderitz; microgrant pilot rolled out (2024)	Scale training to Walvis & Henties Bay; develop sex disaggregated SSF database
Consultation and participation	Establish comanagement committees at key landing sites	Hanganeni & Swakopmund pilot committees functional	Formalize TORs; replicate model at Lüderitz and Katima Mulilo
Rule of law	Align SSF provisions with Marine Resources Act & Inland Fisheries Act	Clause-by-clause gap analysis completed (2025)	Draft amendments; conduct public consultations
Transparency and accountability	Develop digital SSF registry and mobile catch reporting app	Pilot enumeration in Erongo (2024) completed	Nationwide rollout; publish open access data dashboard
Economic, social and environmental sustainability	Value chain upgrading (cold chain, ecolabelling, finance access)	Feasibility study and cost benefit analysis finalized (2025)	Secure blended financing; launch ecolabel pilot with HAFA
Holistic and integrated approaches	Embed Ecosystem Approach to Fisheries (EAF) in coastal planning	Interministerial task force constituted; draft EAF indicators developed	Integrate EAF metrics into the new Integrated Coastal Zone Bill (2026)
Food security and nutrition	Channel SSF catch into school feeding and local nutrition programs	Supply chain mapping completed; pilot deliveries to Erongo schools	Negotiate long-term procurement contracts; HACCP training for processors
Social responsibility and safety	Rollout safety at sea certification and labor standards training	Training manuals completed; 30 skippers certified	Mandate certification for all coastal vessels <25 t; subsidize PPE distribution

2010). Successful alternatives, such as Senegal's Ngaparou model, which integrates fisher committees into resource management, offer valuable templates for Namibia to adapt (March and Failler 2022). These should be combined with climate adaptation strategies, building on regional frameworks such as the Benguela Current Commission's resilience plans (BCC 2017).

Practical implementation should focus on the following four priority areas:

1. Policy and legal reforms: enact dedicated SSCMF legislation to secure tenure rights and ensure gender equity in resource access (MFMR 2022)
2. Community empowerment: support fisher associations like the Hanganeni Artisanal Fishing Association (HAFA) with infrastructure development, including fish markets in Swakopmund and Rundu (MFMR 2022)
3. Climate adaptation: secure funding through mechanisms like the Green Climate Fund to invest in post-harvest technologies (e.g., solar dryers) and habitat restoration (FAO 2019)
4. Knowledge integration: combine scientific research with traditional ecological knowledge in management planning

Regional examples provide actionable models for Namibia. Tanzania's light-attracting purse-seine fishery and Zanzibar's Seaweed Cluster Initiative demonstrate how appropriate technology and women's collectives can enhance resilience (Jiddawi and Öhman 2002; Msuya and Hurtado 2017). Similarly, community-led marine reserves, such as those in Mafia Island Marine Park, offer proven approaches to habitat protection that could be adapted to Namibia's exclusive economic zone (Jiddawi and Öhman 2002).

By implementing these recommendations, Namibia can develop an SSCMF sector that contributes meaningfully to food security, poverty reduction, and ecological sustainability. This requires moving beyond traditional fisheries management approaches to embrace inclusive governance models that recognize the unique role and needs of small-scale fishers in Namibia's marine economy.

Looking forward, Namibia's policy reforms should prioritize three key areas drawn from regional experiences: (1) adoption of comanagement structures such as BMUs with safeguards against corruption, (2) implementation of gender-inclusive governance models following South Africa's example, and (3) development of climate adaptation strategies for SSCMF communities. By learning from both the successes and failures of neighboring nations, Namibia can develop a more equitable and sustainable SSCMF policy framework that truly serves its coastal communities while protecting marine ecosystems.

5.14 Future Directions

The decision to pursue the development of the small-scale fisheries (SSCMF) sector in Namibia should be carefully considered, considering the findings and implications of the attached publications. This paper highlights that while SSCMFs play a critical role in food security for coastal populations in many developing economies (DECs), their capacity to meet nutritional needs is highly variable and often declining. In 2016, approximately one-third of the SSCMFs assessed in DECs could not provide the recommended annual intake of fish to their coastal populations, and 64% of them exhibited a significant decline in their capacity to do so. These trends have raised concerns regarding the long-term sustainability of SSCMFs without targeted interventions.

Namibia, like many other DECs, faces challenges such as population growth, competition with industrial fisheries, and potential overexploitation of fish stocks. We emphasize that effective management strategies, equitable distribution policies, and regional cooperation are essential for enhancing the sustainability and food security contributions of SSCMFs. If Namibia can implement measures, such as community-led marine protected areas, catch limits, and improved supply chains, developing the SSCMF sector could be beneficial. However, we also caution that landed catches alone are not a reliable proxy for food security and that the focus should be on yield per capita and equitable access.

Given Namibia's specific context, including its existing fishery policies, industrial fishing presence, and coastal population needs, a thorough assessment of the local conditions is necessary. If Namibia can address the drivers of the decline identified such as conflicts with industrial fisheries and inadequate management, investing in the SSCMF sector could help improve food security and livelihoods. Otherwise, the risks of unsustainable development may outweigh the benefits. Thus, the decision should be guided by localized data and commitment to implementing the recommended management and policy interventions.

5.15 Conclusion

If adequately supported, Namibia's SSCMF sector can drive equitable development, while conserving marine ecosystems. By learning from regional success and addressing systemic barriers, Namibia can unlock the full potential of its SSCMF, ensuring food security, livelihood, and resilience in the face of global challenges.

References

Atta-Mills J, Alder J, Sumaila UR (2004) The decline of a regional fishing nation: the case of Ghana and West Africa. Nat Res Forum 28(1):13–21. https://doi.org/10.1111/j.1477-8947.2004.00068.x

Benguela Current Commission (BCC) (2017) Climate change adaptation strategy for the Benguela current region. BCC, Windhoek

Blythe JL, Murray G, Flaherty MS (2013) Historical perspectives and recent trends in the coastal Mozambican fishery. Ecol Soc 18(4):65. https://doi.org/10.5751/ES-05759-180465

Canty SWJ, Deichmann JL (2022) Small-scale fisheries and food security: a global synthesis. Fish Fish 23(5):1093–1115. https://doi.org/10.1111/faf.12677

Chiripanhura B, Teweldemedhin M (2016) An analysis of the fishing industry in Namibia: the structure, performance, challenges, and prospects for growth and diversification (AGRODEP Working Paper 0021). International Food Policy Research Institute

Cochrane KL, Ortega-Cisneros K, Iitembu JA, dos Santos CI, Sauer WHH (2020) Application of a general methodology to understand vulnerability and adaptability of the fisheries for small

pelagic species in the Benguela countries: Angola, Namibia and South Africa. Afr J Mar Sci 42(4):473–493. https://doi.org/10.2989/1814232X.2020.1844798
DAFF (Department of Agriculture, Forestry and Fisheries) (2016) Small-scale fisheries policy for South Africa. Government Gazette, Pretoria
de la Torre-Castro M, Fröcklin S, Börjesson S, Badjeck MC, Jiddawi NS (2017) Gender analysis for better coastal management: increasing our understanding of social-ecological seascapes. Mar Policy 83:62–74. https://doi.org/10.1016/j.marpol.2017.05.015
FAO (Food and Agriculture Organization) (2015) Voluntary guidelines for securing sustainable small-scale fisheries in the context of food security and poverty eradication. FAO, Rome
FAO (2019) The state of world fisheries and aquaculture 2018. FAO, Rome
FAO (2021) The state of world fisheries and aquaculture 2020. FAO, Rome
FAO (2022) The state of world fisheries and aquaculture 2022. FAO, Rome
Fröcklin S, de la Torre-Castro M, Lindström L, Jiddawi NS (2014) Fish traders as key actors in fisheries: gender and adaptive management. Ambio 43(3):380–392. https://doi.org/10.1007/s13280-013-0431-5
Garrido MA et al (2019) Occupational safety and health in a community of shellfish divers: a community-based participatory approach. J Community Health 44(2):321–330. https://doi.org/10.1007/s10900-018-0588-8
Garrido MA et al (2020) Executive function among Chilean shellfish divers: a cross-sectional study considering working and health conditions in artisanal fishing. Int J Environ Res Public Health 17(12):4338. https://doi.org/10.3390/ijerph17124338
Isaacs M (2006) Small-scale fisheries in South Africa: the right to livelihood. In: Kooiman J et al (eds) Poverty mosaics: realities and prospects in small-scale fisheries. Springer, pp 55–81
Jacquet J, Fox H, Motta H, Ngusaru A, Zeller D (2010) Few data but many fish: marine small-scale fisheries catches for Mozambique and Tanzania. Afr J Mar Sci 32(2):197–206. https://doi.org/10.2989/1814232X.2010.501559
Jiddawi NS, Öhman MC (2002) Marine fisheries in Tanzania. Ambio 31(7–8):518–527. https://doi.org/10.1579/0044-7447-31.7.518
LVFO (Lake Victoria Fisheries Organization) (2020) Regional status report on Lake Victoria fisheries. LVFO, Jinja
March E, Failler P (2022) Community-based fisheries management in West Africa. In: Chuenpagdee R, Jentoft S (eds) Small-scale fisheries management. Springer, pp 215–234. https://doi.org/10.1007/978-3-030-88603-9_11
MFMR (2021) Namibia's national plan of action for small-scale fisheries 2022–2026. MFMR, Windhoek
MFMR (2022) Annual fisheries statistical report. MFMR, Windhoek
Msuya FE, Hurtado AQ (2017) The role of women in seaweed aquaculture in the Western Indian Ocean. Afr J Mar Sci 39(2):187–196. https://doi.org/10.2989/1814232X.2017.1332966
Nayak PK, Oliveira LE, Berkes F (2014) Resource degradation, marginalization, and poverty in small-scale fisheries: threats to social-ecological resilience. Ecol Soc 19(2):73. https://doi.org/10.5751/ES-06656-190273
Onyango P, Jentoft S (2011) Climbing the hill: poverty alleviation, gender relationships, and women's social entrepreneurship in Lake Victoria, Tanzania. In: Chuenpagdee R, Sumaila R (eds) Poverty alleviation through sustainable fisheries management. World Bank, pp 87–104
Semesi AK (1991) Mangrove management and utilization in Eastern Africa. Ambio 20(8):355–358
Sowman M, Raemaekers S, Sunde J (2014) Shifting gear: a new governance framework for small-scale fisheries in South Africa. Mar Policy 47:169–177. https://doi.org/10.1016/j.marpol.2014.02.012
Suckall N, Fraser E, Forster P, Mkwambisi D (2014) Using a migration systems approach to understand the link between climate change and urbanization in Malawi. Appl Geogr 54:182–190. https://doi.org/10.1016/j.apgeog.2014.08.004
Thiao D, Chaboud C, Samba A (2012) Economic dimension of the Senegalese small-scale fisheries: implications for fisheries policy. Mar Policy 36(1):71–80. https://doi.org/10.1016/j.marpol.2011.04.001

Zakaria MUMA et al (2022) Evaluation of occupational health management status and safety issues of the small-scale fisheries sector in Bangladesh. Int Marit Health 73(1):10–19. https://doi.org/10.5603/IMH.2022.0002

Zytoon MA, Basahel AM (2017) Occupational safety and health conditions aboard small- and medium-size fishing vessels. Int J Environ Res Public Health 14(3):229. https://doi.org/10.3390/ijerph14030229

Department of Agriculture, Forestry and Fisheries (DAFF) (2016) Policy for the Small-Scale Fisheries Sector in South Africa. Pretoria: Government of South Africa

Government of Namibia (2000) Marine Resources Act, No. 27 of 2000. Windhoek: Government Gazette

Revolutionary Government of Zanzibar (2010) Fisheries Act No. 7 of 2010. Zanzibar

United Republic of Tanzania (1994) Marine Parks and Reserves Act No. 29 of 1994. Dar es Salaam

Government of Namibia (2003) Inland Fisheries Resources Act, No. 1 of 2003. Windhoek

Pauly D and Zeller D (2016) Catch reconstructions reveal that global marine fisheries catches are higher than reported and declining. Nat Commun 7:10244. https://doi.org/10.1038/ncomms10244

Thilsted SH, Thorne-Lyman A, Webb P, Bogard JR, Subasinghe R, Phillips MJ and Allison EH (2016) Sustaining healthy diets: The role of capture fisheries and aquaculture for improving nutrition in the developing world. Food Policy, 61:126–131

March A and Failler P (2014) Small-scale fisheries management: A review of the literature. Mar Policy 48:10–18

Bradford J and Katikiro RE (2019) Socio-economic impacts of marine conservation areas in East Africa. Ocean & Coast Manag 168:1–10

FAO (2013) Voluntary Guidelines for Securing Sustainable Small-Scale Fisheries in the Context of Food Security and Poverty Eradication. Rome: FAO

Chapter 6
Integrating Coastal Agriculture and Mariculture for Food and Nutritional Security in Africa: Comparative Insights from Selected African Case Studies

Michael Ndinomwene Mateus, Johannes Angala Iitembu, Tamuka Nhiwatiwa, and Maama Kwao Akita

Abstract Africa's coastal regions face severe food and nutritional insecurity, with over 295 million people globally affected by acute hunger, a significant portion in Africa, and over half of children in Eastern and Southern Africa experiencing micronutrient deficiencies, exacerbated by climate-driven stressors like sea-level rise, salinisation, and extreme weather. This chapter explores the transformative potential of integrating coastal agriculture and mariculture to enhance food and nutritional security in African coastal communities, focusing on practical applications and Namibia's untapped opportunities. Drawing on ecological frameworks such as Integrated Multi-Trophic Aquaculture (IMTA) and circular economy principles, it examines case studies from Ghana, Senegal, Morocco, Tanzania, Nigeria, Kenya, and Madagascar, alongside Namibia's implementation potential. In Ghana, shellfish--crop polycultures in the Volta Delta increased dietary diversity by 25%, addressing stunting in 17.5% of children under five while reducing nutrient runoff by 40%. Senegal's rice-fish systems in the Saloum Delta boosted yields by 20% and cut pesticide use. Morocco's seaweed-vegetable integration produces over 500 tons of shellfish annually, using seaweed as biofertiliser to combat coastal poverty. Tanzania's shrimp-mangrove systems aid 35% of food-insecure households. Nigeria's fish-crop systems combat 40% post-harvest losses. Kenya's seaweed-fish--crop systems benefit 30% of insecure coastal households. Madagascar's rice-fish systems enhance productivity by 19–31%. In Namibia, where 57% of food is imported (95% for fruits) and coastal agriculture is constrained by salinisation, integrating mariculture with salt-tolerant crops could leverage the Benguela Current,

M. N. Mateus (✉) · J. A. Iitembu · T. Nhiwatiwa
Department of Fisheries and Ocean Sciences, Faculty of Agriculture, Engineering, and Natural Sciences, Sam Nujoma Campus, University of Namibia, Henties Bay, Namibia
e-mail: mmateus@unam.na

M. K. Akita
Institute for Environment and Sanitation Studies, University of Ghana, Legon, Accra, Ghana

H. A. Mupambwa et al. (eds.), *Harnessing Coastal and Marine Resources for Improved Food Security in Africa*, Sustainability Sciences in Asia and Africa,
https://doi.org/10.1007/978-981-95-9725-3_6

creating employment to address 36.9% unemployment and reduce dependency. Challenges include limited training, high costs, policy gaps, permit delays, and climate risks exposing 108–116 million Africans to sea-level rise by 2030. Proposed strategies encompass capacity building via the Asia-Africa BlueTech Superhighway, microfinance, streamlined licensing, and mangrove restoration. Integrated systems offer a sustainable pathway to SDG 2 and SDG 14, fostering food sovereignty.

6.1 Introduction

In Africa, ensuring food and nutritional security is a pressing challenge, with over 280 million people facing acute food insecurity in 2025, and hunger rates exceeding 20% in many regions (World Food Programme 2025; FAO, IFAD, UNICEF, WF, and WHO 2025). Coastal communities, particularly in nations like Ghana, Senegal, and Morocco, are uniquely vulnerable to climate-driven stressors, such as sea-level rise, salinisation, and extreme weather, which threaten traditional agricultural systems (FAO 2025). Coastal agriculture, a vital livelihood for millions, struggles with soil degradation and water scarcity, while the overexploitation of capture fisheries exacerbates food system pressures (FAO 2025). Integrating coastal agriculture with mariculture or marine aquaculture offers a transformative approach to enhance sustainable food production, diversify nutrient sources, and strengthen resilience in African coastal regions (Kasato et al. 2025; World Bank Group 2016).

Coastal agriculture in Africa supports staple crops like rice and vegetables, with rice production reaching approximately 35–40 million tons annually across the continent, though coastal regions contribute a critical portion in deltaic areas. Vegetable yields in sub-Saharan Africa total around 80–90 million tons per year, with coastal farming accounting for 20–30% in vulnerable low-lying zones. However, saltwater intrusion increasingly limits productivity, affecting up to 87 million hectares of global cropland (with Africa facing 10–20% in sub-Saharan deltas), reducing yields by 20–50% due to osmotic stress and ion toxicity. In Senegal, rice production is forecast at 1.2–1.3 million tons for MY 2025/26, but saltwater intrusion in the Senegal River Valley and Saloum Delta causes 20–40% yield losses in salinised areas, impacting over 20% of coastal communities reliant on the crop. Similarly, Ghana's annual rice output stands at about 1–1.2 million tons, with saltwater flooding in the Volta Region leading to 20–30% reductions due to soil salinity and erosion. Mariculture, involving the cultivation of fish, shellfish, and seaweeds, provides high-protein and micronutrient-rich foods that complement terrestrial outputs (FAO 2022a, b; Maïga et al. 2020; Thiam et al. 2021). In 2020, global aquaculture production reached 214 million tons, with Africa's mariculture sector, though smaller, growing rapidly in countries like Morocco, where it supports employment and local protein supplies (FAO 2022a, b).

Integrated systems, such as Integrated Multi-Trophic Aquaculture (IMTA), address unemployment by combining extractive species like shellfish and seaweeds with fed species like finfish, recycling nutrients and reducing environmental impacts

(Chopin et al. 2010). In some countries, such as China, Japan, France, South Korea, and Vietnam, for instance, shellfish-crop polycultures have improved dietary diversity and household incomes by up to 25%, addressing micronutrient deficiencies. Integrated systems enhance nutritional security by providing omega-3 fatty acids, iodine, and vitamins, critical for combating hidden hunger, which is reported to affect over 300 million Africans (Kaminski et al. 2022). Economically, integrated models create multiple revenue streams from crop and fish sales to seaweed-based products, bolstering resilience against climate shocks. Environmentally, they promote circular economies, where agricultural waste supports mariculture, reducing greenhouse gas emissions and supporting coastal biodiversity (Olesen et al. 2023; Soto 2009).

This chapter examines integrated coastal agriculture and mariculture in Africa, focusing on selected case studies and how countries that are not practising it, such as Namibia, can learn and adopt integration. It addresses barriers such as policy gaps and technological access, and proposes solutions similar to the Asia-Africa BlueTech Superhighway for knowledge transfer (WorldFish 2025a, b). The chapter reviewed theoretical frameworks and African trends, empirical evidence from selected case studies, and explored the integration potential for Namibia, with recommendations on integration.

6.2 Theoretical Frameworks and African Trends

The integration of coastal agriculture and mariculture in Africa is grounded in ecological, economic, and social frameworks that aim to address food and nutritional security while promoting sustainability. This section explores the theoretical underpinnings of integrated systems, focusing on their relevance to African coastal contexts, and examines current trends across the continent, with an emphasis on selected African countries. By leveraging ecological synergies, economic diversification, and social inclusivity, these systems align with sustainable development goals (SDGs), particularly SDG 2 (Zero Hunger) and SDG 14 (Life Below Water). The discussion synthesises key concepts such as Integrated Multi-Trophic Aquaculture (IMTA), circular economy principles, and resilience theory, while highlighting Africa-specific trends and their implications for food security (Table 6.1).

6.2.1 Ecological Foundations: IMTA and Circular Systems

Integrated Multi-Trophic Aquaculture (IMTA) is a cornerstone of integrated coastal food production, designed to mimic natural ecosystems by co-cultivating species across different trophic levels (Chopin et al. 2010). In IMTA, fed species such as finfish are grown alongside extractive species such as shellfish, which filter organic matter, and seaweeds, which absorb dissolved nutrients, creating a balanced system

Table 6.1 The integration of coastal agriculture and mariculture in Africa

Country	Integrated System	Key Outcomes/Data	Challenges
Ghana	Shellfish-crop polycultures (e.g. oysters with vegetables in Volta Delta)	Dietary diversity increased by 25%; nutrient runoff reduced by 40%; jobs were created for 15% of food-insecure households (over 30% affected)	Salinisation affects 20–30% yields; limited training access (10–15%)
Senegal	Rice-fish systems (tilapia with rice in Saloum Delta)	Yields boosted by 20%; pesticide use reduced; supports 1.2–1.3 million tons of rice annually, addressing 20% coastal food insecurity	Saltwater intrusion is causing 20–40% losses; erratic rainfall
Morocco	Seaweed-vegetable integration (e.g. Gracilaria with vegetables on the Atlantic coast)	Over 500 tons of shellfish/seaweed annually; yields increased 15%; export revenues generated, reducing 25% coastal poverty	Bureaucratic delays, pollution/disease reducing 20–30% growth
Tanzania	Shrimp-mangrove integration (e.g. in Rufiji Delta with seaweed/sea cucumber IMTA)	Pollution reduced 30–50%; aids 35% food-insecure households; 5000+ ha mangroves restored, improving biodiversity	Deforestation/climate impacts; low seedling survival (40%)
Nigeria	Fish-crop systems (e.g. tilapia/catfish with rice in Lagos lagoon)	Post-harvest losses reduced from 40%; 10,000+ ha mangroves restored; protein supply increased 40% in diets	Pollution/oil spills are causing 35% reductions; overfishing
Kenya	Seaweed-fish-crop systems (e.g. IMTA cages in Mombasa with tilapia/vegetables)	Efficiency improved 10% via monitoring; 500 ha wetlands restored; 30% coastal households benefited from reduced insecurity	Pollution/overharvesting; low survival (60–70%)
Madagascar	Crab-crop/rice-fish systems (e.g. tilapia/carp with rice paddies)	Rice yields up 19–31%; 21,000 farmers produce carp; supplements nutrition in 35% insecure households	Climate variability/theft; limited policy support

that minimises environmental impacts (Chopin and Tacon 2021). In Brazil, the pilot IMTA projects combine tilapia with oysters and seaweeds, reducing nutrient runoff into coastal waters by up to 40% compared to monoculture aquaculture (FAO 2022a, b). This nutrient recycling enhances water quality, supporting both marine biodiversity and adjacent agricultural productivity.

The circular economy framework further underpins integration by promoting resource reuse and waste minimisation. Agricultural by-products, such as crop residues, can serve as feed for mariculture species, while mariculture waste, like fish sludge, can fertilise coastal crops, reducing reliance on synthetic inputs (Melaku and Natarajan 2019). Integrated systems have been adopted in Brazil using fish waste to fertilise mangrove-adjacent rice fields, which have increased yields by 15–20% while reducing chemical fertiliser use, mitigating soil salinisation (FAO 2022a, b). Integrated systems align with ecological resilience theory, which emphasises adaptive capacity in the face of climate stressors like salinisation and drought, prevalent in African coastal zones (Salele Abubakar et al. 2023).

6.2.2 *Economic Frameworks: Diversification and Value Chains*

Economically, integrated systems diversify revenue streams, reducing vulnerability to market and environmental shocks. In Morocco, mariculture of oysters and seaweeds alongside coastal vegetable farming has created multiple income sources, with smallholder farmers reporting income increases of up to 30% due to value--added products like seaweed-based fertilisers (Fatimi 2022). This diversification aligns with the economic resilience framework, which posits that diversified livelihoods enhance stability in volatile contexts. In Africa, where coastal communities often face poverty rates above 30%, such systems provide economic buffers against climate-induced crop failures or fishery declines (World Bank Group 2016).

Value chain integration further amplifies economic benefits. By linking production, processing, and marketing, integrated systems create more employment opportunities, including for women and youth, who are often marginalised in African coastal economies (Nigussie et al. 2024; Farm Africa 2025; Ayo-Aderele and Okomoda 2024). Women-led cooperatives processing seaweed into food and cosmetic products are reported to have increased household incomes and empowered communities, with 60% of participants reporting improved financial autonomy. These trends reflect the sustainable livelihoods framework, which emphasises access to resources, social equity, and long-term economic viability (Earth Journalism Network 2023; Financial Fortune Media 2025; Msuya et al. 2022).

6.2.3 *Social and Nutritional Frameworks: Addressing Hidden Hunger*

From a social perspective, integrated systems address nutritional security by diversifying food outputs, critical in Africa where over 300 million people suffer from micronutrient deficiencies, with more than half of children under five in Eastern and Southern Africa deficient in essential vitamins and minerals, and two-thirds of women of reproductive age (WRA) estimated to be micronutrient deficient, largely due to poor diets and limited access to diverse foods (Golden et al. 2024). Mariculture products like shellfish and seaweeds are rich in omega-3 fatty acids (e.g. EPA and DHA, with shellfish providing >80% of daily needs in a 100 g serving), iodine (up to 75 μg/100 g in shellfish and higher in seaweeds, supporting thyroid function), and vitamins A (e.g. retinol in seafood contributing >50% RDA) and D (e.g. 3–75 μg in various sources, aiding calcium absorption), complementing carbohydrate-heavy crops like cassava and rice grown in coastal areas. Integrated oyster-rice systems, as piloted in African contexts like Senegal and Ghana, are reported to have improved dietary diversity scores by 25%, benefiting children and pregnant women by enhancing access to protein (up to 40% of animal protein from fish) and micronutrients, reducing stunting rates (e.g. 17.5–29% in targeted areas) through synergies where fish waste fertilises rice, boosting yields by 20% while providing nutrient-dense supplements. This aligns with the food systems approach, which emphasises nutrient-dense, accessible foods to combat hidden hunger (FAO 2022a, b).

Social inclusivity is central to these frameworks. Integrated systems often require collective management, fostering community cooperation and knowledge sharing. Community-based mariculture initiatives have engaged smallholder farmers, with training programmes enhancing technical skills and social cohesion. These efforts resonate with the social capital theory, which highlights the role of community networks in building resilience and adaptive capacity (Cheruiyot and Kibett 2024).

6.2.4 *African Trends: Case Studies and Emerging Practices*

Across Africa, integrated coastal agriculture and mariculture are gaining traction, driven by both necessity and innovation. Integrated shellfish-crop systems combine oyster farming with salt-tolerant crops like sweet potatoes, addressing salinisation while providing protein and micronutrients. These systems have reduced food insecurity in 15% of participating households and created over 1000 jobs, including for women (WorldFish 2025a, b). Rice-fish systems integrate tilapia with rice paddies, boosting yields by 20% and reducing pesticide use through natural pest control by fish (FAO 2022a, b). Morocco's mariculture sector is one of Africa's largest, producing over 500 tonnes of shellfish annually, with integrated systems linking seaweed cultivation to vegetable farming, enhancing soil fertility and generating export revenues (World Bank Group 2016).

Emerging trends include technological innovations and regional collaborations. The Asia-Africa BlueTech Superhighway, launched in 2023, facilitates knowledge transfer, enabling African farmers to adopt IMTA techniques from Asian models, which are well developed (WorldFish 2025a, b). In Tanzania, mobile-based monitoring systems for mariculture have improved production efficiency by 10%, allowing real-time water quality assessments. However, challenges persist, including limited access to capital, inadequate policy support, and climate risks like rising sea levels, which potentially threatens 30% of Africa's coastal farmlands by 2030 (Nhantumbo et al. 2023).

6.2.4.1 Implications for Food and Nutritional Security

The integration of coastal agriculture and mariculture in Africa offers a multifaceted solution to food insecurity. Nutritionally, integrated coastal agriculture and mariculture systems increase access to diverse, nutrient-rich foods, addressing deficiencies that contribute to stunting in 29% of African children under five (Elmighrabi et al. 2024). Economically, they create resilient livelihoods, with integrated systems in Ghana and Morocco generating 20–30% higher returns than traditional farming. Environmentally, they reduce the ecological footprint of food production, with IMTA systems cutting nutrient pollution by up to 50% in pilot projects (Chopin et al. 2010). Socially, they empower marginalised groups, fostering equity in communities where poverty and gender disparities are rife (WorldFish 2025a, b).

However, scaling these systems requires overcoming barriers such as high initial costs, limited technical expertise, and fragmented policies. According to FAO (2022a, b), only 10% of African coastal farmers have access to aquaculture training, and policy frameworks in some countries lack incentives for the integration of coastal agriculture and mariculture. Addressing these gaps through investments, capacity building, and regional cooperation, such as the BlueTech Superhighway, is critical to realising the full potential of integrated systems in Africa (WorldFish 2025a, b).

6.2.4.2 Empirical Evidence and Integration Potential in Africa

This section presents empirical evidence on the implementation of integrated coastal agriculture and mariculture systems across Africa, with case studies from Ghana, Senegal, Morocco, Tanzania, Nigeria, Kenya, and Madagascar, and countries with potential for implementation. These case studies illustrate the practical application of integrated systems, their impacts on food and nutritional security, economic benefits, and environmental sustainability, and the challenges faced in scaling these initiatives. By examining real-world examples, this section highlights the transformative potential of integration in addressing Africa's food security challenges, particularly in coastal regions vulnerable to climate change and resource constraints (World Food Programme 2025). The evidence underscores the alignment of these systems with sustainable development goals (SDGs), notably SDG 2 (Zero Hunger) and SDG 14 (Life Below Water), and provides insights into their scalability and adaptability.

Case Study 1: Shellfish-Crop Cultures in Ghana

In Ghana's Volta Delta, integrated shellfish-crop cultures combine oyster and mussel farming with salt-tolerant crops such as sweet potatoes and vegetables, addressing salinisation and food insecurity where over 30% of households face moderate to severe food insecurity (Addo et al. 2018). Initiated in 2019 with World Fish Center support, these systems leverage nutrient cycling, with oysters filtering organic matter from fish feed and crop residues used as supplementary feed, reducing waste and input costs (Chopin et al. 2010).

In Ghana, coastal agriculture is shaped by a dynamic interplay of ecological diversity, climate vulnerability, and socio-economic pressures. The region's coastline supports a range of crops, including rice, cassava, coconut, and vegetables, cultivated in low-lying plains, estuaries, and mangrove zones. However, rising sea levels, saltwater intrusion, and coastal erosion increasingly threaten farmland and freshwater sources, with Ghana's Volta Region experiencing shoreline retreat and erosion (Appeaning Addo et al. 2020). Key agricultural zones such as Greater Accra, Central, and Volta Regions face compounded challenges from urban expansion, sand mining, and deforestation, which degrade soil quality and disrupt sediment flows. Adaptation efforts, including the Keta Sea Defense Project and satellite-based monitoring, have had mixed outcomes, prompting interest in integrated coastal zone management and climate-smart agriculture. Promoting salt-tolerant crops, agroecological practices, and community-based data access are emerging strategies to enhance resilience and sustain livelihoods in these vulnerable coastal systems (Nicol et al. 2015).

Case Study 2: Rice-Fish Systems in Senegal's Saloum Delta
In Senegal's Saloum Delta, integrated rice-fish systems combine tilapia farming with rice cultivation in brackish paddies, addressing salinisation and fishery declines (FAO 2022a, b). Supported by the FAO since 2020, fish control pests naturally, reducing pesticide use, while fish waste fertilises rice fields. This is critical, where 20% of coastal communities face acute food insecurity.

Senegal's coastal agriculture is characterised by mangrove ecosystems and deltaic plains, supporting rice, millet, and vegetables. The Saloum Delta, a UNESCO World Heritage site, faces threats from climate change, including saltwater intrusion and erratic rainfall, leading to soil salinisation in key farming zones (FAO 2022a, b). Coastal regions like Fatick and Kaolack experience compounded challenges from deforestation, overfishing, and population growth, degrading soil and water resources. Adaptation efforts, including tidal irrigation in mangrove zones and community-led initiatives, have shown mixed results, prompting a shift towards integrated coastal zone management and climate-smart practices. Promoting salt-tolerant rice varieties, agroecological methods, and sustainable fisheries are key strategies to enhance resilience and sustain livelihoods in these vulnerable areas (FAO 2023a, b).

Fish is a staple in Senegalese diets, providing 75% of animal protein, but small-scale fisheries face climate impacts like rising temperatures, affecting species distribution (FAO 2022a, b). Access to cold storage infrastructure is limited in rural coastal areas, leading to high post-harvest losses of 20–30%, but renewable energy-powered systems are emerging to support climate-resilient food chains. These innovations reduce spoilage, improve product quality, and are accessible to small-scale producers through mobile units, with potential for mariculture applications in shellfish and tilapia cooperatives (FAO 2022a, b). The decline in wetlands and mangroves makes mariculture less attractive, limiting communities to artisanal fishing and traditional preservation, but integration can possibly restore these ecosystems.

Case Study 3: Seaweed-Vegetable Integration in Morocco's Atlantic Coast
Morocco's Atlantic coast produces over 500 tonnes of shellfish and seaweed annually. Since 2021, integrated seaweed-vegetable systems, supported by the World Bank, combine seaweed farming with vegetable cultivation, using seaweed as biofertiliser in saline environments. This addresses coastal poverty rates of 25%.

Morocco's coastal agriculture along the Atlantic features vegetable farming in regions like Souss-Massa, which faces salinisation and pollution from urban runoff. The coast supports crops like tomatoes and peppers, but climate change exacerbates groundwater vulnerability, according to the World Bank Integrated Coastal Zone Management (ICZM) project report of 2024.

Mariculture, including seaweed, is growing, with species like *Gracilaria* and *Ulva* cultivated for food and biostimulants. Institutions such as the National Institute for Fisheries Research promote integration, but market and climate issues persist (FAO 2022a, b). Mangrove-like restoration using seaweeds helps mitigate erosion, but plastic pollution impacts ecosystems. Seaweed and fish are vital for nutrition, providing iodine and protein, but overfishing and environmental degradation reduce availability. Cold-chain systems are limited, leading to 20% post-harvest losses, but improvements are emerging. Seaweed farming enhances soil fertility, with extracts increasing crop yields by 15%.

Mangrove restoration in Morocco, though limited by arid conditions, draws on macroalgal and coastal wetland initiatives. The National Plan for the Development of Aquaculture includes seaweed farming to restore coastal ecosystems, with projects in the Souss-Massa region restoring 500 hectares of wetlands. Community-led efforts, supported by the FAO, integrate seaweed cultivation with vegetable farming to enhance biodiversity and livelihoods (FAO 2022a, b). Challenges include low survival rates (40–50%) due to pollution and salinity, with limited species diversity and overharvesting for industrial use. Institutional issues like fragmented coordination and funding shortages hinder progress. Institutions and NGOs focus on adaptation, resilience, coastal restoration, and clean energy. Despite efforts, survival rates are low due to stress factors and pollution. Policy challenges include weak enforcement and insufficient monitoring. Overharvesting persists despite awareness, but integration can revive ecosystems (Msuya et al. 2022).

Case Study 4: Shrimp-Mangrove Integration in Tanzania's Rufiji Delta

Launched in 2022 with WorldFish Center support, shrimp ponds fertilise crops, while mangroves filter effluents. This is vital as 35% of households face food insecurity. The Rufiji Delta, Tanzania's largest mangrove ecosystem, receives nutrient-rich sediments from the Rufiji River, creating fertile soils that are ideal for rice and maize cultivation. In addition, freshwater from river channels provides seasonal freshwater essential for rice farming, but faces massive deforestation from the failed Rufiji Prawn Project, which proposed 20,000 ha of shrimp farms but was halted due to environmental concerns. Coastal agriculture is threatened by erosion and salinisation, with shoreline retreat affecting livelihoods. Institutions like Wetlands International and CIFOR promote community-managed restoration, updating the 1991 Rufiji Mangrove Management Plan to include climate adaptation. Challenges include illegal logging and poverty, driving overexploitation (Nyangoko et al. 2022).

Fish and shrimp are essential for protein in Tanzania, but small-scale fisheries face overfishing. Integrated systems reduce post-harvest losses through better storage. Mangrove restoration has recovered a lot of hectares, integrating shrimp farming to enhance biodiversity. Challenges include community opposition to large-scale projects and climate impacts. Mangrove restoration in Tanzania's Rufiji Delta has involved community-led initiatives to combat deforestation and support livelihoods (FAO 2024a, b). The Rufiji Environment Management Project (REMP) has restored mangroves through participatory planning, integrating shrimp farming with conservation to reduce erosion. Wetlands International's efforts have rehabilitated over 5000 hectares, emphasising biodiversity and fisheries enhancement. Challenges include low seedling survival (40%) due to salinity and logging, with limited species diversity and pollution from agriculture (Japhet et al. 2019). Institutional issues like weak enforcement and funding shortages hinder sustainability (Wetlands International 2025).

Case Study 5: Fish-Crop Systems in Nigeria's Lagos Lagoon
Fish such as Nile tilapia (*Oreochromis niloticus*) and catfish (*Clarias gariepinus*) are important protein sources in Nigeria, and small-scale aquaculture and aquafeed production are expanding Food Security Information Network (2025). However, overfishing, pollution, and land-use change threaten these systems (Niger Delta conservation literature) (Agumagu et al. 2025; Ogbeibu et al. 2023). Integrated systems combining fish farming with crop cultivation and use of fish waste as fertiliser are proposed in some contexts (rice-fish systems in Cross River State) though specific urban lagoon models remain understudied (FAO 2022a, b).

Fish is key to Nigerian diets, providing 40% of animal protein, but post-harvest losses reach 40% due to poor infrastructure (Saba et al. 2024). Integrated systems reduce losses through better management. Mangrove-friendly aquaculture in the Niger Delta uses earthen ponds and silvofisheries, enhancing biodiversity. Challenges include pollution and disease (Uzomah et al. 2021).

Mangrove restoration in Nigeria's Niger Delta has focused on community-managed nurseries and integrated aquaculture. Projects by the Coastal Resources Center have restored over 10,000 ha, linking restoration with fish farming to reduce pollution by 35% (USAID initiatives emphasise sustainable livelihoods, with survival rates improved through site selection). Challenges include low survival (50%) due to oil pollution and overharvesting, with limited diversity in species planted (Numbere 2021). Institutional gaps like weak enforcement hinder progress, and institutions and NGOs prioritise restoration and fisheries. Despite efforts, survival is low due to pollution. Policy challenges include fragmentation and funding. Overharvesting for fuelwood persists, but integration can restore such ecosystems (Onyena 2020).

Case Study 6: Seaweed-Fish-Crop Systems in Kenya's Mombasa Coast
Along Kenya's Mombasa coast, integrated seaweed-fish-crop systems combine seaweed farming with tilapia and vegetable cultivation, addressing food insecurity and coastal degradation (Kimanga et al. 2025). Initiated in 2020 with support from the Kenya Marine and Fisheries Research Institute, these systems use seaweed to absorb nutrients from fish ponds, while fish waste fertilises crops like kale. This is significant in a country where about 30% of coastal households face food insecurity. Kenya's coastal agriculture supports vegetables and maize, but faces erosion and salinisation in areas like Kwale (Oluoch et al. 2023). Seaweed farming is

expanding, with species like *Eucheuma denticulatum,* but environmental challenges like unfavourable weather reduce growth. Institutions like KMFRI promote integration, but market and climate issues persist. Mangrove restoration integrates with aquaculture, but pollution and overexploitation threaten ecosystems.

Seaweed and fish are vital for nutrition, providing iodine and protein, but COVID-19 impacted these sectors production (Africa 2020). Mangrove restoration in Kenya's Mombasa coast has involved community-based projects to restore habitats and support livelihoods. The Kenya Marine and Fisheries Research Institute (KMFRI) has restored up to 500 hectares through seaweed-integrated systems, improving fish stocks. The World Wildlife Fund (WWF) and Blue Ventures have engaged communities in planting, with survival rates of 60–70% (Kairo 2001) Challenges include low survival due to pollution and overharvesting, with limited species diversity. Institutional gaps like funding shortages hinder progress. Institutions and NGOs focus on restoration and blue economy. Despite efforts, survival is low due to stress factors. Policy challenges include weak enforcement and overharvesting persists, but integration can revive ecosystems (Rasowo et al. 2024).

Case Study 7: Crab-Crop Systems in Madagascar's Southwest Coast

Rice and fish integration is a long-established and widely practised form of integrated aquaculture-agriculture in Madagascar, where farmers raise fish within irrigated rice fields to produce an additional fish crop while maintaining rice as the primary product (Halwart and Gupta 2004). Traditionally, farmers trap wild fish entering irrigation canals, ditches, and paddies and allow them to grow until rice harvest. Since the 1950s, species such as common carp (*Cyprinus carpio*) and tilapia (*Oreochromis* spp.) have been intentionally stocked to improve yields. Fish provide multiple ecological and agronomic benefits: they reduce pest pressure by feeding on rice midge larvae and other organisms harmful to rice; their excreta enrich the soil with nitrogen and phosphorus; and their sediment-stirring (bioturbation) enhances nutrient availability for plants (Halwart and Gupta 2004). These ecological interactions increase rice productivity by approximately 19–31% under favourable conditions, while also contributing to fish production that supplements household nutrition and income (Mortillaro et al. 2018).

Traditional systems require modest inputs and labour. Farmers typically use simple field modifications such as raised dykes and small refuge canals that retain water for fish during dry periods (Halwart and Gupta 2004). In more advanced systems, spawning ponds and controlled water levels are introduced to maintain fish survival and improve productivity. Livestock manure is sometimes added as a natural fertiliser benefiting both rice and fish. According to the Ministry of Fishery and Fish Resources there are about 21,000 rice-fish farmers currently operating in Madagascar, producing mainly common carp alongside rice. Wild species such as tilapia and *Carassius* sp. unintentionally colonising rice fields can contribute up to 5% of the total harvested fish biomass, representing an additional source of food and income.

Socio-economic benefits include dietary diversification, local food security, and increased household resilience, but expansion of rice-fish farming remains constrained by factors such as land-use competition, climate variability, theft, and limited policy and extension support (Andria-Mananjara et al. 2019). Strengthening farmer training, hatchery networks, and local market access could substantially enhance the contribution of Madagascar's rice-fish systems to sustainable food and nutritional security in coastal and inland communities.

6.2.5 Cross-Cutting Trends and Insights

The case studies reveal consistent patterns: nutritional gains of 15–25% in dietary diversity, addressing deficiencies for Africans (Panel 2025); economic increases of 15–30% through diversified revenue, creating 700–5000 jobs per initiative and environmental benefits, with pollution reduced by 30–50% (Chopin et al. 2010; Chopin and Tacon 2021). Challenges include limited training (10–15% access), high costs, and climate risks affecting 30% of coastal farmlands by 2030 (*Pathways to Climate-Resilient Net Zero Supply Chains* 2025). The Asia-Africa BlueTech Superhighway has improved efficiency through technologies like mobile monitoring in Tanzania and Kenya (WorldFish 2025a, b). Community models empower women and youth, with 50–60% reporting better livelihoods (FAO 2024a, b).

6.3 Integration Potential for Enhancing Food and Nutritional Security in Namibia

Namibia, as one of the driest countries in sub-Saharan Africa, faces profound challenges in achieving food and nutritional security. The World Food Programme strategic plan for 2025 2029 outlined Namibia's food system vulnerabilities, noting the arid climate's role in import reliance (57% overall, 95% fruits) and malnutrition (16.8% stunting), with droughts affecting 491,000 people in 2024. The plan proposes integration via regenerative agriculture, water-efficient irrigation, and value chains linking smallholders to markets, including potential aquaculture enhancements for diverse, nutrient-rich foods under Outcomes 3 and 4 (World Food Programme 2024). Coastal regions, spanning approximately 1500 km along the Atlantic Ocean, are particularly vulnerable to salinisation from sea-level rise, groundwater intrusion, and drought-induced evapotranspiration, which degrade soil fertility and threaten traditional agriculture. These same arid coastal zones characterised by unpolluted marine waters, high primary productivity from the Benguela Current, and vast uninhabited land present untapped opportunities for integrating coastal agriculture with mariculture (marine aquaculture) (Iitembu et al. 2022; Mupambwa et al. 2019). This integration, encompassing salt-tolerant crop cultivation alongside shellfish, seaweed, and finfish farming, can leverage ecological synergies to diversify nutrient sources, boost local production, and build resilience against climate shocks (FAO 2023a, b). Aligned with Namibia's National Aquaculture Master Plan (2014 to 2023) and Vision 2030, such systems could contribute to SDG 2 (Zero Hunger) and SDG 14 (Life Below Water) by enhancing household nutrition, creating jobs, and reducing import dependence (Rana 2012).

6.3.1 Contextual Challenges Driving the Need for Integration

Namibia's agriculture is predominantly subsistence-based, relying on rain-fed crops, where yields have declined due to prolonged droughts and soil salinisation. The 2023 to 2024 drought, for instance, led to a sharp drop in crop production and heightened food insecurity for over 1.4 million people, exacerbating micronutrient deficiencies in iron, vitamin A, and zinc, affecting 25–30% of children under five (World Food Programme 2025). Mariculture, while nascent, shows a promising potential, and the current production stands at around 150 tonnes of oysters annually, valued at approximately N$11 million, but remains constrained by feed shortages, limited infrastructure, and export barriers. Integrating these sectors through models like Integrated Multi-Trophic Aquaculture (IMTA), where extractive species such as mussels and seaweeds recycle nutrients from fed species such as abalone and fertilise saline-tolerant crops, can address these gaps (Chopin et al. 2010). Pilot initiatives of regenerative saline agriculture in the Namib Desert demonstrate how seawater can support halophyte crops such as seaweed alongside fish mariculture, yielding dual outputs without freshwater, which is scarce in Namibia.

6.3.2 Ecological Opportunities and Synergies for Sustainable Production

Namibia's coastal ecosystems offer unique advantages for integrated systems. The Benguela Current Upwelling System provides nutrient-rich, cold waters ideal for high-yield mariculture of species like Pacific oysters (*Crassostrea gigas*), abalone (*Haliotis midae*), and Mediterranean mussels (*Mytilus galloprovincialis*), which reach market size in half the time compared to other regions (). These can be co-cultured with seaweeds such as kelp in IMTA set-ups, reducing eutrophication risks by 30–50% while providing biofertilisers for coastal crops (Chopin et al. 2010; Chopin and Tacon 2021). Salinised soils, affecting 40% of coastal farmlands, can be reclaimed using seaweed extracts and shellfish waste, boosting crop yields by 15–20% for salt-tolerant varieties like quinoa or date palms (Crespi and Lovatelli 2011). In the Erongo Region, for example, integrating oyster farming with saline vegetable production could sequester carbon that can reach up to 200,000 credits per 500 hectares by 2030 while enhancing biodiversity through restored wetlands, contributing to achieving the United Nations goals and the vision 2030. Climate projections indicate a 13.57% rise in soil salinity by 2071 under RCP 8.5, underscoring the urgency of such adaptive integrations to maintain productivity and prevent salinity rises (Pandey et al. 2024).

6.3.3 Nutritional Opportunities for Combating Hidden Hunger

Integrated systems can significantly improve dietary diversity in Namibia, where over 20% of the population faces micronutrient malnutrition. Mariculture products like oysters and mussels are rich in zinc, iron, and omega-3 fatty acids, complementing carbohydrate-heavy staples from coastal crops (Walakira et al. 2023). In coastal communities, where stunting affects 29% of children, combining these with nutrient-dense halophytes could reduce deficiencies by 15–25%, as evidenced by similar systems in South Africa. Freshwater linkages such as using mariculture waste to irrigate inland tilapia-rice polycultures could extend benefits to inland regions and increase protein access for over 50% of rural households (FAO 2024a, b). The Namibian Standards Institute's ongoing certification for exports could also enable value-added products like fortified seaweed snacks to be exported addressing iodine deficiencies prevalent in 15% of coastal populations.

6.3.4 Economic Opportunities: Job Creation and Value Chain Development

With unemployment at 36.9% and coastal poverty rates exceeding 27%, integrated systems offer substantial economic potential (Iitembu et al. 2022). Mariculture alone could create 150 jobs per 500 hectares through farming, processing, and eco-tourism, while localised feed production from fish by-products reduces import costs by 20 to 30%. The National Aquaculture Master Plan targets 4000 MT of production by 2030, requiring 6000 MT of feed opportunities for private-sector plants using drought-resilient crops like pearl millet to make the feed. In Walvis Bay and Lüderitz, expanding oyster and abalone farms could generate more income annually, with about 70% value retention through downstream processing (Mupambwa et al. 2019).

6.3.5 Social and Policy Opportunities for Inclusivity and Governance

Integration promotes equity, empowering women and youth, who comprise 60% of smallholder farmers in coastal areas (FAO 2024a, b). The Aquaculture Act (2002) provides a robust framework, but opportunities lie in streamlining permits, which currently take a while to be processed and incentives for IMTA. The Asia-Africa BlueTech Superhighway could facilitate knowledge transfer, such as from the Spanish feed factories, accelerating adoption by 10 to 20% in other countries (WorldFish 2025a, b). Public–private partnerships, with other companies and stakeholders such as Oceana and NovaNam, could localise 50% of the value chain, aligning with Vision 2030's goals for poverty alleviation to ensure that everyone in Namibia is not threatened by food insecurity and lack of adequate nutrition (Markets 2025).

6.3.6 *Challenges and Pathways Forward*

Despite employment opportunities, waste reduction, and high nutritional value of food, barriers such as drought-induced food shortages, constraints on feed production, and high start-up costs deter 60% of smallholders (World Bank Group 2016). Policy gaps in mariculture licensing and climate adaptation funding limit scaling (Mupambwa et al. 2019). To realise potential, Namibia should prioritise investments in desalination for irrigation, R&D for indigenous species, and climate-smart subsidies that could potentially increase output by 25% by 2030 (Iitembu et al. 2022). Pilot expansions in Erongo could serve as models, transforming Namibia's coastal deserts into hubs of nutritional resilience.

Table 6.2 presents a Namibia-specific Integration Opportunity Matrix, outlining challenges and opportunities across key categories.

Table 6.2 Namibia-specific integration opportunity matrix

Category	Challenges	Opportunities
Ecological	Soil salinisation (40% of farmlands affected); drought reducing water availability; pollution from mining runoff (Mupambwa et al. 2019)	Benguela Current for high-yield mariculture; IMTA for nutrient recycling (30–50% eutrophication reduction); wetland restoration for biodiversity (Rawson et al. 2001)
Nutritional	Micronutrient deficiencies (25–30% in children); high food imports (70%) leading to "hidden hunger" (World Food Programme 2025)	Oysters/mussels for zinc/iron; seaweed for iodine; halophytes for vitamins, reducing deficiencies by 15–25% (Muunda et al. 2024)
Economic	High start-up costs (deterring 60% of smallholders) and feed shortages; unemployment (33%)	150 jobs/500 ha; N$50-100 M annual revenue from farms; value-added products reducing imports by 20–30% (Simasiku et al. 2024)
Policy	Gaps in licensing; limited funding for adaptation; outdated Master Plan targets not met (Iitembu et al. 2022; Mupambwa et al. 2019)	Streamline permits; incentives for IMTA; partnerships like BlueTech Superhighway for tech transfer (Markets 2025)

6.4 Conclusion

Integrated coastal agriculture and mariculture systems present a powerful opportunity to enhance nutritional security in Africa by diversifying diets, optimising nutrient productivity, improving accessibility, and fostering resilience against climate and market shocks. These systems have demonstrated tangible impacts, increasing dietary diversity by 15–25% and reducing micronutrient deficiencies in coastal communities across countries like Senegal, Morocco, Tanzania, Nigeria, Kenya, and Madagascar. Their socioeconomic benefits, including job creation and income stability, further supports nutritional outcomes by enabling households to access diverse foods. However, realising their full potential requires overcoming multifaceted challenges, including technological and knowledge gaps, economic constraints, policy deficiencies, environmental risks, social barriers, and infrastructure limitations. These challenges, if unaddressed, limit the scalability and reliability of nutrient-rich food production, particularly in regions where food insecurity affects 20 to 45% of populations.

Strategic interventions can unlock the transformative potential of these systems. Capacity building through localised training and technology transfer, as demonstrated by the Asia-Africa BlueTech Superhighway, can enhance adoption and efficiency. Financial support via subsidies and microfinance, coupled with policy reforms like streamlined licensing and integrated incentives, can address economic and regulatory barriers. Climate adaptation measures, such as seaweed, mangrove restoration, and pollution control, are critical to safeguarding production systems. Community engagement, awareness campaigns, and integration with national nutrition programmes can boost consumption and equitable access, including for women and youth. Infrastructure improvements and market development can ensure the availability and affordability of nutrient-rich foods that can reach remote areas.

By addressing these challenges through coordinated, multi-stakeholder efforts, integrated coastal agriculture and mariculture systems can play a pivotal role in transforming Africa's food systems. They offer a sustainable, resilient, and equitable pathway to combat malnutrition, supporting the continent's journey towards achieving SDG 2 and SDG 14 by 2030. In summary, integrating coastal agriculture and mariculture in Namibia and other African countries harnesses the desert–sea interface to forge a pathway towards food sovereignty. By capitalising on ecological synergies, nutritional complementarities, and economic multipliers, these systems can safeguard against climate threats, foster inclusive growth, and secure a nourished future for coastal communities.

References

Addo KA, Nicholls RJ, Codjoe SNA, Abu M (2018) A biophysical and socioeconomic review of the Volta Delta, Ghana. J Coastal Res 34(5):1216–1226. https://doi.org/10.2112/JCOASTRES-D-17-00129.1

Africa S (2020) Seaweed farming: recovery efforts amid losses | Science Africa. https://news.scienceafrica.co.ke/seaweed-farming-recovery-efforts-amid-losses/

Agumagu OO, Marchant R, Stringer LC (2025) Land use and land cover change dynamics in the Niger Delta region of Nigeria from 1986 to 2024. Land 14(4):765. https://doi.org/10.3390/land14040765

Andria-Mananjara DE, Martel P, Raliniaina M, Andriamarolaza R, Rakotomanana D, Mikolasek O, Mortillaro JM (2019) Malagasy integrated aquaculture-agriculture. In: FAO. 2019. Report of the special session on advancing integrated agriculture aquaculture through agroecology, Montpellier, France, 25 August 2018. FAO Fisheries and Aquaculture Report No. 1286. Rome

Appeaning Addo K, Brempong E, Jayson-Quashigah P-N (2020) Assessment of the dynamics of the Volta river estuary shorelines in Ghana. Geoenviron Disasters 7. https://doi.org/10.1186/s40677-020-00151-1

Aquaculture Act 18 of 2002 (2002) Government Gazette of the Republic of Namibia, No. 2888, 30 December 2002. https://www.npc.gov.na/wp-content/uploads/2022/06/Aquaculture-Act-2002.pdf

Ayo-Aderele E, Okomoda V (2024) Transforming Nigerian women participation in coastal aquaculture through AABS-IMTA intervention. WorldFish. https://worldfishcenter.org/impact-story/transforming-nigerian-women-participation-coastal-aquaculture-through-aabs-imta

Cheruiyot J, Kibett J (2024) Evaluation of social capital attributes as predictors of collective actions among smallholder farmers in Tinderet Sub-County, Kenya. East Afr J Interdisc Stud 7:252–267. https://doi.org/10.37284/eajis.7.1.1932

Chopin T, Tacon A (2021) Importance of seaweeds and extractive species in global aquaculture production. Rev Fish Sci Aquac 29:139–148. https://doi.org/10.1080/23308249.2020.1810626

Chopin T, Troell M, Reid GK, Knowler D, Robinson S, Buschmann A, Pang S, Fang J (2010) Integrated multi-trophic aquaculture. In: Policies to ensure a sustainable aquaculture sector, pp 195–217

Crespi V, Lovatelli A (2011) Aquaculture in desert and arid lands: development constraints and opportunities. Food and Agriculture Organization of the United Nations

Earth Journalism Network (2023) Empowering women, boosting economies: seaweed farming's coastal resilience solution in Kenya and Tanzania. https://earthjournalism.net/stories/empowering-women-boosting-economies-seaweed-farmings-coastal-resilience-solution-in-kenya

Elmighrabi NF, Fleming CAK, Agho KE (2024) Factors associated with childhood stunting in four North African countries: evidence from multiple indicator cluster surveys, 2014–2019. Nutrients 16(4):473. https://doi.org/10.3390/nu16040473

FAO (2022a) The State of World Fisheries and Aquaculture 2022. https://openknowledge.fao.org/handle/20.500.14283/cc0461en

FAO (2022b) Blue transformation—roadmap 2022–2030. https://doi.org/10.4060/cc0459en

FAO (2023a) Aquaculture growth potential in Southern Africa. WAPI factsheet

FAO (2023b) FAO in Africa: highlights in 2023. (2024). https://doi.org/10.4060/cd0291en

FAO (2024a) The contribution of small-scale fisheries to healthy food systems and sustainable livelihoods in the United Republic of Tanzania. Food and Agriculture Organization. https://doi.org/10.4060/cc7606en

FAO (2024b) The State of World Fisheries and Aquaculture 2024. FAO. https://doi.org/10.4060/cd0683en

FAO (2025) Pathways to climate-resilient net zero supply chains. https://doi.org/10.4060/cd5934en

FAO, IFAD, UNICEF, WF and WHO (2025) The state of food security and nutrition in the world 2025. FAO; IFAD; UNICEF; WFP & WHO. https://doi.org/10.4060/cd6008en

Farm Africa (2025) Sustainable mariculture creates jobs for young people in Kenya. Retrieved 4 Oct 2025, from https://www.farmafrica.org/story/sustainable-mariculture-creates-jobs-for-young-people-in-kenya/

Fatimi A (2022) Seaweed-based biofertilizers: a patent analysis. Recent Pat Biotechnol 16(2):144–154. https://doi.org/10.2174/1872208316666220128105056

Financial Fortune Media (2025) Seaweed turns the financial tide for Zanzibari women. Retrieved 4 Oct 2025, from https://africasolutionsmediahub.org/2025/05/24/seaweed-turns-the-financial-tide-for-zanzibari-women

Food Security Information Network (2025) GRFC 2025—Country profile—Nigeria. https://www.fsinplatform.org/grfc-2025-country-profile-nigeria

Golden CD, Zamborain-Mason J, Levis A, Rice BL, Allen LH, Hampel D, Hazen J, Metcalf CJE, Randriamady HJ, Shahab-Ferdows S, Wu SM, Haneuse S (2024) Prevalence of micronutrient deficiencies across diverse environments in rural Madagascar. Front Nutr 11:1389080. https://doi.org/10.3389/fnut.2024.1389080

Halwart M, Gupta MV (2004) In: WorldFish Center, & FAO (eds) Culture of fish in rice fields. Worldfish Center

Iitembu J, Gabriel N, Tjipute M, Asino H, Hamukwaya J (2022) The governance of aquaculture in Namibia as a vehicle for food security and economic growth. In: Food security for African smallholder farmers, pp 391–403. https://doi.org/10.1007/978-981-16-6771-8_23

Japhet E, Mangora MM, Trettin CC, Okello JA (2019) Natural recovery of mangroves in abandoned rice farming areas of the Rufiji Delta, Tanzania. WIO J Mar Sci 18:25–36. https://doi.org/10.4314/wiojms.v18i2.3

Kairo JG (2001) Ecology and restoration of mangrove systems in Kenya. PhD Dissertation, Free University of Brussels

Kaminski AM, Little DC, Middleton L, Syapwaya M, Lundeba M, Johnson J, Huchzermeyer C, Thilsted SH (2022) The role of aquaculture and capture fisheries in meeting food and nutrition security: testing a nutrition-sensitive pond polyculture intervention in rural Zambia. Foods 11(9):1334. https://doi.org/10.3390/foods11091334

Kasato Y, Mwansisya A, Islam S, Juma NS, Massawe A, Shuaibu A, Ying X (2025) Seaweed in Africa: a mini review of industry, benefits and safety concerns. Int J Res Stud Biosci 11(1):1–8

Kimanga F, Ladan L, Mirera D, Maundu A, Moyoni H, Bironga C, Onyango J (2025) Gendered-value chain opportunities and challenges in seaweed aquaculture: the changing gender and socio economic dynamics in Mwazaro and Kibuyuni Villages, South Coast Kenya. Multidiscip J Tech Univ Mombasa 4(1):1–17. https://doi.org/10.48039/mjtum.v4i1.85

Maïga Y, Mawussi G, Omar Ndaw F, Fall A (2020) Screening of rice lines (Oryza spp L. 1753) for salinity tolerance at vegetative stage under Senegal River valley conditions. J Exp Agric Int. https://doi.org/10.9734/JEAI/2020/v42i430501

Markets R (2025) Namibia ocean & coastal fishing and aquaculture industry report 2025, featuring local and international players—Oceana, NovaNam, Fishcor, Tunacor, Etosha fishing, and Hangana. GlobeNewswire News Room. https://www.globenewswire.com/news-release/2025/03/13/3042061/28124/en/Namibia-Ocean-Coastal-Fishing-and-Aquaculture--Industry-Report-2025-Featuring-Local-and-International-Players-Oceana-NovaNam-Fishcor--Tunacor-Etosha-Fishing-and-Hangana.html

Melaku S, Natarajan P (2019) Status of integrated aquaculture: agriculture systems in Africa. Int J Fish Aquatic Stud 7(4):263–269

Mortillaro J, Raminoharisoa EA, Randriamihanta TH (2018) La rizipisciculture: Un modèle agroécologique innovant et efficient

Msuya F, Bolton J, Pascal F, Narrain K, Nyonje B, Cottier E (2022) Seaweed farming in Africa: current status and future potential. J Appl Phycol 34:1–21. https://doi.org/10.1007/s10811-021-02676-w

Mupambwa HA, Hausiku MK, Nciizah AD, Dube E (2019) The unique Namib desert-coastal region and its opportunities for climate smart agriculture: a review. Cogent Food Agric 5(1):1645258. https://doi.org/10.1080/23311932.2019.1645258

Muunda V, Jacobs F, Naesje T, Hay C (2024) Freshwater fish as a potential nutritional gap-filler in combating malnutrition in Namibia. Namibian J Environ 9

Nhantumbo B, Dada O, Ghomsi F (2023) Sea level rise and climate change—impacts on African coastal systems and cities. https://doi.org/10.5772/intechopen.113083

Nicol A, Langan S, Victor M, Gonslaves J, Majule AE, Tarimo A, Mary AL, Waldorf A, Kebede A, Wamba A, Teshome A, MacDonald A, Alterra Wageningen UR, Ndawula A, Ibrahim A, Anteneh A, Lucerna A, Geleta A, Zeyege A, Karelina Z (2015) Water-smart agriculture in East Africa. International Water Management Institute (IWMI). https://doi.org/10.5337/2015.203

Nigussie L, Minh TT, Senaratna Sellamuttu S (2024) Youth inclusion in value chain development: a case of the aquaculture in Nigeria. CABI Agric Biosci 5(1):44. https://doi.org/10.1186/s43170-024-00243-0

Numbere AO (2021) Mangrove restoration under different disturbances regime in The Niger Delta, Nigeria. In: Mangrove ecosystem restoration. IntechOpen

Nyangoko BP, Shalli MS, Mangora MM, Gullström M, Berg H (2022) Socioeconomic determinants of mangrove exploitation and management in the Pangani River Estuary, Tanzania. Ecol Soc 27(2). https://doi.org/10.5751/ES-13227-270232

Ogbeibu AE, Oribhabor BJ, Ogbeibu AE, Oribhabor BJ (2023) The Niger delta mangrove ecosystem and its conservation challenges. In: Mangrove biology, ecosystem, and conservation. https://doi.org/10.5772/intechopen.112543

Olesen I, Bonaldo A, Farina R, Gonera A, Hughes AD, Navrud S, Orsini F, Parma L, Zornoza R (2023) Moving beyond agriculture and aquaculture to integrated sustainable food systems as part of a circular bioeconomy. Front Mar Sci 10:1178014. https://doi.org/10.3389/fmars.2023.1178014

Oluoch OB, Muindi EM, Gogo EO (2023) Farmers' perception on the effects and management of saline water on soil and crops within Kilifi and Kwale counties. Afr J Food Agric Nutr Dev. https://doi.org/10.22004/ag.econ.340701

Onyena A (2020) A review of the threat of oil exploitation to mangrove ecosystem: insights from Niger Delta, Nigeria. Glob Ecol Conserv. https://doi.org/10.1016/J.GECCO.2020.E00961

Pandey K, De Bruijn JA, De Moel H, Botzen W, Aerts JCJH (2024) Simulating the effects of sea level rise and soil salinization on adaptation and migration decisions in Mozambique. In: Risk assessment, mitigation and adaptation strategies, socioeconomic and management aspects. https://doi.org/10.5194/egusphere-2024-17

Panel G (2025) Building resilience and enhancing nutrition in Africa's food systems

Rana KJ (2012) National aquaculture master plan for Namibia part 2: freshwater aquaculture. Section 1: The master plan for freshwater Aquaculture (2013-2023)

Rasowo JO, Nyonje B, Olendi R, Orina P, Odongo S (2024) Towards environmental sustainability: further evidences from decarbonization projects in Kenya's blue economy. Front Mar Sci 11:1239862. https://doi.org/10.3389/fmars.2024.1239862

Rawson MV, Chen C, Ji R, Zhu M, Daoru W, Wang L, Yarish C, Sullivan J, Chopin T, Carmona R (2001) Integration of fed and extractive aquaculture

Saba AO, Eyo VO, Elegbede IO, Fakoya KA, Ojewole AE, Dawodu FO, Adewale RA, Amal MNA, Saba AO, Eyo VO, Elegbede IO, Fakoya KA, Ojewole AE, Dawodu FO, Adewale RA, Amal MNA (2024) Sustaining the blue bounty: fish food and nutrition security in Nigeria's evolving blue economy. AIMS Agric Food 9(2):500–530. https://doi.org/10.3934/agrfood.2024029

Salele Abubakar H, Abdurrasheed N, Dauda A (2023) Integrated rice-cum-fish farming in Nigeria: prospects, status and challenges. J Aquac Res Dev 14:1–6. https://doi.org/10.35248/2155-9546.23.14.778

Simasiku E, Hove K, Kanyeme M, Awala S, Matengu K, Iita T (2024) Comparative effect of fish feeds on the growth and survival rate of the Redbreast Tilapia (Coptodon rendalli) under intensive culture conditions. https://doi.org/10.20944/preprints202407.1808.v1

Soto D (2009) Integrated mariculture: a global review. Food and Agriculture Organization

Thiam S, Villamor GB, Faye LC, Sène JHB, Diwediga B, Kyei-Baffour N (2021) Monitoring land use and soil salinity changes in coastal landscape: a case study from Senegal. Environ Monit Assess 193(5):259. https://doi.org/10.1007/s10661-021-08958-7

Uzomah A, Lundebye A-K, Kjellevold M, Chuku FA, Stephen OA (2021) A review of chemical contaminants in marine and fresh water fish in Nigeria. Foods 10(9). https://doi.org/10.3390/foods10092013

Walakira JK, Hinrichsen E, Tarus V, Langi S, Ibrahim NA, Badmus O, Aziz A, Baumüller H (2023) Scaling aquaculture for food security and employment in Africa: insights from Egypt, Kenya and Nigeria

Wetlands International (2025) Tanzania Mangrove Track Record. https://www.wetlands.org/publication/tanzania-mangrove-track-record/

World Bank Group (2016) Seaweed aquaculture for food security, income generation and environmental health in tropical developing countries. World Bank, Washington, DC. https://doi.org/10.1596/24919

World Food Programme (2024) Draft Namibia country strategic plan (2025–2029). World Food Programme Executive Board

World Food Programme (2025) Global Report on Food Crises (GRFC) 2025. https://www.wfp.org/publications/global-report-food-crises-grfc

WorldFish (2025a) Asia-Africa BlueTech Superhighway. Retrieved 4 Oct 2025, from https://worldfishcenter.org/project/asia-africa-bluetech-superhighway

WorldFish (2025b) Social and economic inclusion. Retrieved 4 Oct 2025, from https://worldfishcenter.org/research-theme/social-and-economic-inclusion

Chapter 7
Examining the Marine Aquaculture's Contribution to Sub-Saharan Africa's Food System Diversification and Nutritional Security

Daniel K. Mbangula

Abstract Sub-Saharan Africa (SSA) is home to about one-third of the world's population that suffers from chronic hunger. Micronutrient deficiencies affect millions of people in addition to the high prevalence of chronic hunger. In the meantime, there is growing agreement among scientists from various fields that fish is essential to enhancing food and nutrition security. Thus, by reviewing the body of research, the current review seeks to analyse the critical role that coastal and marine ecosystems play in ensuring food security across the African continent. Furthermore, this study illustrates how fish and the aquatic industry as a whole contribute to ensuring food and nutrition security in SSA. Fish is a great source of animal protein and vital micronutrients like calcium, iodine, zinc, and vitamins. These nutrients are crucial for human nutrition and have been shown to lower the risk of non-communicable diseases and malnutrition. The role that the fisheries industry can play in preventing hunger and undernutrition, particularly for the impoverished and marginalized in SSA, should be acknowledged by policymakers, development organizations, and the general public. The production of aquatic food has a lot of potential as a workable way to address the urgent problem of providing the African population with the protein they need. It is essential for both economic expansion and food security, particularly in tropical Africa. Aquatic food sources are becoming more and more necessary as the population grows quickly, as they offer affordable protein and vital nutrients. A number of obstacles prevent aquatic food from being fully utilized in the tropical African environment, despite Africa's abundant aquatic food supplies and the significance of aquatic food in sustainable and healthful diets. The results show how different countries consume different amounts of aquatic protein, highlighting the significance of taking local factors into account when developing nutritional interventions and regulations. A consistent pattern of lower aquatic protein availability is found when the supply of protein from aquatic sources and land animals was compared across African countries. Animal protein consumption

D. K. Mbangula (✉)
Faculty of Education and Human Sciences, University of Namibia, Windhoek, Namibia

H. A. Mupambwa et al. (eds.), *Harnessing Coastal and Marine Resources for Improved Food Security in Africa*, Sustainability Sciences in Asia and Africa,
https://doi.org/10.1007/978-981-95-9725-3_7

is comparatively lower per capita, especially in Africa. The average amount of land animal feed on the continent was far more abundant than the average amount of aquatic food. This implies that the majority of dietary protein requirements in African nations are satisfied by land animal sources. Policymakers, dietitians, and other stakeholders should take note of these findings since it emphasizes the necessity of focused approaches to increase protein accessibility in African nations. Additionally, it urges that any disparities in the region's use of land-based and aquatic protein sources be addressed.

Keywords Marine aquaculture • Sub-Saharan Africa • Food System Diversification and Nutritional Security • Sustainability

7.1 Introduction

As the second of the Sustainable Development Goals 2030, the issue of food and nutrition security is at the top of the global political agenda. By 2030, the objective seeks to eradicate hunger, attain food security, and enhance nutrition. In order to meet their dietary needs and food preferences for an active, healthy life, as well as a hygienic environment and sufficient health, education, and care, all people must always have physical, social, and economic access to food that is sufficient in quantity and quality in terms of variety, diversity, nutrient content, and safety, according to the Committee on World Food Security (2021). Aquatic foods play a vital role in the nutrition and food security of billions of people because invertebrates, algae, and fish contribute 15 and 6 percent of all protein and total animal protein consumed worldwide, respectively (FAO 2024). Foods of aquatic origin contain certain significant traces of micronutrients such as iron, zinc, calcium, and vitamins, as well as the most common source of essential omega-3 fatty acids such as docosahexaenoic (DHA) and eicosapentaenoic acid (EPA), on top of being an excellent source of healthy proteins (Golden et al. 2021). They are highly diverse, with more than 600 species or species types grown and more than 2370 species grown in the wild, with each species possessing a distinct nutritional pattern (Golden et al. 2021). Aquatic food species have a better land-water use rate and environmental stressors such as greenhouse gas, nitrogen, and phosphorus emissions compared to terrestrial animals (Gephart et al. 2021). Aquatic food assists the way to healthy and sustainable diet. Additionally, it offers more nutritious sources of food that have less environmental impact, besides being more nutritious (Gephart and Golden 2022). Nevertheless, since the primary subjects of the conversation and decision-making in the food system are agriculture and livestock, the aquatic foods are quite often overlooked, although their role in human health and well-being is undeniable (Bennett et al. 2021).

Historically, aquatic food is primarily based on capture fisheries. However, in the last 30 years, the growth of aquaculture has been very high to meet the world's demand for aquatic food, whereas the fishery catches have remained relatively stable. In 2022, aquaculture surpassed capture fisheries and became the largest source of aquatic animals in history, with aquatic food production reaching 223 million tonnes (FAO 2024). Aquatic food systems must change towards resilient, inclusive, and sustainable models. Also, in order to fully realize the potential of aquatic foods and maximize their contribution to feeding the world's population in an equitable and sustainable manner (FAO 2024, 2022). In order to provide wholesome, reasonably priced, and nutritious diets—especially for the most vulnerable populations, such as those in Sub-Saharan Africa. While fostering equitable growth, this transformation aims to expand aquatic food systems and increase contributions from capture fisheries, aquaculture, and associated supply chains (Agostini et al. 2024). The radical change that food systems require relies on innovation that is considered to be one of the key forces in the systems (Herrero et al. 2020). The innovations may be either a technology-driven product or service, or they may be new forms of doing things, new business models, or other groundbreaking advances in the provision of the necessary services in solving a complex problem (IDIA 2021). Innovations can be described as a novel solution that has the transformative capability to speed up the impact. Disconnected innovations, however, are rarely yielding positive results on a number of dimensions of sustainability. Research on food dimensions are necessary to develop innovative approaches to sustain food systems and nutritional security. Current research and innovation frameworks are deemed insufficient to fully drive these transformations (Herrero et al. 2021). Thus, to achieve significant change in food systems, we must reconsider how we approach innovation itself (Kok et al. 2019). This makes this study more relevant to the debate on this topic under review.

7.1.1 Chapter Aims/Objectives

By the end of this chapter, this study aims to achieve the following objectives:

- To analyse the critical role that coastal and marine ecosystems play in ensuring food security across the African continent.
- To discuss the challenges faced in coastal and marine ecosystems in sustainable management, and innovative solutions being implemented to overcome these challenges.
- To elaborate on the effects of climate change, its impact on coastal productivity, and its implications for food production.
- To describe findings of the study about Marine Aquaculture's Contribution to Sub-Saharan Africa's Food System Diversification and Nutritional Security.
- To draw a conclusion from the study findings

7.2 Critical Role that Coastal and Marine Ecosystems Play in Ensuring Food Security Across the African Continent

The nutritional advantages of eating aquatic foods, such as proteins, vital amino acids, vitamins, minerals, and heart-healthy omega-3 fatty acids, have been shown in studies (Tacon 2023). Consuming them is associated with a number of health advantages, such as improving eye and brain function and offering a more nutritious substitute for processed and red meats, which can have detrimental effects on well-being (Golden et al. 2021). However, aquatic species differ in their nutrient composition, which is mostly determined by their fat content. As an illustration of their importance in human nutrition, carp species like *Amblypharyngodon mola* and *Puntius sophore* have high vitamin K contents, ranging between 402.9 and 786.9 IU per 100 g of flesh (Bianchi et al. 2022). Although the amount of high-quality protein, vitamins D, and B12 that Mackerel offers is still quite low, it is notable that 11 nations have exceeded the global average (Chan et al. 2021). These nations, which include Gabon, Congo, the Gambia, Sierra Leone, Ghana, Egypt, and Côte d'Ivoire, as well as a few small island developing states (SIDS), are known for their comparatively higher rates of aquatic food consumption. On the other hand, for a variety of reasons, including social, economic, and geographic ones, the majority of African nations struggle with comparatively low levels of aquatic food consumption (Rocha et al. 2022). Nevertheless, despite these obstacles, many African countries are beginning to recognize the economic potential and nutritional advantages of increasing their consumption of aquatic foods.

Furthermore, aquatic foods offer more than omega-3 fatty acids; they serve as an exceptional reservoir of high-quality protein, crucial for constructing and mending body tissues. They play a vital role in maintaining a robust immune system and facilitating the production of enzymes and hormones (Tacon et al. 2020). Moreover, they encompass all the essential amino acids the body cannot synthesize independently, making them excellent protein sources for plant-based alternatives. In contrast to land-based animals and plant-based substitutes, aquatic foods are especially rich sources of vitamin D and iodine (Craig et al. 2021). Aquatic foods also have lower levels of saturated fat, which makes them a healthier option than land-based animals, which frequently have high levels of saturated fat linked to an increased chance of developing heart disease (Aldaya et al. 2021). Aquatic foods, especially shellfish, that are sourced sustainably are therefore better for the environment and require less energy and resources to produce than animals raised on land (Laso et al. 2022). Africa's varied ecosystems and cultural customs make it crucial to consider the variations in consumption patterns by region.

7.3 Challenges Faced in Coastal and Marine Ecosystems in Sustainable Management, and Innovative Solutions Being Implemented to Overcome These Challenges

In Africa, sustainable aquatic food production is essential for maintaining food security, stimulating economic growth, and advancing environmental preservation (El Bilali et al. 2019). Rivers, lakes, and coastal regions make up the vast water resource base of the African continent, offering a wealth of chances to produce a variety of aquatic foods, such as fish, crustaceans, and molluscs (Zihao et al. 2020). However, a number of issues, such as pollution, climate change, inadequate governance, and overfishing, present serious risks to the sustainability of aquatic food production in Africa (Ahern et al. 2021). Among these issues, overfishing is one of the most urgent ones, endangering the long-term viability of aquatic food production in Africa (Lucrezi 2022). Unsustainable fishing methods have significantly diminished many local fisheries (Haque et al. 2021). The sustainability of fish stocks and the livelihoods of millions of people who depend on them for income and sustenance are at risk due to this depletion (Bi et al. 2022). Numerous fish stocks have been reduced in tropical Africa due to overfishing, which is typified by unsustainable fishing methods like utilizing destructive gear and pursuing juvenile fish (Srinivasan et al. 2012). With significant repercussions for nearby communities, the overfishing-induced decline in fish stocks poses a serious threat to Africa's aquatic food production sustainability (Ikram et al. 2023). Fish availability and affordability are directly impacted by this circumstance, restricting their access to essential food sources (Ansari et al. 2021). Omega-3 fatty acids, which are vital for brain development, cardiovascular health, and cognitive function, are abundant in fish (Sanjay Panda and Diwan 2023). The risk of cardiovascular disorders, malnutrition, and other disorders associated with omega-3 deficiency may increase if the supply of fish declines. The essential institutional and legal frameworks for managing aquatic resources are absent from a number of tropical African nations (DiNicolantonio and O'Keefe 2020).

In particular, inadequate governance is a major obstacle to the sustainable production of aquaculture food in Africa and the effective management of fisheries (Wilson and McCay 2019). Numerous nations lack the institutional and legal frameworks that are required in the area, which leads to insufficient enforcement of rules, a high frequency of illicit fishing, and disputes between different user groups (Begum et al. 2022). This lack of governance also affects aquaculture management, as poor oversight and control worsen pollution and disease transmission (Edokpayi et al. 2017). Furthermore, the aforementioned problems are made worse by the effects of climate change. Ocean acidification, rising sea temperatures, and changing weather patterns have an impact on the distribution and productivity of (Lubembe et al. 2022) fish stocks. Droughts and floods are examples of extreme weather events that damage infrastructure and interfere with production processes. Additionally, the sustainability of coastal aquaculture is threatened by rising sea levels (Gitz et al. 2016), which has an effect on the accessibility and cost of fish resources (Katikiro and Macusi 2012). The quality and safety of aquatic foods are also impacted by variations in water temperature and salinity because warmer waters encourage the growth of toxic bacteria (Gomez-Zavaglia et al. 2020). Additionally, the availability and safety of aquatic foods are directly impacted by climate change-induced extreme weather events like droughts and floods, which lead to malnutrition and other health problems (Neetoo and Duchenne-Moutien 2021). The sustainability of aquatic food production in Africa is seriously threatened by aquatic pollution in addition to climate change (Mpumi et al. 2020). Aquatic food safety and nutritional quality are negatively impacted by untreated sewage, waste disposal, and contamination of water bodies brought on by industrial and agricultural operations (Mateo-Sagasta et al. 2017).

As a result, fish and other aquatic organisms have accumulated dangerous substances like antibiotics and heavy metals due to aquatic pollution, which could be harmful to human health (Qadri and Faiq 2020; Okon et al. 2022). Accordingly, eating fish tainted with mercury may cause developmental problems and neurological harm. According to Mitra et al. (2022), exposure to polycyclic aromatic hydrocarbons (PAHs) may also increase the risk of cancer, necessitating a more aggressive strategy to address aquatic organisms and human exposure. To address these issues and advance the sustainability of aquatic food production in Africa, a number of initiatives have been put into place. These initiatives include enhancing governance. The African Union's Policy Framework and Reform Strategy for Fisheries and Aquaculture in Africa serves as an example of this. With an emphasis on improved governance, ethical fishing methods, and greater investment, this framework offers a thorough framework for the sector's sustainable development (NEPAD Planning and Coordinating Agency 2016). Other projects focus on promoting environmentally friendly fishing methods. For instance, the World Bank-supported Sustainable Fisheries Management Project improves the livelihoods of fishermen, promotes sustainable fishing methods, and fortifies institutional and legal frameworks (Guenard 2021). Maximizing the potential of the region's aquatic resources is typically the shared goal of these initiatives.

Africa has enormous unrealized potential for improving the production of aquatic food (De Angelis et al. 2021). Numerous sustainable practices across the continent may guarantee the long-term safety and nutritional value of fish supplies (Khan et al. 2021). This can be achieved by implementing laws and policies, like quotas and gear limits, that encourage sustainable fishing methods. Additionally, initiatives are needed to control pollution and lessen the negative effects of climate change on aquatic ecosystems (Odeku and Paulos 2017). Overfishing, pollution, and climate change can all affect aquatic ecosystems, but they can still benefit from aquatic foods (Zhang et al. 2023). Millions of Africans' health and well-being will eventually improve with the implementation of suitable management measures (Andrews et al. 2021). Increasing local communities' understanding of the significance of consuming aquatic foods in a safe and healthful manner will be necessary for these initiatives. Additionally, encouraging studies on the safety and nutritional value of these foods can ensure that the required standards are met while opening up new opportunities (Khan et al. 2021). The sustainability and prospects for aquatic food production vary between landlocked and coastal African nations. Access to the ocean gives coastal nations the advantage of being able to use a variety of marine resources, including fish and shellfish (Bennett et al. 2020). The marine biodiversity that includes valuable fish species is generally accessible in coastal African nations. Seafood is a vital component of their diets because of the ocean's high protein content, significant fatty acids, and vital micronutrients (Aakre et al. 2020). However, landlocked nations depend on inland fisheries, lakes, and rivers to meet their food needs (Kolding et al. 2019).

7.4 Effects of Climate Change, Its Impact on Coastal Productivity, and Its Implications for Food Production

Climate change is the biggest challenge facing the sufficiency of tropical Africa in consuming aquatic foods. The fluctuation of the weather conditions and increase in the sea temperatures negatively affect fish stocks' distribution and productivity, so that communities cannot access and use these most important food sources (Barange et al. 2018). Thus, the complexity of the challenges hindering the access and consumption of aquatic foods has to be addressed by means of coordinated activities and comprehensive solutions that would acknowledge the complexity of the problems associated with the consumption of aquatic foods (Elkin and Katz 2019). So, the benefits of using the vast potential of aquatic food production can be enjoyed by the people, economies, and environment of tropical Africa. Implementing a number of initiatives to advance ethical practices in aquaculture is an interesting activity in the area. In order to ensure that all aquaculture activities are met with ethical standards, the Aquaculture Stewardship Council (ASC) has emerged as one of the well-known organizations that certify (Kruk and Peters 2018). The ASC certification is needed in order to make sure that the aquaculture industry is sustainable and a steward of the environment. The challenge that needs to be overcome in order to encourage the widespread implementation of ethical aquaculture practices is the high ASC certification rates, particularly in the Global South and Africa (Bush et al. 2013).

Moreover, the mentioned issues are aggravated by the impact of climate change. The alteration of the weather pattern, acidification of the ocean, and the increase in sea temperatures influence the distribution and productivity of fish stocks (Maulu et al. 2021; Lubembe et al. 2022). Examples of extreme weather include droughts and floods, which cause the destruction of infrastructure and disrupt the production process. There is also the risk that rising sea levels can cause a threat to the sustainability of coastal aquaculture (Gitz et al. 2016), which also impacts the availability and prices of fish resources (Katikiro and Macusi 2012). Water temperature and salinity also affect the quality and safety of aquatic foods due to the increased population of the toxic bacteria whose growth is promoted by warmer waters (Gomez-Zavaglia et al. 2020). Also, the occurrence and safety of aquatic foods are directly influenced by extreme weather conditions such as droughts and floods, resulting in climate change that directly causes malnutrition and other medical issues (Duchenne-Moutien and Neetoo 2021). Aquatic pollution poses a significant threat to the sustainability of the production of aquatic food in Africa, other than climate change (Mpumi et al. 2020). The untreated sewage, waste dumping, and water pollution caused by industries and agriculture adversely affect the quality and safety of aquatic food, as well as food nutritional quality (Mateo-Sagasta et al. 2017). Subsequently, aquatic organisms such as fish and others have stored harmful chemicals such as antibiotics and heavy metals courtesy of aquatic pollution, which may pose a threat to human health (Okocha et al. 2018; Okon et al. 2022).

7.5 Chapter Findings

The following are some of the chapter's findings that were found based on these themes:

7.5.1 Coastal and Marine Ecosystem's Role on Food Security

The study results revealed that the African countries are beginning to recognize the economic potential and nutritional advantages of increasing their consumption of aquatic foods. Also, the aquatic foods are not only sources of omega-3 fatty acids, they are also an outstanding store of high-quality protein which maintains and repairs body tissues. They are critical in ensuring that the immune system is strong and aids the production of enzymes and hormones (Tacon et al. 2020). In addition, they contain all the necessary amino acids that the body cannot produce on its own, which makes them good protein sources of plant-based substitutes. Unlike terrestrial animal-based and plant-based alternatives, aquatic food is particularly highly endowed with vitamin D and iodine (Craig et al. 2021). Saturated fat is also less prevalent in aquatic foods, which is why they are a healthier choice in comparison with terrestrial animals, which often contain high amounts of saturated fats tied to a high risk of developing heart disease (Aldaya et al. 2021). Sustainably sourced aquatic foods, particularly shellfish, are thus more environmentally friendly and consume less energy and resources to produce compared to animals kept on the farmland (Laso et al. 2022). The diversity of the African ecosystem and cultural practices is important to consider the differences in regional consumption patterns. This implies that the marine ecosystems contribute greatly to food security.

7.5.2 Challenges in Coastal and Marine Ecosystems

Results indicate that a number of issues, such as pollution, climate change, inadequate governance, and overfishing, present serious risks to the sustainability of aquatic food production in Africa (Ahern et al. 2021). Overfishing is among them; unfortunately, this threatens the sustainability of aquatic food production in Africa in the long term (Lucrezi 2022). Because of unsustainable fishing practices, a number of local fisheries have been reduced (Haque et al. 2021). This depletion threatens the sustainability of fish stocks and the livelihoods of millions of people, who either rely on fish income and sustenance (Bi et al. 2022). Overfishing has also led to the depletion of a variety of fish stocks in tropical Africa, characterized by unsustainable fishing practices such as the use of destructive fishing equipment and juvenile capture (Srinivasan et al. 2012). The depletion in fish stocks caused by overfishing has great consequences for the surrounding populations, and its impact threatens the sustainability of the aquatic food production in Africa (Ikram et al. 2023). This situation directly affects fish availability and affordability and limits their access to necessary sources of food (Ansari et al. 2021). Such conclusions demonstrate that there is a great necessity to have improved and sustainable practices in fishing and aquatic food management systems.

7.5.3 Impact of Climate Change on Marine Ecosystems

The study findings revealed that changing weather patterns and rising sea temperatures have a negative impact on the distribution and productivity of fish stocks, making it difficult for communities to obtain and consume these essential food sources (Barange et al. 2018). Thus, to adequately deal with the multifaceted concerns connected with aquatic food consumption, it is necessary to organize the actions and comprehensive solutions that take into account the multidimensional character of the barriers to access to and consumption of aquatic foods (Elkin and Katz 2019). Also, the temperature and salinity level of waters influence the quality and safety of aquatic foods, as warm water promotes the development of harmful bacteria (Gomez-Zavaglia et al. 2020). Also, the existing conditions and safety of aquatic foods are directly influenced by extreme weather conditions such as droughts and floods caused by climate change, which contribute to malnutrition and other health issues (Duchenne-Moutien and Neetoo 2021). Aquatic pollution is also a serious threat to the sustainability of the aquatic production of food in Africa, alongside climate change (Mpumi et al. 2020). These findings show that there is an urgent intervention required to combat climate change in Africa and the world at large in order to salvage the marine and coastal food security, particularly in the developing nations.

7.6 Discussion

With the trends discussed in this chapter, the data indicated that the coastal and marine ecosystems play a major role in ensuring food security across the African continent. This can be attributed to the fact that coastal and marine ecosystems have been the greatest contributors of fish and other aquatic food and hence play a significant role in food security in the African continent. Also worth noting is the fact that the sustainability of both coastal and marine ecosystems will be a long-term benefit to food supply to the African nations that are densely populated. Nevertheless, other challenges are also associated with the sustainability of coastal and marine environments. Some of these issues are the unsustainable fishing techniques, untreated sewage, dumping of waste materials, and water pollutants occasioned by the industrial and agricultural activities. Secondly, lack of legal frameworks of the issue in the African countries also threatens the endeavours to address the challenges encountered by the coastal and marine ecosystem environments. Therefore, it is necessary to come up with a powerful sustainable management strategy to address these issues. These involve, the promotion of research on the safety and nutritional quality of such foods that can indeed result in the achievement of the desired standards, besides creating new opportunities. Also, implementing different programmes to foster ethical aquaculture. As it has been stated above in this chapter, one such initiative is that of improving governance, The African Union Policy Framework and reform strategy in fisheries and aquaculture in Africa. This framework can provide comprehensive guidelines on the sustainable development of the sector due to its focus on enhanced governance, ethical fishing practices, and increased investments. The climate change problem is another burning issue as far as the coastal and marine ecosystems are concerned, in the context of ensuring food security. Climate change, e.g. altering weather patterns and increasing sea temperatures have adverse effect on the location and productivity of fish stocks, and communities find it hard to access and consume these important food resources. Examples of extreme weather events that cause the destruction of infrastructure and disruption of production processes are droughts and floods. There is therefore a need to develop climate-resilient approach in order to prevent these effects.

7.7 Conclusion

It is a challenging but necessary task to convert aquatic food systems into inclusive, resilient, and sustainable models. The important aspects of innovation ecosystems that must be reinforced in order to support this transition have been emphasized in this chapter. In order to ensure that a variety of perspectives are heard and that the advantages of innovation are distributed widely, inclusivity and equity are not only moral requirements but also pragmatic requirements for the success of innovation ecosystems. This study provided practical suggestions for designing interventions to maximize efficacy within diverse African communities by illustrating the dynamics of local culture, socioeconomic status, environmental conditions, and individual dietary behaviours. It is advised that future studies look into the precise sources of aquatic protein and how it functions in these countries' overall diets in order to enable focused strategies to improve public health and food security. The pattern of an aquatic protein supply, which displayed lower values than the land animal protein supply across African countries, is further illustrated by this study. These results demonstrate the importance of land-based animal sources in supplying the population's protein requirements. In particular, imports of aquatic food had a major impact on regional consumption during the study year, indicating the impact of trade economics on aquatic food intake. In order to improve protein availability and address imbalances between land and aquatic protein sources in African nations, policymakers, nutritionists, and stakeholders must have a better understanding of these factors influencing the differences and their dynamics. Thus, more research is needed to examine the fundamental causes of these gaps and possible remedies for the production and consumption of sustainable aquatic source food in the area.

In addition, millions of people in tropical Africa depend on the complex and varied aquatic food production industry for their livelihood. A range of fishing, aquaculture, and fish farming methods are used in this industry. Nevertheless, the industry has faced difficulties and impediments, such as inadequate laws, inadequate infrastructure, and limited financial access. Various strategies, like improved financial accessibility and ecologically friendly aquaculture methods, have been put forth to address these problems. To guarantee successful results for a healthier African population, these strategies must be put into practice with greater pragmatism and continuous observation.

References

Aakre I, Bøkevoll A, Chaira J, Bouthir FZ, Frantzen S, Kausland A et al (2020) Variation in nutrient composition of seafood from Northwest Africa: implications for food and nutrition security. Foods 9:1516

Agostini V, Olsen E, Tiffay C, Alison E, Coetzee J, Cojocaru A-L, Costello C, Darias MJ, Fabinyi M, Fulton E, Gelcich S, Haraksingh Thilsted S, Hassouni FZ, Mills K, Msuya F, Obura D, Tigchelaar M (2024) Ocean decade vision 2030 white papers—challenge 3: ocean contributions to nourishing the world's population. UNESCO-IOC. https://doi.org/10.25607/r7qs-a228

Ahern M, Thilsted S, Oenema S, Kühnhold H (2021) The role of aquatic foods in sustainable healthy diets. UN Nutrition Discussion Paper

Aldaya MM, Ibañez FC, Domínguez-Lacueva P, Murillo-Arbizu MT, Rubio-Varas M, Soret B et al (2021) Indicators and recommendations for assessing sustainable healthy diets. Food Secur 10:999. https://doi.org/10.3390/foods10050999

Andrews N, Bennett NJ, Le Billon P, Green SJ, Cisneros-Montemayor AM, Amongin S et al (2021) Oil, fisheries and coastal communities: a review of impacts on the environment, livelihoods, space and governance. Energy Res Soc Sci 75:102009. https://doi.org/10.1016/j.erss.2021.102009

Ansari FA, Guldhe A, Gupta SK, Rawat I, Bux F (2021) Improving the feasibility of aquaculture feed by using microalgae. Environ Sci Pollut Res 28:43234–43257. https://doi.org/10.1007/s11356-021-14989-x

Barange M, Bahri T, Beveridge MC, Cochrane KL, Funge-Smith S, Poulain F (2018) Impacts of climate change on fisheries and aquaculture. United Nations' Food Agric Organ 12:628–635

Begum A, Uddin MK, Rahman MM, Shamsuzzaman MM, Islam MM (2022) Assessing policy, legal and institutional frameworks of marine megafauna conservation in Bangladesh. Mar Policy 143:105187. https://doi.org/10.1016/j.marpol.2022.105187

Bennett NJ, Finkbeiner EM, Ban NC, Belhabib D, Jupiter SD, Kittinger JN et al (2020) The COVID-19 pandemic, small-scale fisheries and coastal fishing communities. Coast Manag 48:336–347

Bennett A, Basurto X, Virdin J, Lin X, Betances SJ, Smith MD, Allison EH, Best BA, Brownell KD, Campbell LM, Golden CD, Havice E, Hicks CC, Jacques PJ, Kleisner K, Lindquist N, Lobo R, Murray GD, Nowlin M, Patil PG, Rader DN, Roady SE, Thilsted SH, Zoubek S (2021) Recognize fish as food in policy discourse and development funding. Ambio 50:981–989. https://doi.org/10.1007/s13280-020-01451-4

Bi M, Zhang Z, Guo X, Wan L (2022) Evaluation of sustainable utilization of African marine fishery resources. Aust Fish 8:4. https://doi.org/10.3390/fishes8010004

Bianchi M, Hallström E, Parker RW, Mifflin K, Tyedmers P, Ziegler F (2022) Assessing seafood nutritional diversity together with climate impacts informs more comprehensive dietary advice. Commun Earth Environ 3:188. https://doi.org/10.1038/s43247-022-00516-4

Bush SR, Belton B, Hall D, Vandergeest P, Murray FJ, Ponte S et al (2013) Certify sustainable aquaculture? Science 341:1067–1068. https://doi.org/10.1126/science.1237314

Chan CY, Tran N, Cheong KC, Sulser TB, Cohen PJ, Wiebe K et al (2021) The future of fish in Africa: employment and investment opportunities. PLoS One 16:e0261615. https://doi.org/10.1371/journal.pone.0261615

Committee on World Food Security (2021) Coming to terms with terminology: food security, nutrition security, food security and nutrition, food and nutrition security 2021. http://www.fao.org/fsnforum/sites/default/files/file/Terminology/MD776(CFS_Coming_to_terms_with_Terminology).pdf. Accessed on 17 Dec 2021

Craig WJ, Mangels AR, Fresán U, Marsh K, Miles FL, Saunders AV et al (2021) The safe and effective use of plant-based diets with guidelines for health professionals. Nutrients 13:4144. https://doi.org/10.3390/nu13114144

De Angelis P, Tuninetti M, Bergamasco L, Calianno L, Asinari P, Laio F et al (2021) Data-driven appraisal of renewable energy potentials for sustainable freshwater production in Africa. Renew Sust Energ Rev 149:111414. https://doi.org/10.1016/j.rser.2021.111414

DiNicolantonio JJ, O'Keefe JH (2020) The importance of marine omega-3s for brain development and the prevention and treatment of behaviour, mood, and other brain disorders. Nutrients 12:2333. https://doi.org/10.3390/nu12082333

Duchenne-Moutien RA, Neetoo H (2021) Climate change and emerging food safety issues: a review. J Food Prot 84:1884–1897. https://doi.org/10.4315/JFP-21-141

Duchenne-Moutien RA, & Neetoo H (2021) Climate Change and Emerging Food Safety Issues: A Review. Journal of Food Protection (or a similar specialised food science/climate review venue based on context)

Edokpayi JN, Odiyo JO, Durowoju OS (2017) Impact of wastewater on surface water quality in developing countries: a case study of South Africa. Water Quality 10:10–5772. https://doi.org/10.5772/66561

El Bilali H, Callenius C, Strassner C, Probst L (2019) Food and nutrition security and sustainability transitions in food systems. Food Energy Secur 8:e00154. https://doi.org/10.1002/fes3.154

Elkin Z, Katz I (2019) Implementation of the sustainable development goals: national review. Global Risk Insights 430:1–429

FAO (2022) Blue transformation - roadmap 2022–2030. FAO

FAO (2024) The state of world fisheries and aquaculture 2024. FAO

Gephart JA, Henriksson PJG, Parker RWR, Shepon A, Gorospe KD, Bergman K, Eshel G, Golden CD, Halpern BS, Hornborg S, Jonell M, Metian M, Mifflin K, Newton R, Tyedmers P, Zhang W, Ziegler F, Troell (2021) Blue foods environmental performance. Nature. https://doi.org/10.1038/s41586-021-03889-2

Gitz V, Meybeck A, Lipper L, Young C, Braatz S (2016) Climate change and food security: risks and responses. Food and Agriculture Organization of the United Nations

Golden CD, Koehn JZ, Shepon A, Passarelli S, Free CM, Viana DF, Matthey H, Eurich JG, Gephart JA, Fluet-Chouinard E, Nyboer EA, Lynch AJ, Kjellevold M, Bromage S, Charlebois P, Barange M, Vannuccini S, Cao L, Kleisner KM, Rimm EB, Danaei G, DeSisto C, Kelahan H, Fiorella KJ, Little DC, Allison EH, Fanzo J, Thilsted SH (2021) Aquatic foods to nourish nations. Nature:1–6. https://doi.org/10.1038/s41586-021-03917-1

Gomez-Zavaglia A, Mejuto JC, Simal-Gandara J (2020) Mitigation of emerging implications of climate change on food production systems. Food Res Int 134:109256. https://doi.org/10.1016/j.foodres.2020.10925

Guenard R (2021) Poisson from a petri dish. Inform 32:229. https://doi.org/10.4060/ca9229en

Gephart JA, Golden CD (2022) Environmental and nutritional double bottom lines in aquaculture. One Earth, 5(4):324–328. https://doi.org/10.1016/j.oneear.2022.03.018

Haque AB, Cavanagh RD, Seddon N (2021) Evaluating artisanal fishing of globally threatened sharks and rays in the Bay of Bengal, Bangladesh. PloS One 16:e0256146. https://doi.org/10.1371/journal.pone.0256146

Herrero M, Thornton PK, Mason-D'Croz D, Palmer J, Benton TG, Bodirsky BL, Bogard JR, Hall A, Lee B, Nyborg K, Pradhan P, Bonnett GD, Bryan BA, Campbell BM, Christensen S, Clark M, Cook MT, de Boer IJM, Downs C, Dizyee K, Folberth C, Godde CM, Gerber JS, Grundy M, Havlik P, Jarvis A, King R, Loboguerrero AM, Lopes MA, McIntyre CL, Naylor R, Navarro J, Obersteiner M, Parodi A, Peoples MB, Pikaar I, Popp A, Rockström J, Robertson MJ, Smith P, Stehfest E, Swain SM, Valin H, van Wijk M, van Zanten HHE, Vermeulen S, Vervoort J, West PC (2020) Innovation can accelerate the transition towards a sustainable food system. Nat Food 1:266–272. https://doi.org/10.1038/s43016-020-0074-1

Herrero M, Thornton PK, Mason-D'Croz D, Palmer J, Bodirsky BL, Pradhan P, Barrett CB, Benton TG, Hall A, Pikaar I, Bogard JR, Bonnett GD, Bryan BA, Campbell BM, Christensen S, Clark M, Fanzo J, Godde CM, Jarvis A, Loboguerrero AM, Mathys A, McIntyre CL, Naylor RL, Nelson R, Obersteiner M, Parodi A, Popp A, Ricketts K, Smith P, Valin H, Vermeulen SJ, Vervoort J, van Wijk M, van Zanten HH, West PC, Wood SA, Rockström J (2021) Articulating the effect of food systems innovation on the sustainable development goals. Lancet Planet Health 5:e50–e62. https://doi.org/10.1016/S25425196(20)30277-1

IDIA (2021) Strengthening innovation ecosystems

Ikram M, Rashid MT, Shaukat M (2023) Climate change: a cataclysmic threat to humanity. J Xi'an Shiyou Univ Nat Sci Ed 9:635–645

Katikiro RE, Macusi ED (2012) Impacts of climate change on west African fisheries and its implications on food production. J Environ Sci Manag 15

Khan A, Ahmed SM, Sarr C, Kabore Y, Kahasha G, Bangwe L et al (2021) Nourishing nations during pandemics: why prioritise fish diets and aquatic foods in Africa. Maritime Stud 20:487–500. https://doi.org/10.1007/s40152-021-00236-z

Kok KPW, den Boer ACL, Cesuroglu T, van der Meij MG, de Wildt-Liesveld R, Regeer BJ, Broerse JEW (2019) Transforming research and innovation for sustainable food systems—a coupled-systems perspective. Sustainability 11:7176. https://doi.org/10.3390/su11247176

Kolding J, Zwieten PA, van Marttin F, Funge-Smith S, Poulain F (2019) Freshwater small pelagic fish and fisheries in the main African great lakes and reservoirs in relation to food security and nutrition. FAO, Rome

Kruk S, Peters H (2018) Towards a socially responsible aquaculture stewardship council. In: Oxford Briefing Notes, pp 1–18

Laso J, Ruiz-Salmón I, Margallo M, Villanueva-Rey P, Poceiro L, Quinteiro P et al (2022) Achieving sustainability of the seafood sector in the European Atlantic area by addressing eco-social challenges: the NEPTUNUS project. Sustain For 14:3054. https://doi.org/10.3390/su14053054

Lubembe SI, Turyasingura B, Chavula P (2022) Reflection on impacts of climate change on fisheries and aquaculture: sub-Sahara Africa. Weather 6:62–72

Lucrezi S (2022) Public perceptions of marine environmental issues: a case study of coastal recreational users in Italy. J Coast Conserv 26:52–14. https://doi.org/10.1007/s11852-022-00900-4

Mateo-Sagasta J, Zadeh SM, Turral H, Burke J (2017) Water pollution from agriculture: a global review. Executive summary. FAO, Rome

Mitra S, Chakraborty AJ, Tareq AM, Emran TB, Nainu F, Khusro A et al (2022) Impact of heavy metals on the environment and human health: novel therapeutic insights to counter the toxicity. J King Saud Univ Sci 34:101865. https://doi.org/10.1016/j.jksus.2022.101865

Mpumi N, Machunda RS, Mtei KM, Ndakidemi PA (2020) Selected insect pests of economic importance to Brassica oleracea, their control strategies and the potential threat to environmental pollution in Africa. Sustain For 12:3824. https://doi.org/10.3390/su12093824

Maulu S, Hasimuna OJ, Haambiya LH, Monde C, Musuka CG & Makorwa TH (2021) Climate Change Effects on Aquaculture Production. Front Sustain Food Syst 5:609097

NEPAD Planning and Coordinating Agency (2016) The Pan-African fisheries and aquaculture policy framework and reform strategy: key policy messages transition to sustainability, profitability and equitable distribution, vol 7. NEPAD Planning and Coordinating Agency

Odeku KO, Paulos BM (2017) Prohibition of pollution of marine environments: challenges and prospects. Environ Econ 8:127–136. https://doi.org/10.21511/ee.08(3-1).2017.05

Okocha RC, Olatoye IO, Adedeji OB (2018) Food safety impacts of antimicrobial use and their residues in aquaculture. Public Health Rev 39:21–22. https://doi.org/10.1186/s40985-018-0099-2

Okon EM, Okocha RC, Adesina BT, Ehigie JO, Alabi OO, Bolanle AM et al (2022) Antimicrobial resistance in fish and poultry: public health implications for animal source food production in Nigeria, Egypt, and South Africa. Front Antibiot 1:1043302. https://doi.org/10.3389/frabi.2022.1043302

Qadri R, Faiq MA (2020) Freshwater pollution: effects on aquatic life and human health. Fresh Water Pollute Dyn Remediat:15–26. https://doi.org/10.1007/978-981-13-8277-2_2

Rocha CP, Cabral HN, Marques JC, Gonçalves AM (2022) A global overview of aquaculture food production with a focus on the activity's development in transitional systems—the case study of a south European country (Portugal). J Marine Sci Eng 10:417. https://doi.org/10.3390/jmse10030417

Sanjay Panda RKM, Diwan AD (2023) Omega fatty acids of fishes and human brain development: a study on naturopathy and homoeopathy. J Surv Fish Sci 10:4204–4213

Srinivasan UT, Watson R, Sumaila UR (2012) Global fisheries losses at the exclusive economic zone level, 1950 to present. Mar Policy 36:544–549. https://doi.org/10.1016/j.marpol.2011.10.001

Tacon AG (2023) Contribution of fish and seafood to global food and feed supply: an analysis of the FAO food balance sheet for 2019. Rev Fish Sci Aquac 31:274–283. https://doi.org/10.1080/23308249.2022.2124364

Tacon AG, Lemos D, Metian M (2020) Fish for health: improved nutritional quality of cultured fish for human consumption. Rev Fish Sci Aquac 28:449–458. https://doi.org/10.1080/23308249.2020.1762163

Wilson DC, McCay BJ (2019) Fishery management, human dimension. In: Encyclopedia of ocean sciences, volume 2. Elsevier Ltd, pp 1018–1023

Zhang K, Zhang F, Zeng Z, Ma S (2023) Impact of extreme climate events on marine ecosystems: adaptation and challenges. Front Mar Sci 10:1187357. https://doi.org/10.3389/fmars.2023.1187357

Zihao J, Angelucci DE, Igreja MA, Arnold LJ, Badal E, Callapez P et al (2020) Last interglacial Iberian Neandertals as fisher-hunter-gatherers. Science 367:eaaz7943

Chapter 8
Renewable Energy as an Enabler of Coastal and Marine Food Security: The Case of Namibia

Marina N. Iyambo and Al-Mas Sendegeya

Abstract This chapter explores the role of renewable energy (RE) in enhancing coastal and marine food security in Namibia. It is particularly relevant to policymakers, government agencies, research institutes, small-scale fisheries, and private sector innovators seeking insights into how RE can be effectively integrated into marine food systems. By highlighting the interconnectedness of energy use and marine-based production, the chapter emphasizes the need for sustainable energy solutions to address challenges such as infrastructure limitations, post-harvest losses, and climate impacts. Namibia's marine food systems are energy-intensive and reliant on fossil fuels, presenting opportunities for RE integration. Various RE technologies and their applications in marine food systems are examined. It also discusses global best practices and their potential to improve efficiency, reduce emissions, and support sustainable livelihoods. Recommendations include expanding RE infrastructure, integrating hybrid systems, supporting small-scale fisheries with solar-powered tools, and fostering innovation through pilot projects and international collaboration. Ultimately, the chapter concludes that transitioning to RE is not merely a technological upgrade but a strategic imperative for Namibia to achieve its blue economy goals, mitigate climate risks, and ensure long-term food security.

Keywords Energy Use · Marine Supply Chains · Technology Intergration

---

M. N. Iyambo (✉)
Sam Nujoma Marine and Coastal Resources Research Centre, Sam Nuujoma Campus, University of Namibia, Henties Bay, Namibia
e-mail: marinaiyambo@gmail.com

A.-M. Sendegeya
Department of Electrical and Mechanical Engineering, Namibia University of Science and Technology, Windhoek, Namibia

H. A. Mupambwa et al. (eds.), *Harnessing Coastal and Marine Resources for Improved Food Security in Africa*, Sustainability Sciences in Asia and Africa,
https://doi.org/10.1007/978-981-95-9725-3_8

8.1 Introduction

8.1.1 *Background*

The global imperative to tackle climate change has markedly expedited the shift towards RE (Karlilar Pata and Balcilar 2024). This transition is particularly vital in sectors that are both energy-intensive and environmentally sensitive, such as food systems. The Food and Agriculture Organization (FAO) states that energy consumption accounts for around one-third of greenhouse gas (GHG) emissions from food systems (Nagpal et al. 2021). This statistic, although not exclusive to maritime and coastal food systems, underscores the escalating carbon footprint associated with food processing, packaging, and distribution stages, which have experienced the most significant emissions increases since 1990.

The energy industry significantly contributes to climate change, with fossil fuels responsible for over 75% of worldwide energy-related emissions (Osman et al. 2023). Ironically, although energy production has contributed to environmental deterioration, it is also crucial for alleviating climate impacts, especially in susceptible sectors such as fisheries and aquaculture, which are increasingly influenced by ocean warming, acidification, and ecosystem stress.

Marine and coastal food systems are intricately connected to energy consumption. Energy is a crucial input throughout the entire value chain, encompassing the powering of fishing vessels, aquaculture activities, refrigeration, processing, and transportation. In modern nations, these systems are generally automated and energy-demanding, heavily dependent on power and fuel. Conversely, developing nations frequently utilise smaller-scale systems reliant on diesel generators or biomass alternatives (Gorjian et al. 2022; Nagpal et al. 2021).

Namibia, with its long coastline and abundant marine species, stands to gain significantly from a transition to renewable energy, which is crucial for both climate mitigation and the enhancement of food security.

8.1.2 *Overview of RE Resources in Namibia*

RE refers to energy derived from natural processes that are continuously replenished within the Earth's ecosystem. These sources include sunlight, wind, rain, tides, biomass, and geothermal heat, each of which is either directly or indirectly powered by the sun or the Earth's internal heat (Mohtasham 2015). They are called renewable because they regenerate naturally and are not depleted when used sustainably, unlike fossil fuels, which are finite and environmentally damaging. As (Turkenburg et al. 2000) explained, RE sources have been integral to human civilization for centuries.

Biomass was historically used for heating and cooking, while wind and hydropower were harnessed for movement and, later, electricity generation.

Fundamentally, most RE flows are driven by solar insolation, the sun's radiation that powers atmospheric and hydrological cycles, or by geothermal energy, which originates from the Earth's molten core (Mohtasham 2015; Turkenburg et al. 2000). These energy flows are part of the Earth's natural balance, making RE not only sustainable but also essential for long-term environmental and economic resilience.

8.1.2.1 Solar Energy

Solar energy technologies use sunlight to produce electricity or heat, offering a clean and renewable alternative to fossil fuels. Two main types of solar technologies are:

- *Solar Panels (Photovoltaic or PV Technology):* These convert sunlight directly into electricity. Modern solar panels are efficient and widely used for powering homes, businesses, and even satellites.
- *Concentrated Solar Power (CSP):* This technology uses mirrors to focus sunlight onto a specific point, generating heat that is used to produce electricity. CSP is ideal for large-scale energy production.

Namibia receives approximately 10 hours of sunlight per day throughout the year, and an average annual solar irradiation of 2700 kWh/m2. Solar photovoltaic (PV) systems in Namibia can generate double the annual electricity of comparable systems in Germany (Frankl and Madeira 2024).

8.1.2.2 Wind

Wind energy harnesses the kinetic energy of moving air masses to generate electricity. This is achieved using wind turbines, which convert wind's motion into mechanical power and then into electrical energy. There are two main types of wind turbines: Horizontal-Axis Wind Turbines (HAWTs) and Vertical-Axis Wind Turbines (VAWTs).

HAWTs are the most widely used and feature a rotor shaft and generator mounted at the top of a tower, with blades facing into the wind. VAWTs, on the other hand, have a vertically oriented rotor shaft and can capture wind from any direction, making them suitable for urban or small-scale applications (Dani et al. 2024).

Key components of a wind turbine include the rotor blades, nacelle (housing the gearbox and generator), tower, yaw system (which orients the turbine to face the wind), and control systems that monitor wind speed and turbine performance.

Namibia also happens to be one of the world's windiest places, with potential concentrated in the southern coast around Lüderitz, where average wind speeds often exceed 7 metres per second (Frankl and Madeira 2024).

8.1.2.3 Biomass

Biomass energy involves a diverse range of technologies and mechanisms that convert organic matter into usable energy in the form of heat, electricity, or biofuels. Biogas production is primarily achieved through anaerobic digestion of feedstocks like animal manure, energy crops, and organic waste, yielding a gas mixture composed of 55% methane (Sharma et al. 2014). The energy output from biomass varies depending on the feedstock, ranging from sugarcane and corn to wood and cassava, and the conversion method used. Common biomass feedstocks include agricultural residues, wood, vegetable oils (e.g. rapeseed, soybean, palm), animal fats, and municipal organic waste (Garba 2020).

These technologies enable the transformation of biomass into solid, liquid, and gaseous energy carriers, supporting RE production while reducing dependence on fossil fuels and contributing to circular economy practices.

Namibia primarily uses biomass through traditional practices, such as burning firewood and charcoal for cooking and heating. Approximately 45 million hectares of Namibia's land are prone to the encroachment of invasive bush species. The sustainable harvesting of biomass from the invasive bush can be used to produce charcoal (Frankl and Madeira 2024). On the other hand, increased animal farming, even among smallholder farmers, generates sizeable quantities of animal manure from goats, cows, chickens, and sheep rearing, which can be used in biogas production.

8.1.2.4 Marine Energy

According to Hecke et al. (2020) marine energy encompasses a range of renewable technologies that harness the kinetic, thermal, and chemical energy of the ocean to generate electricity. These include tidal range, tidal stream, wave energy, ocean thermal energy conversion (OTEC), and salinity gradient (osmotic power) systems. Each technology operates on distinct physical principles and is suited to specific marine environments.

These technologies are still in various stages of development, with tidal and wave energy systems being the most mature. Namibia has a coastline stretching approximately 1572 kilometres along the South Atlantic Ocean. It is strongly influenced by the Benguela Current, and the tidal ranges along the Namibian coast are generally modest. Nonetheless, feasibility studies are necessary to assess the potential for marine energy systems to be adapted to Namibia's coastal environment.

8.1.2.5 Geothermal

Geothermal energy is a renewable resource that harnesses the Earth's internal heat for electricity generation, heating, and cooling. The heat is generated in the Earth's core, which is at a temperature of about 4000 °C. Geothermal energy is the only form of RE that is independent of the sun, having its ultimate source from within the earth (Anatolitis et al. 2024).

Namibia contains numerous geothermal springs, such as Ganigobes, Ai-Ais, and Rehoboth, characterized by high discharge temperatures indicative of subsurface thermal anomalies (Sracek et al. 2015). Although these surface indicators suggest geothermal activity, no comprehensive subsurface investigation has been undertaken to evaluate the viability of energy generation. Consequently, these springs function as first indicators rather than verified energy sources, necessitating more geophysical and geological examination.

8.1.3 Importance of Marine Resources and RE

The Benguela Current Upwelling System makes Namibia's marine ecosystems some of the most productive in the world, with a lot of demersal and pelagic species living there (MFMR 2017). These habitats are the most important parts of the country's fisheries, aquaculture, and mariculture sectors, which are key to its blue economy goals. But these maritime resources still are not being used to their full potential because of ongoing problems with infrastructure, especially in electricity and water systems (Remmert 2018).

RE offers a transformative opportunity to unlock this potential. The coastal environment of Namibia is conducive to wind and ocean energy production. Ombepo Wind Farm near Lüderitz is a prime example of how clean energy can be harnessed in coastal zones (Remmert 2018). When integrated with desalination technologies and cold storage systems, RE can power aquaculture, fish processing, and distribution chains, reducing post-harvest losses and improving food system efficiency.

Offshore RE technologies, such as wave and tidal power, can also be co-located with marine infrastructure to support aquaculture and other blue economy sectors (Roesch et al. 2020). These technologies not only reduce emissions but also enhance marine food systems to climate shocks. According to IRENA, ocean energy is well positioned to decarbonize aquaculture, desalination, and ocean observation, while enabling emerging sectors like green hydrogen and underwater vehicle charging (Roesch et al. 2020).

Moreover, integrating RE into marine food systems supports broader sustainability goals. It reduces reliance on fossil fuels, lowers operational costs, and enables environmentally sound practices such as aquaponics and hydroponics (Gorjian et al. 2022). For Namibia, this transition is not just a technological upgrade, it is a strategic imperative for building a resilient, inclusive, and climate-smart marine economy.

8.2 Systems and Logistics Involved in Marine-Based Food Chains in Namibia

To understand the role renewable energy can play in Namibia's marine-based food systems, it is essential to look at the systems and logistics involved in the marine food value chain. Namibia's coastal and marine food economy involves capturing fish, aquaculture (both freshwater and mariculture), processing, transportation, and export logistics (MET 2012). Each stage presents unique energy demands and operational challenges, which RE can help address.

8.2.1 Marine Capture Fisheries and Harvesting Systems

Namibia's marine capture fisheries are among the most productive globally (MET 2012). The sector targets species such as hake, horse mackerel, sardine, monk, and rock lobster. Fishing operations are conducted using wet trawlers, longliners, and factory vessels (MFMR 2020), which are energy-intensive and rely heavily on fossil fuels for propulsion, refrigeration, and onboard processing. The Ministry of Fisheries and Marine Resources (MFMR) regulates harvesting through Total Allowable Catches (TACs) and vessel monitoring systems, with over 2000 vessel trips monitored in a single reporting period (MFMR 2020).

In addition to industrial fishing, Namibia's small-scale fisheries (SSF) play a vital role in food security and livelihoods. Coastal SSF are concentrated in towns like Walvis Bay, Henties Bay, Swakopmund, and Lüderitz, target species such as kabeljou, galjoen, steenbras, snoek, mullets, and rock lobster using traditional gear like ski-boats, nets, and hook-and-line (FAO 2022). The manual labour involved in this kind of fishing principle can be reduced with RE initiatives. Although ambitious, specific recommendations will be discussed later on in the chapter.

8.2.2 Aquaculture and Mariculture Systems

Namibia's aquaculture sector is branched into freshwater and marine (mariculture) systems. Freshwater aquaculture focuses on species like tilapia and African catfish, while mariculture is dominated by oyster and abalone farming, particularly in Walvis Bay, Swakopmund, and Lüderitz (MFMR 2004, 2020).

Mariculture operations use a variety of systems, including:

- Suspended baskets and longlines for oyster farming
- Onshore raceways and ponds for abalone

The inland aquaculture sector is also expanding, as seen in the Neckartal Dam Inland Fisheries and Aquaculture Research Project launched in 2020. This initiative aimed to assess the development potential of inland fisheries and aquaculture in the //Karas Region showed breeding success (FAO 2022).

These systems require consistent energy inputs for water circulation, aeration, temperature regulation, and monitoring areas where RE technologies like solar-powered pumps and sensors can be highly beneficial.

8.2.3 Processing and Value Addition

Fish processing occurs both offshore (on factory vessels) and onshore in processing hubs like Walvis Bay and Lüderitz. The implementation of the 70:30 policy in the hake fishery has shifted more processing onshore, increasing energy demands at land-based facilities (MFMR 2020).

Value-added products such as canned horse mackerel, dried fish, and fish sausages are produced to enhance economic returns (MFMR 2020). Notably, fish drying is a traditional processing method still widely practiced, especially by women in inland communities, contributing to household income and food preservation (FAO 2022).

For coastal and industrial processes, cold storage, packaging, and quality control systems are required all of which are energy-intensive and present opportunities for RE integration, particularly solar and wind energy.

8.2.4 Transportation, Cold Chain, and Export Logistics

Namibia's marine food products are transported via road and sea, with cold chain logistics playing a critical role in maintaining product quality. Ports in Walvis Bay and Lüderitz serve as key nodes for both domestic distribution and international exports (NPC 2017).

The development of a fish processing hub and cruise port in Walvis Bay, as outlined in the NDP5, further underscores the importance of integrated logistics infrastructure (NPC 2017). These facilities require reliable energy for refrigeration, lighting, and operations areas, where RE can reduce operational costs and carbon footprints.

8.2.5 Monitoring, Compliance, and Research Systems

The MFMR also oversees port inspections, transshipment monitoring, and research activities related to stock assessments and ecosystem health (MFMR 2020). These functions rely on data collection, communication systems, and mobile operations, which can benefit from decentralized RE solutions such as solar-powered monitoring stations.

Additionally, Namibia's marine and coastal tourism sector, including recreational fishing, adds another layer to the logistics network. Thousands of anglers visit the coast annually for rock-and-surf or ski-boat angling, further emphasizing the need for sustainable infrastructure and energy systems (Chiripanhura 2016).

It is indisputable that Namibia's food security depends on its inland and marine fisheries. The nation's marine-based food systems, from the ports to the inland aquaculture projects, involve a lot of human labour and some energy-intensive processes. At this point, RE stops being only an add-on solution and starts to play a vital role in ensuring food security.

8.3 Current Energy Landscape in Namibia's Coastal Region

Namibia's coastal region, stretching from the southern town of Lüderitz to the central hub of Walvis Bay, is emerging as a focal point for RE development. With abundant solar irradiation and strong coastal winds, the region holds immense potential to support the country's energy transition and food security goals, particularly in marine-based sectors.

The government's vision for RE deployment across the northern, central, and southern coastal corridors includes large-scale solar PV and wind projects, desalination plants, and sustainable biomass sources. While some of these developments are linked to green hydrogen production, the underlying renewable infrastructure such as solar farms, wind turbines, and desalinated water systems can directly benefit marine aquaculture, fisheries, and coastal food processing by providing clean, reliable, and decentralized energy.

The Namibian energy infrastructure presents a mix of existing and planned transmission lines, substations, and generation sites. The presence of planned substations suggests an ongoing effort to increase energy access. Some existing renewable energy power plants, such as Omburu, which is located closer to the coastline, are also highlighted.

Overall, Namibia has made significant strides in deploying RE infrastructure, particularly solar and wind power. As of 2025, the country has installed approximately 610 MW of grid-connected generation capacity, with 150 MW operated by Independent Power Producers (IPPs), mostly solar photovoltaic (PV) systems (Neshila 2024).

The country's electricity grid is undergoing progressive liberalization and expansion. The Modified Single Buyer (MSB) model, introduced in 2019, has allowed IPPs to sell electricity directly to contestable buyers, enhancing market flexibility (Morks 2024).

International case studies have provided valuable insights into the energy dynamics of the mariculture or aquaculture systems. For instance, in the U.S. catfish farming sector, electricity accounts for up to 75–83% of direct energy use, primarily for aeration and water quality control, with diesel and petrol making up the remainder (Scroggins et al. 2022).

Similarly, a Mediterranean aquaculture facility reported daily energy consumption exceeding 13,000 kWh, with the hatchery and weaning units being the most energy-intensive due to their need for temperature regulation, lighting, and mechanical systems (Ioakeimidis and Polatidis 2013). In Recirculating Aquaculture Systems (RAS), energy audits revealed that heat pumps alone consumed over 54% of total energy, followed by skimmers and UV systems (Badiola 2014).

8.4 Global Practices

8.4.1 Solar Energy Case Studies

In areas, where arid conditions and limited freshwater resources challenge traditional food production, solar power offers transformative solutions and has the capability to enhance coastal and marine-based food security. Drawing from global best practices, solar energy has been successfully integrated into agri-food systems to improve productivity, reduce post-harvest losses, and support sustainable livelihoods.

Table 8.1 summarizes several initiatives that have emerged from integrating solar technologies into coastal and agricultural systems

Solar-powered irrigation has reduced reliance on diesel pumps and manual labour, enabling more efficient water extraction for farming, particularly in arid and semi-arid regions.

In tropical and coastal areas, solar-powered cold storage facilities have helped preserve perishable goods, reduce post-harvest losses, and improve access to markets for small-scale producers. Aquavoltaics, which combines solar energy generation with aquaculture, involves installing photovoltaic systems over fish ponds to support sustainable seafood production while generating clean energy.

Table 8.1 Global solar energy practices

Sector	Location	Application
Solar Irrigation	India, Bangladesh, Rwanda, Senegal	Solar irrigation pumps initiatives (Nagpal et al. 2021).
Cold Storages	Kenya, India, China, the Philippines	Solar cold storage incorporating PCM and pay as you go initiates to reduce spoilages (Amjad et al. 2023; Igor and Dietram 2021).
Desalination plants	Somalia, Libya, Namibia	Solar-powered desalination initiatives such as community--scale solar stills, solar PV-powered reverse-osmosis systems, and solar stills for runoff water (Hilarydoss et al. 2024; Muftah et al. 2025).
Acquavoltaics	China, Taiwan	Photovoltaic (PV) systems installed over aquaculture ponds (Song et al. 2024)

Additionally, solar-powered desalination technologies have been implemented to provide potable water in coastal communities, ranging from small-scale solar stills to larger solar desalination plants, directly addressing water scarcity and improving livelihoods.

8.4.2 Geothermal

Geothermal energy is increasingly being harnessed to enhance marine food-based security, particularly through innovative applications in aquaculture and related industries. By providing stable, temperature-controlled environments, geothermal systems support the cultivation of aquatic organisms such as fish, crustaceans, mollusks, and algae, thereby improving growth rates and productivity. This is achieved by using geothermal fluids directly or in combination with freshwater, often with heat exchangers to ensure safety and sustainability.

Notable examples are summarized in Table 8.2.

8.4.3 Marine RE

Offshore wind and marine renewable energy (MRE) technologies are well established and have been generating electricity in offshore environments for years. Numerous operational projects around the world are producing clean energy and undergoing environmental assessments, as highlighted in Table 8.3.

Projects like AquaWind in the Canary Islands are testing floating wind turbines combined with aquaculture systems, while earlier European initiatives such as Roter Sand, AquaLast, MytiFit, and MUSES have contributed foundational knowledge on spatial planning and risk governance (Schupp et al. 2021; Van den Burg et al. 2020). These efforts represent important steps towards understanding how offshore energy infrastructure might one day support marine-based food production.

Table 8.2 Global geothermal energy practices

Location	Initiative/ Technology	Application
Iceland	SagaNatura	Utilization of geothermal heat and industrial CO_2 for microalgae cultivation
Iceland & Netherlands	GEOFOOD Project	Integrated fish and vegetable production systems utilizing geothermal and waste heat
Iceland	Geothermal fish drying	Use of geothermal heat for fish drying and preservation
Iceland	Seawater fish farming	Application of geothermal heating in aquaculture to maintain optimal rearing temperatures

Source: Tryggvadottir et al. (2022)

Table 8.3 Global marine energy practices

Location	Technology	Application
Scotland	Tidal turbines (4 × 1.5 MW)	Grid-connected; environmental monitoring (Copping and Hemery 2024)
Spain	Wave energy	Grid-connected; environmental monitoring (Copping and Hemery 2024)
Japan	Tidal systems	Focus on fisheries' interactions (Copping and Hemery 2024)
Australia	Wave energy	Environmental studies (Copping and Hemery 2024)
Canary Islands	Floating wind + aquaculture	Co-location with aquaculture systems (Schupp et al. 2021; Van den Burg et al. 2020)
Europe	Offshore wind farms	Co-location with *Saccharina latissima*, *Mytilus edulis*, and finfish (Pardo et al. 2025)

8.5 Recommendations

To enhance marine food security and address challenges within Namibia's coastal food systems, several strategic interventions are proposed, drawing from global best practices and the country's renewable energy (RE) potential.

Namibia should prioritize the development of robust RE infrastructure by expanding solar and wind projects, leveraging its high solar irradiation and strong coastal winds. These resources can sustainably power aquaculture farms, fish processing plants, and cold storage facilities. Although solar irradiance along the coast may vary compared to inland regions, this can be effectively offset through hybrid systems combining solar and wind power to ensure an uninterrupted energy supply for operations such as refrigeration and water circulation.

The transportation sector within the marine food value chain relies heavily on refrigeration, often using specialized ice during transit. Producing ice using RE can significantly reduce operational costs and carbon emissions.

Supporting small-scale fisheries remains essential, as fishing is the primary livelihood for many coastal communities. Innovative technologies such as solar-powered e-bikes equipped with portable fridges or ice makers can help fishers preserve their catch while minimizing physical strain and post-harvest losses. Additionally, integrating low-power automation tools, such as fishing lines with embedded micro-batteries, can enhance efficiency and productivity.

Currently, Namibia lacks pilot projects and feasibility studies that demonstrate the integration of RE into marine-based food systems. Initiatives such as coupling offshore wind farms with aquaculture operations and conducting targeted energy audits could identify high-impact opportunities for renewable deployment.

Given the high initial capital requirements for RE projects, strengthening policy frameworks is crucial. This includes providing incentives, streamlining approval processes, and encouraging private sector participation. International collaboration is equally important, Namibia can adopt proven RE models from countries like India, China, and Kenya while leveraging global expertise and financing mechanisms.

8.6 Conclusions

Namibia's coastline and marine biodiversity offer a unique opportunity to strengthen national food security through the strategic integration of RE. The country's marine food systems, including fisheries, aquaculture, processing, transportation, and export logistics, are highly energy-intensive and constrained by infrastructure limitations.

RE presents a transformative pathway to overcome these challenges, reduce dependence on fossil fuels, and build a resilient, low-carbon marine economy. By harnessing its abundant solar and wind resources, along with emerging offshore energy technologies, Namibia can sustainably power critical operations such as cold storage, desalination, aquaculture, and logistics.

Global best practices such as solar-powered cold storage, aquavoltaics, and renewable-powered desalination demonstrate how RE can be effectively embedded into marine food systems. Moreover, fostering innovation through pilot projects, conducting energy audits, and supporting small-scale fishers with solar-powered tools and mobility solutions like e-bikes with portable fridges can significantly enhance productivity and reduce labour burdens.

Ultimately, the shift to RE is not just a technological upgrade but a strategic necessity for achieving Namibia's blue economy ambitions, mitigating climate risks, and ensuring long-term food security. By embracing RE, Namibia can unlock the full potential of its marine resources.

References

Amjad W, Munir A, Akram F, Parmar A, Precoppe M, Asghar F, Mahmood F (2023) Decentralized solar-powered cooling systems for fresh fruit and vegetables to reduce post-harvest losses in developing regions: a review. Clean Energy. Oxford University Press 7(3):635–653. https://doi.org/10.1093/ce/zkad015

Anatolitis V, Bojek P, Briens F, Buisson E, Criswell T, Dallard J (2024) The future of geothermal energy. www.iea.org

Badiola M (2014) Energy audits in Recirculating Aquaculture Systems (RAS): as a way forward to guarantee sustainability. https://doi.org/10.13140/2.1.1218.5604

Chiripanhura B (2016) AGRODEP Working Paper 0021: an analysis of the Fishing Industry in Namibia: The structure, performance, challenges, and prospects for growth and diversification. https://doi.org/10.13140/RG.2.2.36280.57609

Copping AE, Hemery LG (2024) Progress in understanding environmental effects of marine renewable energy. https://doi.org/10.2172/2438587

Dani A, Benlamlih M, Mekrini Z, El Mrabet M, Boulaala M (2024) Wind energy conversion technologies and control strategies: a review. Int J Renew Energy Res 14(1):140–154

FAO (2022) Namibia's National Plan of action for small-scale fisheries. https://www.google.com/url?sa=t&source=web&rct=j&opi=89978449&url=https://faolex.fao.org/docs/pdf/nam213210.pdf&ved=2ahUKEwidrs796aCQAxXkdUEAHbNZBgMQFnoECBkQAQ&usg=AOvVaw3eKgauSd1rO3EjgsTS2lni

Frankl P, Madeira R (2024) Renewable energy opportunities for Namibia. http://hdl.handle.net/11159/703245

Garba A (2020) Biomass conversion technologies for bioenergy generation: an introduction. www.intechopen.com

Gorjian S, Fakhraei O, Gorjian A, Sharafkhani A, Aziznejad A (2022) Sustainable food and agriculture: employment of renewable energy technologies. Curr Robot Rep 3(3):153–163. https://doi.org/10.1007/s43154-022-00080-x

Hecke J, Salgado A, Abunofal M, Van Velzen N (2020) Innovation outlook: ocean energy technologies. International Renewable Energy Agency

Hilarydoss S, Nishant K, Nahak SK (2024) Economics, environmental foot print and sustainability of community scale solar desalination plant in water scarce Somalia coastal regions. Clean Techn Environ Policy 26(9):2819–2847. https://doi.org/10.1007/s10098-024-02741-1

Igor CC, Dietram O (2021) Construction of a fish cold store in Kenya Guidelines for the installation, design and calculation of a highly efficient solar-powered cold store using natural refrigerants. www.bmu.de

Ioakeimidis C, Polatidis H (2013) Use of renewable energy in aquaculture: An energy audit case-study analysis. Glob NEST J 15(3). http://www.retscreen.net

Karlilar Pata S, Balcilar M (2024) Decarbonizing energy: Evaluating fossil fuel displacement by renewables in OECD countries. Environ Sci Pollut Res 31(21):31304–31313. https://doi.org/10.1007/s11356-024-33324-8

MET (2012) The National Policy on Coastal Management for Namibia 3. https://www.google.com/url?sa=t&source=web&rct=j&opi=89978449&url=https://www.npc.gov.na/publicpolicy/the-national-policy-on-coastal-management-for-namibia/&ved=2ahUKEwiZ4Y_z56CQAxVpWkEAHbWGHI0QFnoECBgQAQ&usg=AOvVaw1MWXdzYa2iBxNu_A9e874d

MFMR (2004) Namibia's aquaculture strategic plan. https://www.google.com/url?sa=t&source=web&rct=j&opi=89978449&url=https://mfmr.gov.na/documents/411764/436209/Aquaculture%2BStrategic%2BPlan.pdf/df025d1d-bc11-9831-5b49-cb11fe249d89%3Fversion%3D1.0%26download%3Dtrue&ved=2ahUKEwjaovux6KCQAxVOaEEAHTPQCmkQFnoECBkQAQ&usg=AOvVaw3dXMMVkcVbyk0nfczIzqkT

MFMR (2017) Ministry of Fisheries and Marine Resources strategic plan. https://www.google.com/url?sa=t&source=web&rct=j&opi=89978449&url=https://mfmr.gov.na/documents/411764/436229/Strategic%2BPlan%2B2017%2B-%2B2021.pdf/76a029eb-8aa3-061b-545b-2d48459b8350&ved=2ahUKEwj8yLf86KCQAxUHV0EAHRB0DrQQFnoECBcQAQ&usg=AOvVaw2rQOl7TL7KKVaxC7OTa_uH

MFMR (2020) Ministry of Fisheries and Marine Resources annual report. https://mfmr.gov.na/downloads/-/document_library/DWqND60MzhTr/view_file/3809730?_com_liferay_document_library_web_portlet_DLPortlet_INSTANCE_DWqND60MzhTr_redirect=https%3A%2F%2Fmfmr.gov.na%2Fdownloads%2F-%2Fdocument_library%2FDWqND60MzhTr%2Fview%2F436205

Mohtasham J (2015) Review article-Renewable energies. Energy Procedia 74:1289–1297. https://doi.org/10.1016/j.egypro.2015.07.774

Morks RG (2024) A just and equitable transition: politics, partnerships and economics of Namibia's electricity and hydrogen sectors. https://www.iddri.org/sites/default/files/PDF/Publications/Catalogue%20Iddri/Rapport/Ukama_NAMIBIA.pdf

Muftah AK, Zili-Ghedira L, Abugderah MM, Hassen W, Becheikh N, Alshammari BM, Kolsi L (2025) Sustainable water production: solar Energy integration in multi-effect desalination plants. Water (Switzerland) 17(5). https://doi.org/10.3390/w17050647

Nagpal D, Dubois O, Maltsoglou I (2021) Renewable energy and agri-food systems: advancing energy and food security towards sustainable development goals. IRENA and FAO. https://doi.org/10.4060/cb7433en

Neshila L (2024) Namibia country window: Energy System Transformation Outlook (ESTO). www.get-transform.eu

NPC (2017) NDP5 implementation plan. https://www.google.com/url?sa=t&source=web&rct=j&opi=89978449&url=https://www.npc.gov.na/wp-content/uploads/2022/09/IP-Layout_30-June-2017-FINAL.pdf&ved=2ahUKEwiTgqCIraGQAxVNUEEAHWHUOhMQFnoECAwQAQ&usg=AOvVaw2FfUa6I6TvoGjBu6F86lIC

Osman AI, Chen L, Yang M, Msigwa G, Farghali M, Fawzy S, Rooney DW, Yap PS (2023) Cost, environmental impact, and resilience of renewable energy under a changing climate: a review. Environ Chem Lett 21(2):741–764. https://doi.org/10.1007/s10311-022-01532-8

Pardo JCF, Aune M, Harman C, Walday M, Skjellum SF (2025) A synthesis review of nature positive approaches and coexistence in the offshore wind industry. ICES J Marine Sci. Oxford University Press 82(4). https://doi.org/10.1093/icesjms/fsad191

Remmert D (2018) Towards a blue economy harnessing Namibia's coastal resources sustainably. http://www.hss.de/namibia/en/home.html

Roesch R, Boshell F, Ocenic E, Salgado A, Hecke J, Castellanos G, Abdel-Latif A (2020) Fostering a blue economy: offshore renewable energy. International Renewable Energy Agency

Schupp MF, Krause G, Onyango V, Buck BH (2021) Dissecting the offshore wind and mariculture multi-use discourse: a new approach using targeted SWOT analysis. https://doi.org/10.1007/s40152-021-00218-1/Published

Scroggins RE, Fry JP, Brown MT, Neff RA, Asche F, Anderson JL, Love DC (2022) Renewable energy in fisheries and aquaculture: Case studies from the United States. J Clean Prod 376. https://doi.org/10.1016/j.jclepro.2022.134153

Sharma S, Meena R, Sharma A, Kumar Goyal P, Prof A (2014) Biomass conversion technologies for renewable energy and fuels: A review note. IOSR J Mech Civil Eng 11(2):28–35. www.iosrjournals.org

Song F, Lu Z, Guo Z, Wang Y, Ma L (2024) The effects of a fishery complementary photovoltaic power plant on the near-surface meteorology and water quality of coastal aquaculture ponds. Water (Switzerland) 16(4). https://doi.org/10.3390/w16040526

Sracek O, Wanke H, Ndakunda NN, Mihaljevič M, Buzek F (2015) Geochemistry and fluoride levels of geothermal springs in Namibia. J Geochem Explor 148:96–104. https://doi.org/10.1016/j.gexplo.2014.08.012

Tryggvadottir L, Chatenay C, Jóhannesson P, Tulinius H (2022) Direct utilization of geothermal resources. https://esmap.org/sites/default/files/esmap-files/16103-WB_ESMAP%20Direct%20Use-WEB.pdf

Turkenburg WC, Beurskens J, Faaij A, Fraenkel P, Fridleifsson I (2000) Renewable energy technologies. In: World energy assessment: energy and the challenge of sustainability https://sdgs.un.org/publications/world-energy-assessment-energy-and-challenge-sustainability-18054

Van den Burg SWK, Röckmann C, Banach JL, van Hoof L (2020) Governing risks of multi-use: seaweed aquaculture at offshore wind farms. Front Mar Sci 7. https://doi.org/10.3389/fmars.2020.00060

Chapter 9
Bycatch in the Marine Capture Fishery, a Potential for Food Security: A Case of Namibia's Hake Bottom Trawl Fishery

Samuel K. Mafwila, Evans Simasiku, Johannes Angala Iitembu, Greg Mbaimbai, Anna-Marie Nambonga, and Kudakwashe Hove

Abstract Bottom trawling, a pervasive demersal fishing technique, is intrinsically non-selective and consequently yields substantial quantities of bycatch, much of which is routinely discarded at sea. The present chapter critically examines the taxonomic composition, spatial heterogeneity, and temporal trends of bycatch in Namibia's hake-directed bottom trawl fishery over the period 1997–2014, with a view to assessing its untapped potential for bolstering national food security programmes. An extensive observer dataset comprising 4256 sampling events was analysed to characterize 23 bycatch taxa, which collectively amounted to 9,031,480 tonnes of incidental catch. Five species, namely, *Trachurus capensis, Trachipterus trachypterus, Helicolenus dactylopterus, Lophius vomerinus, and Genypterus capensis*, dominated, accounting for 63.09% of the total bycatch biomass. Multivariate statistical techniques were employed to elucidate spatial patterns. Notably, central and southern fishing grounds exhibited consistently elevated densities of key bycatch species, indicating robust, geographically stable biomass 'hotspots'. These findings underscore the dual ecological and socio-economic value of bycatch taxa. By formalizing the retention and landing of a portion of currently discarded bycatch, Namibia could generate a reliable, low-cost source of marine

S. K. Mafwila (✉)
Office of the Director, Katima Mulilo Campus, University of Namibia, Katima Mulilo, Namibia
e-mail: smafwila@unam.na

E. Simasiku · G. Mbaimbai · K. Hove
Department of Wildlife Management and Tourism Studies, Katima Mulilo Campus, University of Namibia, Katima Mulilo, Namibia

J. A. Iitembu
Department of Fisheries and Ocean Sciences, Sam Nujoma Campus, University of Namibia, Henties Bay, Namibia

A.-M. Nambonga
National Marine Information and Research Centre, Ministry of Fisheries and Marine Resources, Swakopmund, Namibia

H. A. Mupambwa et al. (eds.), *Harnessing Coastal and Marine Resources for Improved Food Security in Africa*, Sustainability Sciences in Asia and Africa,
https://doi.org/10.1007/978-981-95-9725-3_9

protein for school-feeding initiatives, drought relief efforts, and orphanage support, without exacerbating fishing pressure. The chapter concludes with policy recommendations to amend retention regulations, protocols for post-harvest handling, and strategies for integrating low-value species into domestic relief and market supply chains, thereby transforming discarded biomass into a tangible contribution to national food security.

Keywords Bycatch · Bottom trawl · Benguela Current · Food security · Fisheries

9.1 Introduction

The marine environment of Namibia falls within the Benguela Current system. The Benguela Current System is one of the four major upwelling systems in the world (Kirkman et al. 2016). Upwelling systems are highly productive, supporting multiple fishery industries (Kirkman et al. 2016). Namibian commercial fisheries are dominated by three species: hake (*Merluccius capensis and Merluccius paradoxus*); horse mackerel (*Trachurus capensis*), and pilchard (*Sardinops ocellatus*) (Lange 2003), with hake being the most dominant in terms of value (Wilhelm et al. 2015). Hake species are caught by bottom trawling off the Namibian coast (NatMIRC 2007). This method is, however, not selective and many species apart from hake are caught during the fishing process. The non-specific nature of many fishing gears has an influence not only on their target species but on many other species referred to as bycatch (Hall et al. 2000).

However, the bycatch issue is significant because it involves the capture of non--target species, which can impact marine ecosystems and affect the sustainability of fisheries. However, managing bycatch effectively can contribute positively to food security. Namibia's fisheries, including the hake demersal fishery, are vital for the country's food security, providing important protein sources and income generation (Engelhard et al. 2024). Yet, these fisheries face challenges related to bycatch, a common issue that needs addressing through sustainable practices and improved management techniques to ensure long-term food security benefits (Crowder and Murawski 1998). Effective management of bycatch can potentially transform these challenges into opportunities. For instance, adopting real-time catch monitoring tools, like the automated system described in the demersal trawl fishery sector, can significantly reduce unwanted bycatch. These systems allow fishers to react quickly to catch compositions that include unwanted species, optimizing fishing activities to increase target catch efficiency while minimizing bycatch (Sokolova et al. 2021).

Addressing bycatch in the Namibian context offers a unique opportunity to reconcile fisheries productivity with ecosystem conservation, thereby enhancing food security. Implementing effective bycatch management measures such as spatial closures, seasonal restrictions, and gear modifications can mitigate adverse impacts and support the resilience of marine populations (Barange and Pillar 1998). Moreover, strategic utilization of bycatch resources can contribute to increasing the availability of protein sources for local communities, especially in a setting where fish is a central component of nutrition and livelihood (Bianchi et al. 1993; Iitembu et al. 2021).

Hence, this chapter explores the state of bycatch in Namibia's hake demersal fishery, emphasizing its implications for food security. By examining the composition, spatial-temporal patterns, and management strategies related to bycatch, we aim to highlight pathways towards sustainable fishery practices that safeguard marine biodiversity while maximizing the contribution of fisheries to national food security goals. As such, understanding and managing bycatch is not only a conservation imperative but also a critical component of building resilient and sustainable marine food systems in Namibia and beyond (Blaber and Wassenberg 1989; Wassenberg and Hill 1987).

9.2 Potential Bycatch Species in Namibian Hake Demersal Trawls for Food Security

Namibia's hake demersal fishery, primarily targeting deep-water hake (*Merluccius paradoxus*) and shallow-water hake (*Merluccius capensis*), is a cornerstone of the nation's economy and food security (Paterson and Kainge 2014). Operating within the highly productive Benguela Current ecosystem, this fishery predominantly employs bottom trawling (Mafwila 2011). While effective for target species, this method is inherently non-selective, leading to the capture of a diverse array of marine organisms alongside the hake (Mafwila 2011). Understanding the scale and composition of this incidental catch is paramount to assessing its potential for food security.

A study analysing observer data from the hake-directed bottom trawl fishery in Namibia from 1997 to 2014 provides significant insights into the quantities and composition of bycatch. During this period, approximately 23 bycatch species were recorded (Table 9.1), collectively weighing about 9,031,480 tonnes (Mbaimbai 2021), 18 of which are commercially important (Mafwila 2011). These include species such as silver kob (*Argyrosomus inodorus*), snoek (*Thyrsites atun*), horse mackerel (*Trachurus capensis*), large-eye dentex (*Dentex macrophthalmus*), and ribbonfish (*Trachipterus trachypterus*). Together, they represent about 78% of the total bycatch volume, indicating a largely untapped resource that could contribute significantly to national food and nutritional security (Alverson et al. 1994; Bianchi et al. 1999).

Despite the market and nutritional value of these species, most are underutilized due to low market demand, poor processing infrastructure, and limited awareness (Pinedo and Polacheck 2004a, b). In a country where food insecurity and malnutrition persist, especially in rural areas, this bycatch represents a critical opportunity to improve fish access for low-income communities (FAO 2020).

Table 9.1 Taxonomic list of bycatch species in the bottom trawl hake fishery (1997–2014) by region (North, Central, and South). W denotes the weight (tonnes), and %W denotes the percentage weight contribution of a species

Species	Station A (North)		Station B (Central)		Station C (South)		Total	
	W	%W	W	%W	W	%W	W	%W
Allocyttus verrucosus	536.83	0.04	680.79	0.03	2018.36	0.04	3236.05	0.04
Diplodus capensis	227.08	0.02	–	–	–	–	227.1	0
Epigonus spp	300.01	0.02	1169.04	0.06	19132.32	0.34	20601.45	0.23
Lepidopus caudatus	500.91	0.03	1158.85	0.06	275536.06	4.96	277195.92	3.07
Macrouridae spp	125.69	0.01	42.58	0	–	–	168.28	0
Isurus oxyrinchus	5254.94	0.36	31777.75	1.56	66132.95	1.19	103167.57	1.14
Argyrosomus inodorus	8820.81	0.61	1.44	0	–	–	8822.86	0.1
Brama brama	13774.57	0.95	132906.7	6.53	214866.74	3.87	361555.49	4
Taractichthys logipinnis	6392.39	0.44	52580.55	2.58	554515.56	9.99	613491.53	6.79
Cheilodactylus capensis	60998.17	4.23	27783.48	1.37	2987.83	0.05	91775.07	1.02
Dentex macrophthalmus	474168.17	**32.85**	22312.4	1.1	7359.92	0.13	503874.44	5.58
Trachurus capensis	214469.55	**14.86**	832024.5	**40.88**	254748.45	4.59	1301298.24	**14.41**
Helicolenus dactylopterus	205521.73	14.24	264626.17	**13**	709196.45	**12.77**	1179371.6	**13.06**
Zeus faber	73660.21	5.1	15520.68	0.76	169334.35	3.05	258521.1	2.86
Genypterus capensis	273.26	0.02	9287.85	0.46	833358.76	**15.01**	842920.34	9.33
Lophius vomerinus	205926.64	**14.27**	367924.82	**18.08**	448784.37	8.08	1022668.17	11.32
Thyrsites atun	132.26	0.01	12564.01	0.62	448548.18	8.08	461245.08	5.11
Austroglosus microlepsis	98069.53	6.8	87933.89	4.32	16424.09	0.3	202438.63	2.24
Thunnus spp	294.97	0.02	–	–	375.23	0.01	670.22	0.01
Trachipterus trachypterus	678.1	0.05	34440.76	1.69	1317225.68	**23.72**	1352346.28	**14.97**
Thunnus albacares	620.14	0.04	3.35	0	–	–	623.54	0.01
Todarodes sagittatus	72136.56	5	139988.22	6.88	198271.8	3.57	410408.46	4.54
Raja species	360	0.02	345.32	0.02	14347.19	0.26	15052.55	0.17
Total	**1443242.5**		**2035073.2**		**5553164.3**		**9,031,480**	

The species composition varies along Namibia's coastline. Ribbonfish (Figs. 9.1 and 9.2), which occur throughout Namibian waters, recorded the highest catch volumes (1,352,346.28 tonnes), possibly due to their wide distribution and overlap with hake habitats (Heemstra and Kannemeyer 1986). Conversely, blacktail (*Diplodus capensis*) and *Macrounidae* spp. showed minimal catch (Table 9.1), the former, reflecting its preference for shallower, rocky habitats beyond the trawl fishery's reach (Bianchi et al. 1999).

Bycatch has the potential to alter trophic structures and food web dynamics by removing non-target species that play key ecological roles (Driscoll et al. 2009). Shared feeding grounds between hake and many bycatch species suggest that trophic competition and cascading effects may occur (Mafwila 2011). This underscores the importance of adopting ecosystem-based fisheries management (EBFM) approaches that balance ecological integrity with food security goals.

With appropriate policy and infrastructure support, bycatch species can be transformed into affordable, nutritious food products. Initiatives could include

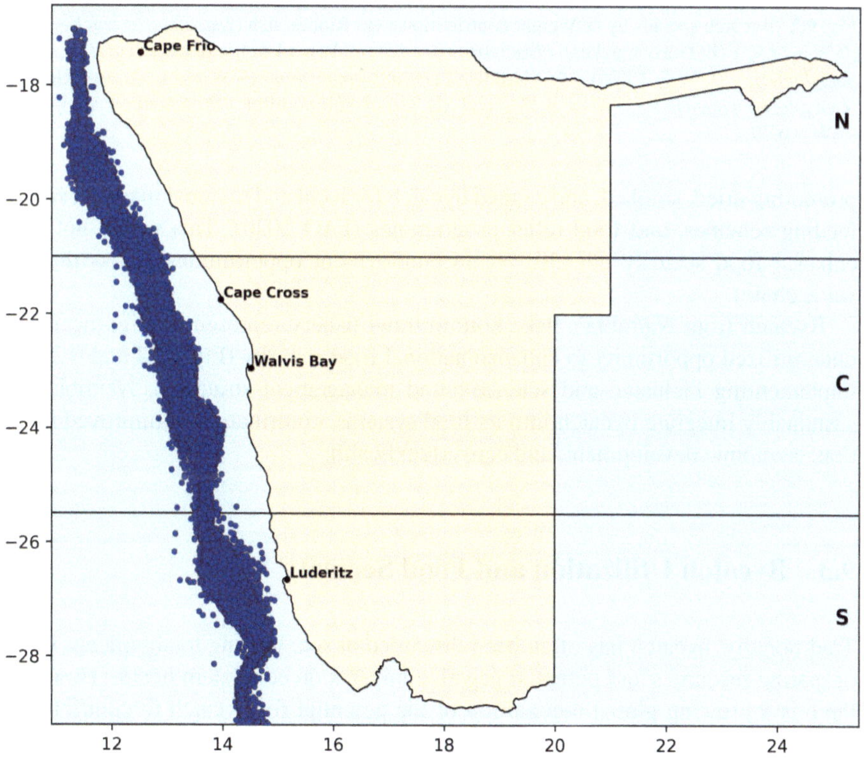

Fig. 9.1 Distribution of some individual trawls covered by the hake bottom-trawl fishery off Namibia. (1997–2014). One dot (blue dot) represents one trawl and the contour lines represent depth. The lines and letters indicate the areas according to sub-regions. (Data source: NatMIRC, MFMR, used with permission). Map drawn by the author using ESRI mapping tools and bathymetry by GEBCO)

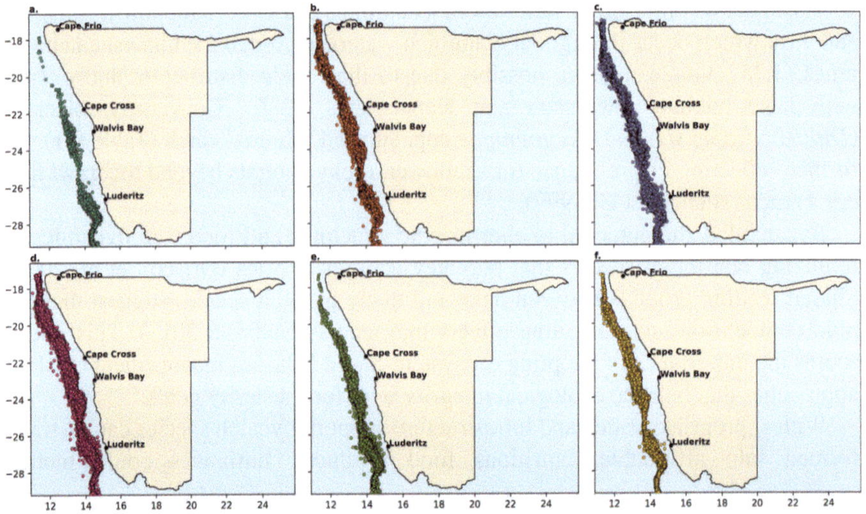

Fig. 9.2 Bycatch species by % Weight contributions: (**a**) Ribbon fish (*Trachipterus trachypterus) (%W = 14.97)*, (**b**) Horse mackerel *(Trachurus capensis)* (%W = 14.41), (**c**) Jacopever *(Helicolenus dactylopterus)* (%W = 13.06), (**d**) Monkfish (*Lophius vomerinus*) (%W = 11.32), (**e**) Kingklip (*Genypterus capensis*) (%W = 9.33), and (**f**) Grootskub-pomfret *(Taractichthys logipinnis)* (%W = 6.79)

promoting dried, smoked, and canned bycatch for local and regional markets, school feeding schemes, and food relief programmes (FAO 2020). This would not only enhance food security but also create employment opportunities across the fish value chain.

Bycatch from Namibia's hake bottom trawl fishery represents a significant yet underutilized opportunity to enhance national food security (Figs. 9.2 and 9.3). By implementing inclusive and science-based management strategies, Namibia can sustainably integrate bycatch into its food systems, contributing to improved nutrition, economic development, and ecosystem health.

9.3 Bycatch Utilization and Food Security

Traditionally, bycatch has often been discarded at sea, leading to significant waste of marine resources and potential negative impacts on ecosystem health. However, there is a growing global recognition of the potential for bycatch to contribute to food security, particularly in regions where protein sources are limited. In Namibia, the utilization of bycatch presents a compelling opportunity to enhance food security, generate economic value, and improve the overall sustainability of the hake demersal fishery.

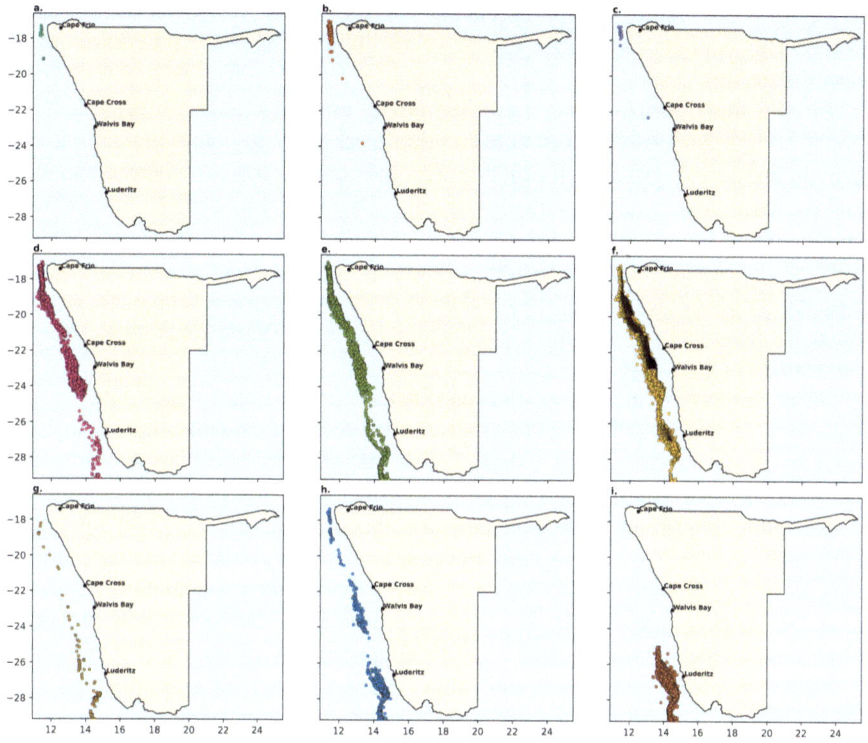

Fig. 9.3 Other potential bycatch species for food security with % weight contribution less than 5: (**a**) Blacktail *(Diplodus capensis),* (**b**) Silver kob *(Argyrosomus inodorus),* (**c**) Yellowfin tuna *(Thunnus albacares),* (**d**) Cape gurnard *(Cheilodactylus capensis),* (**e**) John Dorry *((Zeus faber),* (**f**) West coast sole *(Austroglosus microlepsis),* (**g**) Cardinal fish *(Epigonus spp),* (**h**) Scabbardfish *(Lepidopus caudatus),* and (**i**) Snoek *(Thyrsites atun)*

The assessment of Namibian seafood waste production highlights the quantities and composition of waste generated throughout the seafood value chain, from harvesting to processing. While the study primarily focuses on waste from processing, it underscores the broader issue of underutilized marine resources (Erasmus et al. 2021). Bycatch, if not discarded, can be a significant component of this underutilized resource. The study noted that improved use of seafood waste, including bycatch, can help meet global seafood demand, reduce negative environmental impacts, and provide benefits to human health and biodiversity (Erasmus et al. 2021).Several bycatch species in the Namibian hake fishery, such as Cape horse mackerel (*Trachurus capensis*), monkfish (*Lophius vomerinus*), and kingklip (*Genypterus capensis*), are commercially valuable and consumed in various markets (Erasmus et al. 2021). The economic value of bycatch in Namibia has been reported to be substantial, with figures reaching millions of US dollars annually (Food Business Africa 2023). This indicates that bycatch is not merely an unwanted catch but a potential source of revenue and food. For instance, pilchard bycatch alone has been valued at approximately N$100 million (Food Business Africa 2023). This economic incentive can drive efforts towards better utilization rather than discarding.

From a food security perspective, the utilization of bycatch can provide additional protein sources for local consumption, especially in communities that rely heavily on marine resources for their nutritional needs. Namibia's fisheries are crucial for food security and protein provisioning. By integrating bycatch into the food supply chain, the country can diversify its protein sources and potentially reduce reliance on targeted species, thereby easing pressure on hake stocks. This aligns with the broader objectives of sustainable fisheries management, which aim to maximize the benefits from marine resources for current and future generations (Marine Stewardship Council 2020).

Strategies for bycatch utilization can include processing bycatch into various value-added products. This could range from direct human consumption in local markets to the production of fishmeal, fish oil, or other aquaculture feeds. The recovery of value-added products from seafood waste is recognized as a significant waste reduction strategy that can generate income and address food insecurity (Erasmus et al. 2021). For example, bycatch can be processed into animal feeds or soil fertilizer components, or even as a direct food source for humans (Erasmus et al. 2021). This approach not only minimizes waste but also creates new economic opportunities and contributes to a more circular economy within the fishing sector.

However, effective bycatch utilization requires robust infrastructure for handling, processing, and distributing these species. It also necessitates appropriate regulatory frameworks that encourage utilization while discouraging illegal or unsustainable practices. The Marine Resources Act of Namibia, by disincentivizing bycatch through a levy, aims to promote responsible fishing practices (MFMR 2000). Further policies could be developed to specifically encourage the landing and utilization of bycatch species that have food security potential, ensuring that these resources are not wasted but instead contribute to national food security objectives.

9.4 Management and Policy Frameworks for Bycatch in Namibia

Namibia has been at the forefront of implementing progressive and effective management and policy frameworks to ensure the sustainability of its marine resources, including addressing the challenge of bycatch. The cornerstone of this framework is the Marine Resources Act, Act No. 27 of 2000, which provides the legal basis for the management of marine fisheries in Namibia (MFMR 2000).

The Marine Resources Act (2000) introduced a rights-based and scientific approach to fisheries management. This comprehensive legislation aligns with international guidelines for sustainable, ecosystem-based fisheries management (MFMR 2000). Key provisions of the Act directly address bycatch by:

1. *Setting clear quotas (Total Allowable Catches—TACs)*: TACs are established based on the best available scientific evidence, aiming to prevent overfishing and indirectly reduce the likelihood of excessive bycatch by controlling overall fishing effort (MFMR 2000).
2. *Banning harmful fishing methods and gear*: The Act empowers the Ministry of Fisheries and Marine Resources to regulate fishing gear and methods to minimize their impact on non-target species and habitats (MFMR 2000).
3. *Minimizing bycatch and discards*: A crucial aspect of the Act is its explicit aim to minimize bycatch and discards. It stipulates that a person may not discard any marine resources harvested or taken as bycatch during commercial fishing operations (MFMR 2000). This provision is a strong incentive for fishers to land and potentially utilize bycatch.
4. *Bycatch levy*: The Act also includes a 'bycatch levy' which disincentivizes bycatch by imposing a charge on incidentally caught species (MFMR 2000). This economic disincentive encourages fishers to adopt more selective fishing practices and to avoid areas with high bycatch concentrations.

Beyond the legislative framework, Namibia has implemented practical measures to reduce bycatch, particularly for vulnerable species like seabirds. The hake fishery has been a leader in this regard, with the introduction of mandatory bird-scaring lines since 2015 (Da Rocha et al. 2021). These lines, along with other mitigation measures such as line weighting and night setting for longline vessels, have demonstrably reduced seabird mortality by a remarkable 98% in the longline fleet (Da Rocha et al. 2021). The success of these measures is a result of a collaborative approach involving government, industry, and non-governmental organizations like BirdLife International's Albatross Task Force (Da Rocha et al. 2021; Marine Stewardship Council 2020). The Marine Stewardship Council (MSC) certification of the Namibian hake fishery further validates these efforts, as bycatch reduction is a key condition for maintaining this certification (Marine Stewardship Council 2020).

Observer coverage on fishing vessels is another critical component of Namibia's bycatch management strategy. The hake fishery benefits from 100% observer coverage on its trawl fleet and nearly 50% on its longline fleets (Marine Stewardship Council 2020). These observers play a vital role in collecting independent data on fishing activities, including bycatch, and monitoring compliance with regulations. These data are essential for scientific assessments, informing management decisions, and tracking progress in bycatch reduction (Marine Stewardship Council 2020).

Despite these robust frameworks, challenges remain. The effective implementation and enforcement of regulations are crucial, as evidenced by the less pronounced reduction in seabird bycatch in the trawl fleet compared to the longline fleet (Da Rocha et al. 2021). Continuous monitoring, adaptive management, and ongoing collaboration among stakeholders are necessary to address emerging bycatch issues and ensure the long-term sustainability of the fishery. The policy framework in Namibia provides a strong foundation for managing bycatch, but its success ultimately depends on consistent application and a commitment to continuous improvement.

9.5 Challenges and Opportunities

Despite Namibia's commendable progress in managing its hake demersal fishery and mitigating bycatch, several challenges persist, while new opportunities emerge for further enhancing sustainability and food security. Addressing these challenges and capitalizing on opportunities will be crucial for the long-term viability of the fishery and its contribution to national development.

9.5.1 Challenges

One of the primary challenges lies in the *incomplete implementation and enforcement of bycatch mitigation measures*, particularly within the trawl fleet. While significant reductions in seabird bycatch have been observed in the longline fishery due to the mandatory use of bird-scaring lines, the trawl fleet has shown less progress (Da Rocha et al. 2021). This disparity highlights the need for more effective monitoring and enforcement mechanisms to ensure full compliance across all fishing sectors. The low selectivity of bottom trawling methods inherently leads to high levels of bycatch, making it a persistent challenge to minimize unwanted catches without compromising the efficiency of hake harvesting (Erasmus et al. 2021).

Another challenge is the *lack of comprehensive data on all bycatch species*. While some studies have identified dominant bycatch species and their quantities, detailed information on the population dynamics, growth rates, and longevity of many bycatch species remains limited (Erasmus et al. 2021). This data gap hinders the ability to accurately assess the long-term ecological impacts of bycatch and to develop targeted management strategies for vulnerable non-target species. The removal of certain bycatch species can alter the trophic structure and stability of marine ecosystems, with potential cascading effects that are not yet fully understood (Erasmus et al. 2021).

Economic pressures can also pose a challenge. While bycatch levies aim to disincentivize unwanted catches, the substantial economic value of some bycatch species can create a perverse incentive for their continued capture, even if incidental (Food Business Africa 2023). Balancing the economic interests of the fishing industry with conservation objectives requires careful policy design and robust regulatory oversight. Furthermore, the infrastructure for handling, processing, and marketing bycatch for human consumption or value-added products may not always be sufficient, leading to continued discarding or underutilization.

9.5.2 Opportunities

Despite these challenges, significant opportunities exist to further improve bycatch management and leverage bycatch for food security. One key opportunity lies in *enhancing bycatch utilization*. Many bycatch species, currently discarded or underutilized, have considerable nutritional and economic value (Erasmus et al. 2021). Developing and promoting markets for these species, both domestically and internationally, can transform bycatch from a waste product into a valuable resource. This could involve investing in processing facilities, developing new product lines (e.g. fishmeal, fish oil, or direct human consumption products), and establishing efficient supply chains. Such initiatives would not only reduce waste but also create new employment opportunities and generate additional revenue for the fishing industry and local communities.

Technological advancements offer another promising avenue. Continued research and development into more selective fishing gear and techniques can significantly reduce bycatch rates. This includes exploring modifications to trawl nets, such as the use of Bycatch Reduction Devices (BRDs), and refining longline fishing practices to minimize interactions with non-target species. The successful implementation of bird-scaring lines demonstrates the effectiveness of technological solutions in mitigating bycatch (Da Rocha et al. 2021).

Strengthening regional and international collaboration is also vital. Namibia is part of the Benguela Current Large Marine Ecosystem, which is shared with Angola and South Africa. Collaborative efforts in research, data sharing, and harmonized management strategies across these countries can lead to more effective bycatch management for shared fish stocks and migratory species. Learning from the experiences and best practices of other fisheries globally can also provide valuable insights and innovative solutions.

Finally, *public awareness and education* campaigns can play a crucial role in fostering a greater appreciation for marine biodiversity and the importance of sustainable fishing practices. Educating consumers about the value of bycatch species and promoting their consumption can help drive market demand and support utilization efforts. By transforming public perception of bycatch from a problem to a potential solution for food security, Namibia can further solidify its position as a leader in sustainable fisheries management.

9.6 Conclusion

The Namibian hake bottom trawling fishery stands as a compelling case study in the complex interplay between marine resource utilization, environmental sustainability, and food security. While historically challenged by overfishing and bycatch, Namibia has demonstrated remarkable commitment and success in rebuilding its hake stocks and implementing robust management frameworks. The Marine Resources Act of 2000, coupled with practical mitigation measures such as bird-scaring lines and comprehensive observer coverage, has significantly reduced the ecological footprint of the fishery, particularly in terms of seabird bycatch.

However, the journey towards fully sustainable and food-secure fisheries is ongoing. Bycatch, though mitigated, remains an inherent aspect of demersal fishing. This chapter has highlighted that bycatch in Namibia's hake fishery is not merely a waste product but a valuable, albeit often underutilized, resource with significant potential to contribute to national food security. Species incidentally caught, such as horse mackerel, monkfish, and kingklip, possess both nutritional and economic value that, if properly harnessed, can diversify protein sources, generate additional income, and reduce overall seafood waste.

Realizing the full food security potential of bycatch requires a multi-pronged approach. This includes continued investment in research to better understand bycatch species and their ecological roles, further development and adoption of selective fishing technologies, and the establishment of robust infrastructure for bycatch handling, processing, and marketing. Policy frameworks should continue to evolve, moving beyond mere disincentives for bycatch to actively incentivize its sustainable utilization for human consumption and value-added products.

Namibia's experience offers valuable lessons for other fishing nations grappling with similar challenges. The collaborative spirit between government, industry, and conservation organizations, coupled with a strong scientific foundation, has been instrumental in achieving the current level of sustainability. By embracing bycatch as a potential contributor to food security, Namibia can further enhance the economic and social benefits derived from its marine resources, ensuring a more resilient and equitable food system for its population while safeguarding the health of its precious marine ecosystems. The future of marine capture fisheries, particularly in the context of global food security, hinges on such integrated and forward-thinking approaches that transform challenges into opportunities.

Acknowledgements The authors greatly thank the the government officials under the Ministry of Fisheries & Marine Resources (NATMIRC) in Swakopmund, Erongo Region, Namibia, for availing the dataset for this study. This research did not receive any specific grant from funding agencies in the public commercial, or not-for-profit sectors.

Authorship Statement *Samuel K. Mafwila*: Conceptualization, Resources, Methodology, Formal analysis, Writing—original draft, Writing—review and editing, Supervision, Project administration. *Evans Simasiku*: Writing—review and editing. *Greg Mbaimbai*: Formal analysis, Validation, Investigation, Writing original draft. *Johannes Iitembu*: Conceptualization, Writing—review and editing, Supervision. *AIM Namboga*: Data curation, Validation. *Kudakwashe Hove*: Formal analysis, Writing—review and editing.

References

Alverson DL, Freeberg MH, Murawski SA, Pope JG (1994) A global assessment of fisheries bycatch and discards. FAO Fisheries Technical Paper No. 339. FAO, Rome

Bianchi G, Carpenter KE, Roux JP, Molloy FJ, Boyer D, Boyer HJ (1993) The living marine resources of Namibia. Curso de análisis sociocultural y demográfico. FAO, Roma (Italia) Norwegian Agency for International Development, Rome

Bianchi G, Carpenter KE, Roux JP, Molloy FJ, Boyer D, Boyer HJ (1999) Field guide to the commercial marine and brackish-water species of Namibia. FAO Species Identification Guide for Fishery Purposes. FAO, Rome

Barange M, Pillar SC, Hampton I (1998) Distribution patterns, stock size and life history strategies of Cape horse mackerel Trachurus trachurus capensis, based on bottom trawl and acoustic surveys. In Benguela Dynamics: Impacts of Variability on Shelf-Sea Environments and their Living Resources. Pillar SC, Moloney CL, Payne AIL. and F. A. Shillington (Eds). South African J Mar Sci 19: 433–447

Blaber SJM, and Wassenberg TJ (1989) Feeding ecology of piscivorous birds at subtropical islands in Moreton Bay, with particular reference to the consumption of trawler discards. Mar Biol 101:1–10

Crowder LB and Murawski SA (1998) Fisheries bycatch: Implications for management. Fish 23: 8–17

Da Rocha N, Oppel S, Prince S, Matjila S, Shaanika TM, Naomab C, Yates O, Paterson JRB, Shimooshili K, Frans E, Kashava S, Crawford R (2021) Reduction in seabird mortality in Namibian fisheries following the introduction of bycatch regulation. Biol Conserv 253:108915. https://www.sciencedirect.com/science/article/abs/pii/S0006320720309733

Driscoll R, Chuenpagdee R, Pauly D (2009) Large-scale capture fisheries. In: McLeod K, Leslie H (eds) Ecosystem-based management for the oceans. Island Press, pp 38–46

Engelhard GH, Bova C, Gusha MNC, Harrod OL, Kadhila T, Kanyimba A et al (2024) Climate risk assessment of the fisheries in Namibia. Rev Fish Biol Fish 34(4):1231–1260

Erasmus VN, Kadhila T, Gabriel NN, Thyberg KL, Ilungu S, Machado T (2021) Assessment and quantification of Namibian seafood waste production. Ocean Coast Manage 199:105402. https://www.sciencedirect.com/science/article/abs/pii/S0964569120303094

FAO (2020) The state of world fisheries and aquaculture 2020: sustainability in action. Food and Agriculture Organization of the United Nations, Rome

Food Business Africa (2023) Sustainable fishing efforts suffers setback in Namibia as bycatch caught hits US$3.2M mark. Media Reference. https://www.foodbusinessmea.com/sustainable-fishing-efforts-suffers-setback-in- namibia-as-bycatch-caught-hits-us3-2m-mark/. Accessed 16 May 2025

Hall MA, Alverson DL, Metuzals KI (2000) By-catch: problems and solutions. Mar Pollut Bull 41(1–6):204–219

Heemstra PC, Kannemeyer L (1986) Marine fish diversity in Namibia. Unpublished report, Ministry of Fisheries and Marine Resources, Namibia

Iitembu JA, Erasmus VN, Uanivi U, Nakwaya D, Horaeb RR, Nangolo E, Nashima FP, Iita TK, Mwandemele O (2021) The hits and misses of Namibia's attempt to implement the ecosystem approach to fisheries (EAF) management. Ecosyst Health Sustain 7(1):1920340

Kirkman SP, Blamey L, Lamont T, Field JG, Bianchi G, Huggett JA, Hutchings L, Jackson-Veitch J, Jarre A, Lett C (2016) Spatial characterisation of the Benguela ecosystem for ecosystem-based management. Afr J Mar Sci 38(1):7–22

Lange GM (2003) The value of Namibia's commercial fisheries. DEA Research Discussion Paper 55

Mafwila SK (2011) Trophic interactions and ecosystem dynamics in the northern Benguela: implications for ecosystem-based fisheries management [PhD dissertation, University of Cape Town]

Marine Stewardship Council (2020) A sustainable and bright future for Namibian hake. https://www.msc.org/media-centre/news-opinion/news/2020/11/17/a-sustainable-and-bright-future-for-namibian-hake. Accessed 20 May 2025

Mbaimbai GL (2021) Assessment of bycatch species in the Namibian hake-directed bottom trawl fishery (1997–2014). MSc Thesis, University of Namibia

MFMR (2000) Namibia's Marine Resources Act. In: Future policy. https://www.futurepolicy.org/healthy-ecosystems/oceans/namibias-marine- resources-act/. Accessed 16 July 2025

National Marine Information and Research Centre (NatMIRC) (2007) Hake sampling guidelines. Biomass survey: Hake research. Windhoek, Namibia. Government printers

Paterson B, Kainge P (2014) Rebuilding the Namibian hake fishery: a case for collaboration between scientists and fishermen. Ecol Soc 19(2)

Pinedo MC, Polacheck T (2004a) Sea turtle by-catch in pelagic longline sets off southern Brazil. Biol Conserv 119(3):335–339
Pinedo MC, Polacheck T (2004b) Incidental catch of marine animals in pelagic longline fisheries. In: Sinclair M, Valdimarsson G (eds) Responsible fisheries in the marine ecosystem. FAO/CABI Publishing, pp 109–122
Sokolova M, Mompó Alepuz A, Thompson F, Mariani P, Galeazzi R, Krag LA (2021) A deep learning approach to assist sustainability of demersal trawling operations. Sustain 13:12362
Wilhelm MR, Kirchner CH, Roux JP, Jarre A, Iitembu JA, Kathena JN, Kainge P (2015) Biology and fisheries of the shallow-water hake (Merluccius capensis) and the deep-water hake (Merluccius paradoxus) in Namibia. In: Hakes: biology and exploitation, pp 70–100
Wassenberg TJ and Hill BJ (1987) Feeding by the sand crab Portunus pelagicus on material discarded from prawn trawlers in Moreton Bay, Australia. Mar Biol 95:387–393

Chapter 10
Harnessing the Potential of Seaweed as a Renewable Resource for Improved Food Security in Africa

Martha K. Hausiku-Ikechukwu and Fimanekeni N. Shivute

Abstract Food security in Africa faces major challenges owing to climate change, land degradation, and limitations in traditional agriculture. Addressing the United Nations Sustainable Development Goal 2 (Zero Hunger) requires shifting to sustainable marine resource use, which also aligns with Goal 14 (Life Below Water), aimed at ocean conservation. Marine and coastal resources, particularly seaweeds, offer a sustainable opportunity to diversify food systems, improve biodiversity, and reduce greenhouse gas emissions from agriculture. Seaweed is a nutrient-rich, underexploited marine superfood. Africa is the third largest global producer of cultivated red eucheumatoid seaweed, with Tanzania, Madagascar, and South Africa leading cultivation across 13 countries. Seaweed farming provides stable income, creates jobs, and empowers women in coastal areas. It requires no arable land, making it an environmentally friendly alternative that reduces pressure on terrestrial ecosystems. This chapter explores the food security potential of seaweed by detailing its physicochemical properties and applications, including human food, animal feed, and organic matter. This chapter reviews Africa's seaweed production and its socioeconomic benefits for coastal communities. It concludes with policy and investment recommendations for overcoming environmental, technical, and market barriers. Through coordinated development, seaweed farming can sustainably diversify diets, enhance climate resilience, and significantly contribute to realizing SDG 2 and SDG 14 in Africa, addressing hunger while promoting ocean sustainability.

M. K. Hausiku-Ikechukwu (✉)
Sam Nujoma Marine and Coastal Resources Research Centre, Sam Nujoma Campus, University of Namibia, Henties Bay, Namibia
e-mail: mikechukwu@unam.na

F. N. Shivute
Science and Technology Division, Multidisciplinary Research (MR), Centre for Research Services (CRS), University of Namibia, Windhoek, Namibia

H. A. Mupambwa et al. (eds.), *Harnessing Coastal and Marine Resources for Improved Food Security in Africa*, Sustainability Sciences in Asia and Africa,
https://doi.org/10.1007/978-981-95-9725-3_10

Keywords Blue Economy • Sustainable Development • Functional Foods • Livelihood Diversification • Integrated Multi-Trophic Aquaculture (IMTA) • African Coastal Communities • Marine Resources

10.1 Introduction

Food security remains one of the most critical challenges facing Africa today, shaped by a convergence of factors such as rapid population growth, climate change, conflicts, overfishing, and heavy dependence on land-based agriculture (Ahmed et al. 2021). These interrelated pressures undermine food production and access, resulting in widespread malnutrition and persistent food insecurity in many African regions. Droughts, economic instability, and conflict exacerbate vulnerabilities in food systems, pushing millions into crisis and emergency hunger situations (Kasato et al. 2025).

Within this context, Africa's marine and coastal ecosystems present a largely untapped and promising avenue for diversifying food sources and improving nutritional security, directly contributing to the global ambitions of the Sustainable Development Goal (SDG) 2: Zero Hunger and SDG 14: Life Below Water. Marine resources, including fisheries and emerging aquaculture sectors, offer vital alternatives that can complement terrestrial agriculture and provide sustainable livelihoods for coastal populations (Haque and Mahmud 2025). Sustainable development of these resources is essential for enhancing food availability and nutritional diversity (Yarish et al. 2016).

Seaweed is a marine resource that stands out as a natural resource for addressing food security challenges in Africa. Seaweed cultivation requires no arable land or freshwater, overcoming the major limitations commonly faced by traditional agriculture. Nutritionally, seaweed is rich in proteins, essential vitamins, minerals, and bioactive compounds, making it a potent source for dietary enhancement. Besides food, seaweed farming provides economic opportunities through its use in animal feed, pharmaceuticals, and environmental benefits such as carbon sequestration and ecosystem restoration (Ghaliaoui et al. 2024). Moreover, seaweed farming aligns well with the livelihoods and cultural practices of many African coastal communities, offering both nutritional and economic benefits (Sultana et al. 2023).

This chapter explores the multifaceted role of seaweed in strengthening food security in Africa by assessing its biochemical properties, economic potential, and environmental sustainability. Subsequent sections will discuss seaweed cultivation trends in Africa relative to global trends, identify challenges and opportunities, and offer recommendations for policies and research to unlock the full potential of seaweed for African food security.

10.2 Chemical Properties of Seaweed and Their Applications

Seaweeds and marine macroalgae are chemically diverse organisms rich in bioactive compounds that contribute to their nutritional, industrial, and pharmaceutical significance (Aakre et al. 2020; Hofmann et al. 2025). Their chemical composition varies widely depending on the taxonomic group, environmental conditions, and geographic distribution. They are generally classified into three major divisions—brown (Phaeophyceae), red (Rhodophyceae), and green (Chlorophyceae)— each exhibiting unique biochemical profiles (Abdel-Kareem and ElSaied 2022). These macroalgae are valued for their high carbohydrate content, moderate protein levels, low lipid concentrations, and abundant minerals and vitamins, making them important in marine-based food systems and biotechnological applications (El-Beltagi et al. 2022; Ghaliaoui et al. 2024). A summary of these seaweed bioactive classes and their properties are presented in Table 10.1.

Polysaccharides are the predominant bioactive compounds in seaweeds, constituting 30–60% of their dry weight (Aakre et al. 2020; Khatri et al. 2019). These complex carbohydrates function as hydrocolloids and soluble dietary fibres and include alginate, carrageenan, agar, fucoidan, laminarin, and ulvan (Giridhar Reddy 2022; Xie et al. 2024). Each polysaccharide exhibits distinct structural and chemical characteristics that influence its rheological properties, bioactivity, and industrial applications.

In brown algae, alginate, a copolymer of β-D-mannuronic and α-L-guluronic acids, forms calcium-crosslinked, hydrophilic, and biodegradable gels, making it valuable in wound dressings, food films, and as a stabilising agent in pharmaceuticals and foods (Giridhar Reddy 2022; Lomartire and Gonçalves 2023). Brown algae also produce fucoidan, a sulphated fucan with notable anticoagulant, antiviral, and

Table 10.1 Summary of major seaweed bioactive classes and their properties

Bioactive classes	Key properties
Carrageenan & Agar, Laminarin, Proteins & Peptides	Gelling/Thickening, Low MW glucan (bioactive), ACE-inhibitory, Antioxidant
Fucoidan, Phlorotannins & Bromophenols, Long-chain PUFAs, Sterols, Vitamins, & Minerals	Anticoagulant, Antiviral, Immunomodulatory, Antioxidant, Antimicrobial, Anti-inflammatory, Cardio-metabolic benefits, Membrane modulation, Nutritional
Alginate, Fucoidan, Sterols	Ca^{2+} crosslinked gels (biodegradable), Anticoagulant, Antiviral, Immunomodulatory, Membrane modulation, Bioactivities
Phlorotannins & Bromophenols, Pigments (Carotenoids, Chlorophylls, Phycobiliproteins), Vitamins, & Minerals	Antioxidant, Antimicrobial, Photoprotective, UV absorption
Alginate	Ca^{2+} crosslinked gels, Hydrophilic, Biodegradable
Ulvan	Functional and bioactive contribution

Amlani and YetgîN (2022), Giridhar Reddy (2022), Lomartire and Gonçalves (2023), Matoju et al. (2022), Peñalver et al. (2020), Subbiah et al. (2023), Xie et al. (2024)

antioxidant properties (Kuznetsova et al. 2021; Xie et al. 2024), as well as laminarin, a low-molecular-weight storage glucan with recognized bioactivity. Red algae are rich in agar and carrageenan, which are strong gelling and thickening agents that are widely used as food additives and pharmaceutical excipients. Their sulphate patterns influence gel texture and biological activity (Jayakody et al. 2022). Green algae contain ulvans and sulphated polysaccharides composed of rhamnose, glucuronic acid, and iduronic acid, which contribute to the functional and bioactive potential of Chlorophyta extracts, exhibiting antioxidant and immunomodulatory effects (Cao et al. 2022; Pradhan et al. 2023).

Proteins make up approximately 5–25% of seaweed dry weight, with some red and green species such as Palmaria palmata and Ulva lactuca containing up to 30–40% (Naseri et al. 2020). These proteins provide all essential amino acids, such as lysine, leucine, and valine, with quality comparable to that of terrestrial legumes and eggs (Thiviya et al. 2022). Seaweeds are also rich in aspartic and glutamic acids, which contribute to their characteristic umami flavour. The hydrolysis of seaweed proteins produces bioactive peptides with antioxidant, antihypertensive (including angiotensin-converting enzyme inhibitory), and antimicrobial properties, underscoring their potential as functional food ingredients and nutraceuticals (Figueiredo and Smith 2010; Nadeeshani et al. 2022).

Seaweeds are also rich in vitamins A, B_3, C, D, E, and K, along with phenolic compounds including phlorotannins, flavonoids, and phenolic acids. Phlorotannins, unique to brown algae, exhibit potent antioxidant, anti-inflammatory, and antimicrobial properties (Subbiah et al. 2023) and help protect seaweeds against herbivory and oxidative stress. These compounds have attracted interest for pharmaceutical and cosmetic applications (Polat et al. 2023; Salehi et al. 2019; Shafie et al. 2022). Other secondary metabolites found in seaweeds include terpenoids, halogenated compounds, alkaloids, and polyphenols, which possess diverse bioactivities, such as antiviral, antifungal, and anticancer effects (Carpena et al. 2024; Ghaliaoui et al. 2024; Habeebullah and Alagarsamy 2023).

Additionally, seaweed sterols exhibit membrane-modulating bioactivities, whereas vitamins such as vitamin C and tocopherols, along with mycosporine-like amino acids, provide the antioxidant and UV-protective effects used in cosmeceuticals (Amlani and YetgīN 2022). Seaweeds' high mineral content, especially iodine, is nutritionally valuable but requires safety assessment to avoid potential risks associated with overexposure (Garcia-Vaquero et al. 2021).

Seaweeds contain 15–40% ash by dry weight, accumulating macro- and microelements at concentrations much higher than those in terrestrial plants (Windarto et al. 2025). Key minerals include calcium, potassium, magnesium, sodium, phosphorus, and trace elements such as iodine, iron, zinc, and selenium. Brown algae can accumulate iodine up to 8000 μg/g dry weight, which is important for thyroid health (Garcia-Vaquero et al. 2021; Beerman and Van Tuinen 2023; Lee et al. 2025), enhancing their nutritional value as mineral supplements for humans and animals.

Lipids constitute 1–5% of seaweed dry weight but offer significant nutritional benefits (Rocha et al. 2021). Seaweeds are rich in polyunsaturated fatty acids (PUFAs), particularly omega-3 and omega-6 fatty acids, such as eicosapentaenoic acid (EPA) and docosahexaenoic acid (DHA), which provide anti-inflammatory and cardiometabolic health benefits. Brown algae contain fucosterol and other phytosterols known for their cholesterol-lowering properties, whereas red algae are abundant in glycolipids and phospholipids with antioxidant and anti-inflammatory activities (Dashtaki et al. 2025). Green algae often contain alpha-linolenic acid (ALA), a precursor of essential omega-3 fatty acids that support cardiovascular health (Ramu and Chavadi 2020).

Seaweed pigments are essential for photosynthesis and possess strong antioxidant potential. Brown algae contain fucoxanthin and chlorophyll a and c, which facilitate efficient light absorption in low-light marine environments (Zhukova and Yakovleva 2021). Red algae are rich in unique phycobiliproteins, such as phycoerythrin and phycocyanin, which act as accessory pigments and natural food colorants (Pereira 2018; Summers et al. 2023; Takaichi 2025). These pigments give seaweeds characteristic colours and enhance their nutraceutical value through potent antioxidant effects (Yadav 2024). Green algae are abundant in chlorophyll, which further contributes to photosynthetic efficiency and bioactivity. In addition to pigments, phenolic compounds, such as phlorotannins and bromophenols, provide antioxidant, antimicrobial, UV-protective, and signalling functions (Subbiah et al. 2023). Their bioactivity varies with species and molecular diversity. The combined diversity of pigments and phenolics contributes not only to the nutritional value of seaweeds, but also to their functional applications in health and industry. Table 10.1 presents an overview of the principal bioactive groups, highlighting their defining characteristics, functional roles, and practical uses across various sectors.

The chemical composition of seaweeds reflects their ecological adaptations and highlights their potential as sustainable bioresources. Their diverse polysaccharides, proteins, essential fatty acids, minerals, vitamins, and bioactive secondary metabolites make seaweeds valuable for improving food security, developing functional foods, and producing renewable bioproducts in Africa and worldwide as illustrated in Table 10.2.

10.3 Status of Seaweed Cultivation: Global Context and African Potential

Globally, seaweed aquaculture is a rapidly expanding sector, producing approximately 36.4 million metric tons (live weight) in 2022 and accounting for over half (51.3%) of all mariculture production (FAO 2024). This industry, which has grown at an average of 6.2% annually, is overwhelmingly dominated by Asia, which contributes over 97% of the total output, led by China (~60%) and Indonesia (~25%) (Liu et al. 2025), as illustrated in Fig. 10.1.

The cultivation focuses on a handful of key genera: *Saccharina* and *Laminaria* (kelps) in temperate zones, *Kappaphycus* and *Eucheuma* (carrageenophytes) in the tropics, and *Pyropia* (nori) in East Asia. Red algae, primarily carrageenophytes,

Table 10.2 Applications and functional roles of seaweed

Application area	Functional roles	Examples/notes
Human nutrition and food additives	Provides essential nutrients; supports gut health, satiety; used as gelling, thickening, stabilizing agents; natural colorants and antioxidants; functional ingredients in health foods; improves texture, shelf-life, and reduces sodium in processed foods.	Direct consumption (salads, soups, snacks, sushi); hydrocolloids as food additives and stabilizers.
Animal feed	Enhances animal nutrition, growth, immune response; acts as prebiotic and antimicrobial agent; improves feed conversion and product quality; reduces methane emission in ruminants.	Applied in livestock, poultry, and aquaculture feed to improve growth, health, product quality.
Pharmaceuticals and nutraceuticals	Exhibits antioxidant, antiviral, anti--inflammatory, anticoagulant, antitumour, and immunomodulatory activities; used in drug formulations and functional health supplements.	Immunomodulation, antiviral, anticancer, and antioxidant benefits in pharmaceuticals and nutraceuticals.
Agriculture	Enhances seed germination, root/shoot growth, stress tolerance, crop yield; improves soil structure, moisture retention, microbial activity, cation exchange capacity; enhances mushroom yield and nutritional quality (iodine enrichment); acts as substrate conditioner.	Liquid or solid biofertilizers; soil conditioners to improve fertility and physical properties; mushroom substrate supplements.

Aakre et al. (2020), Brown and Saa (2015), Fitton et al. (2015), Ghaliaoui et al. (2024), Hagan and Anyangwe (2023), Ismail et al. (2023), Kamleshbhai et al. (2024), Liu et al. (2025), Mensah et al. (2023), Mupambwa and Hausiku-Ikechukwu (2018), Olsson et al. (2020), Pereira (2018), Reynolds et al. (2025)

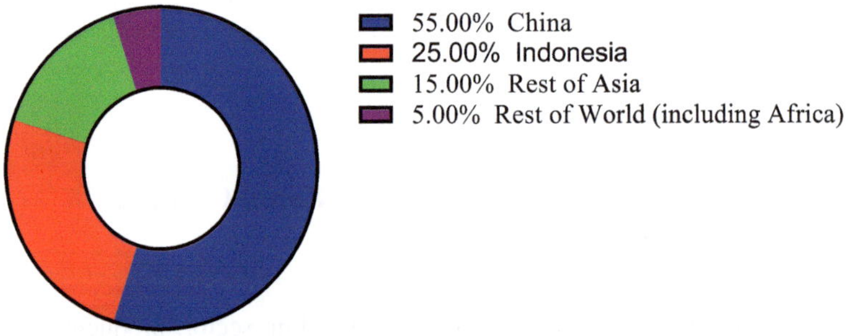

Fig. 10.1 Global Seaweed Production Share (2022 Estimates)

constitute the majority (55.8%) of global production. Despite a vast potential ocean area of 33.7 million km^2 deemed suitable for cultivation, the industry faces a significant threat from climate change, which may reduce suitable farming areas in lower latitudes by up to 50% by mid-century (Liu et al. 2025). Within this global landscape, Africa's seaweed sector represents a growing opportunity to enhance food security and socioeconomic resilience (Table 10.3). The continent's contribution is

Table 10.3 Seaweed production in Africa: Current status and global context

Region/ country	Main species	Production volume	Trend	Key characteristics
Zanzibar (Tanzania)	*Kappaphycus alvarezii, Eucheuma denticulatum*	8785 t FW (2020)	Declining	Largest African producer: yield decline due to disease & temperature stress
Mainland Tanzania	*Kappaphycus alvarezii, Eucheuma denticulatum*	~25,000 t FW (2022)	Stable/Slight Decline	Smallholder farms dominate; high female participation (~80%)
Kenya	*Eucheuma denticulatum, Kappaphycus alvarezii*	~1500–2000 t FW (2023)	Increasing	Emerging farms with NGO support; low export volume
Madagascar	*Kappaphycus alvarezii, Eucheuma spinosum*	~3000–23,000 t FW (2020)	Expanding	Commercial farms with strong Asian export links
South Africa	*Ecklonia maxima, Laminaria pallida*	~7000–10,000 t FW (2024)	Stable	Wild kelp harvest for abalone feed; aquaculture trials ongoing
Namibia	*Macrocystis pyrifera*	Pilot phase	Developmental	Offshore kelp cultivation for blue economy & carbon capture
Other Islands	Mixed species	< 1000 t FW combined	Emerging	Small pilot farms for livelihood diversification
Africa Total	Eucheumatoids & kelps	~120,000 t FW (2022)	Moderate growth	<1% of global production; concentrated in Western Indian Ocean
Global Total	Various	35.8 million t FW (2020)	Steady growth (2–3%)	Asia dominates (>97% of global output)

Sources: Souza (2024), Dogeje et al. (2024), Islam et al. (2015), Msuya (2020), Mutjida (2024), Spillias et al. (2023)

currently less than 1% of the world's total, with an estimated annual production of 120,000 tons of fresh weight (World Bank 2023; Msuya et al. 2022). As illustrated in Fig. 10.1, African production is highly concentrated, with Zanzibar (Tanzania) as the leading producer (~103,000 t), followed by Madagascar (~53,000 t), and South Africa (~17,000 t) (Msuya et al. 2022).

The industry is geographically and technically segmented. Along the East African coast, including Tanzania, Kenya, and Madagascar, cultivation for carrageenan extraction is dominated by the red seaweeds Eucheuma and Kappaphycus, whereby farming relies primarily on shallow-water techniques, such as the off--bottom method, where seaweed-laden ropes are stretched between stakes in sheltered bays. In Tanzania, commercial farming since the 1980s has solidified *Eucheuma denticulatum* as the predominant species due to its resilience, overcoming disease-related die-offs that have plagued *Kappaphycus alvarezii* (Msuya et al. 2022). This system depends on vegetative propagation, which enables sustainable production cycles.

In contrast, Southern Africa pioneered more technologically advanced cultivation methods. South Africa is developing land-based tank cultures and pilot offshore farms for species such as *Gracilaria* (for agar), while also harvesting wild kelp (*Ecklonia maxima, Laminaria pallida*) for abalone feed and alginate extraction (Fleischman et al. 2021). Namibia is advancing its sector, focusing on *Gracilaria* cultivation in lagoon systems, and developing large-scale offshore projects targeting the giant kelp *Macrocystis pyrifera* for commercial and environmental benefits within its blue economic strategy (Souza 2024; Mutjida 2024).

This regional diversity in species and methods, from tropical carrageenophytes to temperate agarophytes and kelps, directly reflects the continent's environmental gradient (Fleischman et al. 2021). Despite facing challenges from pests and diseases, production in key areas, such as Zanzibar, has remained stable to slightly increasing. The seaweed sector's socioeconomic impact on Zanzibar is significant; it ranks as the third-largest foreign exchange earner and employs approximately 30,000 farmers, predominantly women (Lodhia 2023). Supported by government policies and investments in deeper-water cultivation and processing infrastructure, Africa's seaweed industry is poised for continued strategic growth.

Overall, African seaweed farming shows considerable growth potential driven by global demand, led by Zanzibar's established sector and expanding *Gracilaria* cultivation in South Africa.

10.4 Socioeconomic Impact of Seaweed Farming

Seaweed farming generates income, diversifies livelihoods, and can empower women and households, but outcomes vary by value-chain position, market access, and local institutions (Table 10.4). Household income and poverty Seaweed farming provides regular short production cycles (e.g. 30–45 days in Indonesia) and can raise rural household incomes. Studies conclude that it can lift some households above national poverty lines, although not universally and often depending on price volatility and value-chain capture by processors (Rimmer et al. 2021). Women also gain both economic and social benefits from seaweed farming. Community surveys in Indonesian seaweed villages (n = 74 women surveyed) reported positive economic and social impacts from participation in seaweed farming, including earnings, greater financial contribution to households, and perceived well-being gains (Larson et al. 2021).

In local development case studies in Kenya's Kibuyuni village, seaweed cultivation has become a major alternative livelihood with observable social and environmental impacts on the community, illustrating how farming can diversify rural

Table 10.4 Socioeconomic impacts of seaweed farming

Aspect	Socioeconomic impact
Income	Regular short cycles provide frequent cash flows for smallholders and farm labourers.
Employment	Seaweed value chains create on-farm and off-farm jobs (drying, packing, processing) that support coastal communities.
Women	Female participation in farm and post-harvest tasks increases household resilience and women's economic agency in many Indonesian contexts.
Community Wellbeing	Seaweed incomes support diet diversity, schooling, and other household investments where markets and extension support are present.

Sources: Larson et al. (2021), Mirera et al. (2020), Rimmer et al. (2021)

economies and generate non-farm employment and services (Mirera et al. 2020). Mechanisms of benefits arise through local wage/self-employment on farms, income from dried/raw exports, seasonal labour opportunities, and spill-over activities (drying, small processing, transport) that create local value-chain employment (Mirera et al. 2020; Rimmer et al. 2021).

10.5 Challenges Hindering Seaweed Adoption for Improved Food Security in Africa

Challenges hindering seaweed adoption for improved food security in Africa are multifaceted, as illustrated in Fig. 10.2. These challenges manifest in distinct ways across the continent's diverse coastal nations, creating a complex barrier to sectoral growth (Table 10.5).

Production systems are highly vulnerable to shock. In Tanzania, which dominates African production, warming seas have triggered widespread "ice-ice" disease and epiphyte infestations, devastating farms, and forcing a shift from higher-yielding *Kappaphycus* to more resilient but lower-value *Eucheuma* species (Msuya 2020). Meanwhile, in the temperate waters of South Africa and Namibia, emerging cultivation projects face the threat of infrastructure destruction from strong wave action and storms, while land-based tank systems struggle to maintain optimal water quality (Souza 2024).

Compounding these production challenges are significant technical and infrastructural shortcomings, particularly in postharvest management. Across key producer nations such as Tanzania, Kenya, and Madagascar, the predominant practice of drying seaweed directly on the ground leads to contamination with sand and pests, severely degrading quality and restricting market access (Msuya et al. 2022). This low-quality biomass fails to meet international standards for direct food consumption or high-value extracts, despite the proven potential of good aquaculture practices to improve outcomes if widely adopted (Diah et al. 2024). These technical limitations are exacerbated by deep-seated economic and market constraints. In

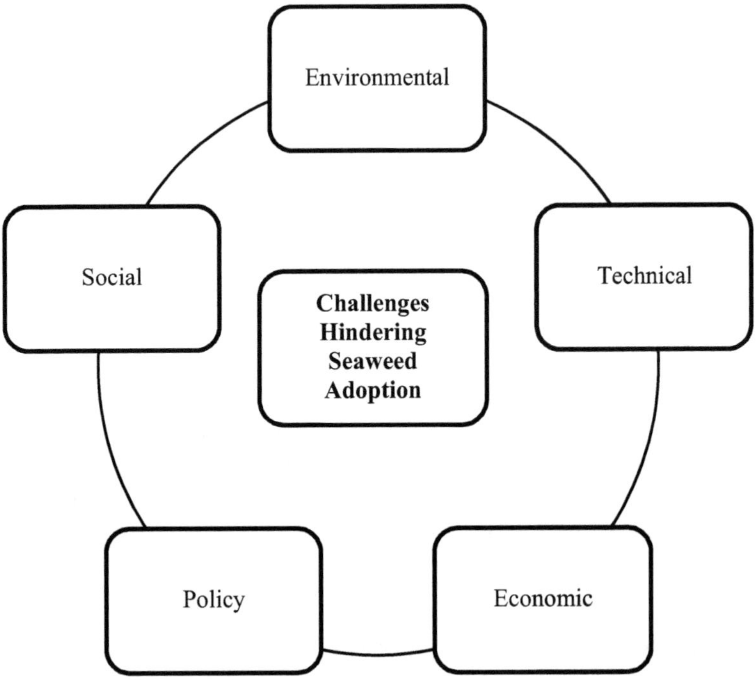

Fig. 10.2 Challenges hindering seaweed adoption for improved food security in Africa

Tanzania and Madagascar, smallholder farmers, who are predominantly women, are largely trapped as price-takers within a commodity model, selling raw dried seaweed for carrageenan extraction, while the high value of processed goods is captured offshore (Mirera et al. 2020). In Southern Africa, nascent ventures in Namibia and South Africa targeting species such as *Gracilaria* and *Macrocystis* face a different economic hurdle: high startup costs and a critical lack of accessible venture capital for blue economy projects, which stifles innovation and scaling (World Bank 2023).

Underpinning these issues is a pervasive policy and governance gap. In Kenya, for instance, significant potential for industry growth remains untapped due to bureaucratic hurdles, a lack of clear maritime spatial plans, and frequent conflicts with the tourism and fishing sectors over coastal access (Mendes et al. 2024). This is symptomatic of a broader pan-African issue, where a few countries have developed dedicated national strategies for seaweed, leaving a vacuum in regulatory guidance, food safety standards for novel products, and incentives for local processing (FAO 2024). Finally, persistent social and gender-specific barriers curtail equitable development in a sector. Although seaweed farming in Tanzania and Zanzibar provides vital income for women, their empowerment is often limited by structural obstacles, including restricted access to finance, difficulties in accessing more productive deeper-water sites, and underrepresentation in cooperative leadership and market negotiations (Mirera et al. 2020; Msuya et al. 2022). Furthermore, the

Table 10.5 Challenges and opportunities for seaweed adoption in African food security

Category	Challenge/barrier	Opportunity/solution	Evidence/location
Biological/ environmental	Disease outbreaks ("ice-ice"), pests, heat stress	Develop resistant strains; implement seasonal cropping	Tanzania: *Kappaphycus* limited by epiphytes; heat stress reduces *Eucheuma* yields
	Extreme weather, wave damage to coastal farms	Invest in robust offshore systems and flexible structures	Namibia: Kelp Blue's offshore farms avoid storm damage and coastal competition
Technical/ infrastructure	Unreliable seed supply; biosecurity risks	Establish hatcheries and quarantine protocols	Tanzania: Seed imports introduce pests due to weak biosecurity
	Poor processing infrastructure and market access	Develop processing clusters and cold chains	Namibia: Poor coastal roads and mining zones limit development
	Slow adoption of improved techniques	Expand participatory training and extension services	Zanzibar: TNC/government programs co-developing best practices
Market/value chain	Low prices for raw seaweed; weak farmer leverage	Develop local value-added products (cosmetics, foods)	Zanzibar: Seaweed Cluster uses <1% of production for value-added goods
	Logistics challenges; export barriers	Establish cooperative marketing and supply chains	Zanzibar: Transport delays and payment issues weaken market links
Institutional/ policy	Weak biosecurity and disease response	Create seaweed--specific regulations and certification	Tanzania: Poor coordination during pest/disease outbreaks
	Complex licensing; marine spatial conflicts	Designate aquaculture zones; streamline permits	South Africa: Kelp value chain roadmap supports policy development
Socioeconomic/ livelihood	Income volatility; seasonal risks	Diversify livelihoods; empower women producers	Zanzibar: 80% of 25,000 farmers are women; income remains unstable
	Health risks and labour issues	Improve safety protocols; provide social protection	Zanzibar: Studies identify sun exposure and injury risks for workers
	Resistance to new methods	Use demonstration farms and peer learning	Zanzibar: Social and cultural constraints limit adoption of improvements

Sources: Biñas et al. (2024), Broch et al. (2019), Charisiadou et al. (2022), Dogeje et al. (2024), Gutierrez-Corley et al. (2025), Hedberg et al. (2018), Lumenyela et al. (2023), Mtemvu and Adebayo (2025), Munguti et al. (2025), Ngajilo et al. (2025), Spagnuolo and Genovese (2024), FAO (2013), Yarish et al. (2016)

general lack of consumer awareness and cultural acceptance of seaweed as a direct food source across many African communities continues to hinder the development of robust local markets, keeping the sector overly reliant on volatile export commodities.

10.6 Strategies for Harnessing Seaweed for Food Security in Africa

Realising the potential of seaweed to enhance food security in Africa necessitates a coordinated, multi-pronged strategy. This approach must be anchored in high-level policy frameworks, such as the African Union's Blue Economy Strategy (FAO 2024), which champions sustainable aquaculture for nutrition and economic growth, and the FAO's Strategic Framework (2022–2031) with its focus on "Blue Transformation" for sustainable aquatic food systems. The strategic pillars summarized in Table 10.6 are critical for guiding stakeholders in building a resilient and impactful seaweed sector.

Strategic investment in research and development (R&D) is fundamental. This should prioritize the identification of nutrient-rich, locally adaptable seaweed species and the development of climate-resilient cultivation techniques such as offshore and integrated multitrophic aquaculture (IMTA) systems. Concurrently, R&D must focus on innovating value-addition processes to improve product quality, shelf life, and consumer appeal in both domestic and international markets (Kasato et al. 2025). Complementing R&D, comprehensive capacity building, and training are essential to translate research into on-the-ground impact. Targeted programs for farmers, processors, and extension officers are needed to strengthen technical and

Table 10.6 Strategic framework for developing the African seaweed sector

Pillar	Key actions	Rationale & policy link
1. R&D	Domesticate resilient native species Advance offshore/IMTA systems Develop value-added products	Creates climate-resilient, diverse sector. Fulfils *AU Blue Economy* innovation mandate.
2. Capacity building	Train farmers/processors in technical & business skills Ensure women's inclusion in programs	Improves yields & livelihoods. Aligns with *FAO* empowerment goals.
3. Market development	Build domestic demand & export channels Invest in local processing & quality control	Creates economic incentives & local value capture. Supports *AU/FAO* trade goals.
4. Policy support	Include seaweed in food security plans Set clear regulations & financial incentives	Creates enabling environment. Implements *AU/FAO* blue economy frameworks.
5. Community engagement	Conduct nutritional/economic outreach Involve communities in management	Ensures social license & equity. Key for *AU/FAO* sustainable development.

Sources: Brakel et al. (2021), Kasato et al. (2025), Khan et al. (2024), Larson et al. (2021), Nakhate and van der Meer (2024), Webb et al. (2023), FAO (2024)

entrepreneurial skills, promote best practices, and ensure the inclusive participation of women, who are pivotal actors in coastal communities (Khan et al. 2024; Nakhate and van der Meer 2024).

To ensure economic viability, a concerted effort must be made towards market development and robust value chains. This involves investing in processing infrastructure, establishing quality control standards, and launching consumer awareness campaigns to stimulate domestic demand for seaweed as a nutritious food source. Developing these local and export markets is crucial for creating stable economic incentives that drive industrial growth (Webb et al. 2023). Underpinning all these efforts is the need for supportive policy and governance. Strong policy frameworks are required to provide clear regulatory guidance, establish food safety standards, and provide financial incentives. Critically, seaweed must be formally integrated into national food and nutrition security plans to unlock public investment and streamline sectoral development, a move strongly advocated by the AU Blue Economy Strategy (Permani et al. 2024).

Finally, the sector's long-term success depends on active community engagement and awareness. Participatory education campaigns that highlight the nutritional and economic benefits of seaweed can foster public acceptance and drive its adoption. Involving local communities in policy dialogue and project design ensures that development is culturally coherent, socially equitable, and ultimately sustainable (Kasato et al. 2025; Webb et al. 2023).

10.7 Conclusion

Harnessing seaweed as a marine and coastal resource offers a promising and sustainable pathway to improve food security across Africa by tapping into both its nutritional benefits and socioeconomic potential. This approach directly leverages ocean productivity to diversify food systems, creating a vital link between sustainable marine resource use and the goal of achieving Zero Hunger (SDG 2). By developing a sustainable aquaculture sector, it simultaneously supports the targets of SDG 14 (Life Below Water). Global and regional examples reveal how Indonesia's large smallholder production, South Korea's rapid industry growth, and Northwest Europe's emerging small-scale operations highlight diverse pathways for development that Africa can adapt to. Seaweed's rich nutrient and bioactive profile support multiple applications in the food, feed, and health sectors, while its farming can generate income, employment, and empower women in coastal areas. To fully realize this potential, coordinated strategies in research and development, capacity building, market expansion, policy support, and community engagement are essential. Addressing environmental, technical, economic, and social challenges will enable sustainable seaweed farming to diversify agriculture, enhance nutrition, build climate resilience, and strengthen livelihoods, thereby playing a key role in advancing food security and contributing to sustainable SDGs across Africa.

References

Aakre I, Tveito Evensen L, Kjellevold M, Dahl L, Henjum S, Alexander J, Madsen L, Markhus MW (2020) Iodine status and thyroid function in a group of seaweed consumers in Norway. Nutrients 12(11):3483. https://doi.org/10.3390/nu12113483

Abdel-Kareem MSM, ElSaied AAF (2022) Chapter 2—Global seaweeds diversity. In: El-Sheekh M, Abomohra AE-F (eds) Handbook of algal biofuels. Elsevier, pp 39–55. https://doi.org/10.1016/B978-0-12-823764-9.00001-7

Ahmed F, Al-Amin AQ, Mohamad ZF (2021) Recent research on climate change and food security. In: Ahmed F, Al-Amin AQ, Mohamad ZF (eds) Climate change and adaptation for food sustainability: implications and scenarios from Malaysia. Springer International Publishing, pp 13–62. https://doi.org/10.1007/978-3-030-85375-4_2

Amlani M, YetgiN S (2022) Seaweeds: bioactive components and properties, potential risk factors, uses, extraction and purification methods. Marine Sci Technol Bull 11(1):9–31. https://doi.org/10.33714/masteb.1021121

Beerman AAA, Van Tuinen ST (2023) Bioavailability of iodine species from various commercially consumed seaweed species. Wageningen Food Safety Res. https://doi.org/10.18174/636611

Brakel J, Sibonga RC, Dumilag RV, Montalescot V, Campbell I, Cottier-Cook EJ, Ward G, Le Masson V, Liu T, Msuya FE, Brodie J, Lim P-E, Gachon CMM (2021) Exploring, harnessing and conserving marine genetic resources towards a sustainable seaweed aquaculture. Plants, People, Planet 3(4):337–349. https://doi.org/10.1002/ppp3.10190

Brown P, Saa S (2015) Biostimulants in agriculture. Front Plant Sci 6. https://doi.org/10.3389/fpls.2015.00671

Biñas JG, Camarista JLA, Cardinal EMSB, Faldas TGA, Gabata JM, Meteoro PMA, Nieles DSL (2024) Lived experiences of seaweed farmers in a coastal municipality of Iloilo, Philippines during COVID-19 pandemic (Doctoral dissertation, Central Philippine University). CPU-IR Repository. https://repository.cpu.edu.ph/handle/20.500.12852/3328

Broch O, Alver M, Bekkby T, Gundersen H, Forbord S, Handå A, Skjermo J, Hancke K (2019) The Kelp Cultivation Potential in Coastal and Offshore Regions of Norway. Front Mar Sci 5. https://doi.org/10.3389/fmars.2018.00529

Cao S, Yang Y, Liu S, Shao Z, Chu X, Mao W (2022) Immunomodulatory activity in vitro and in vivo of a sulfated polysaccharide with novel structure from the green alga Ulvaconglobata Kjellman. Mar Drugs 20(7):447. https://doi.org/10.3390/md20070447

Carpena M, Pereira CSGP, Silva A, Barciela P, Jorge AOS, Perez-Vazquez A, Pereira AG, Barreira JCM, Oliveira MBPP, Prieto MA (2024) Metabolite profiling of macroalgae: biosynthesis and beneficial biological properties of active compounds. Mar Drugs 22(10):478. https://doi.org/10.3390/md22100478

Charisiadou S, Halling C, Jiddawi N, von Schreeb K, Gullström M, Larsson T, Nordlund LM (2022) Coastal aquaculture in Zanzibar, Tanzania. Aquaculture 546: 737331. https://doi.org/10.1016/j.aquaculture.2021.737331

Dashtaki E, Esteki M, Mehdinia A, Simal-Gandara J (2025) Evaluation of the fatty acid composition of representative brown and red seaweeds from the Persian Gulf: an analysis from the nutritional perspective. J Appl Phycol. https://doi.org/10.1007/s10811-025-03683-x

Diah A, Isharyadi F, Sinurat E, Damayanti N, Susanto D, Kristiningrum E, Suroto D, Setyaningsih W (2024) Improving the quality of dried seaweed (Gracilaria sp.) through the implementation of Good Aquaculture Practices (GAP) in Brebes, Central Java. Int J Design Nat Ecodynamics 19:1019–1031. https://doi.org/10.18280/ijdne.190331

Dogeje F, Msuya FE, Ngaruko D, Mohamed AS (2024) Blue economy dynamics: does women's participation in seaweed aquaculture influence their income levels? Empirical evidence from Zanzibar. Aquac Int 33(1):27. https://doi.org/10.1007/s10499-024-01731-w

El-Beltagi HS, Mohamed AA, Mohamed HI, Ramadan KMA, Barqawi AA, Mansour AT (2022) Phytochemical and potential properties of seaweeds and their recent applications: a review. Mar Drugs 20(6):342. https://doi.org/10.3390/md20060342

FAO (2024) The State of World Fisheries and Aquaculture (SOFIA) 2024. Food and Agriculture Organization of the United Nations. - Google Search. (n.d.). Retrieved 19 Nov 2025

Figueiredo E, Smith GF (2010) What's in a name: epithets in Aloe L. (Asphodelaceae) and what to call the next new species. Bradleya 28(28):79–102. https://doi.org/10.25223/brad.n28.2010.a9

Fitton JH, Stringer D, Karpiniec S (2015) Therapies from Fucoidan: an update. Mar Drugs 13:5920–5946. https://doi.org/10.3390/md13095920

Fleischman MJ, Bolton JJ, Rothman MD (2021) Macrocystis (Laminariales, Phaeophyceae) in South Africa: potential for cultivation through holdfast fragmentation and use as feed for the aquacultured abalone, Haliotis midae. J Appl Phycol 33(3):1731–1740. https://doi.org/10.1007/s10811-021-02389-0

Food and Agriculture Organization of the United Nations (2013) Social and economic dimensions of carrageenan seaweed farming (FAO Fisheries and Aquaculture Technical Paper No. 580). FAO. http://www.fao.org/3/i3344e/i3344e.pdf

Garcia-Vaquero M, Rajauria G, Miranda M, Sweeney T, Lopez-Alonso M, O'Doherty J (2021) Seasonal variation of the proximate composition, mineral content, fatty acid profiles and other phytochemical constituents of selected Brown macroalgae. Mar Drugs 19(4):204. https://doi.org/10.3390/md19040204

Ghaliaoui N, Hazzit M, Mokrane H (2024) Seaweeds as a potential source of bioactive compounds. Res Biotechnol Environ Sci 3(1):1–8. https://doi.org/10.58803/rbes.v3i1.19

Giridhar Reddy S (2022) Alginates - a seaweed product: its properties and applications. In: Deniz I, Imamoglu E, Keskin-Gundogdu T (eds) Properties and applications of alginates. IntechOpen. https://doi.org/10.5772/intechopen.98831

Gutierrez-Corley R, Moncada S, Deidun A (2025) Sustainable blue economy practices in seaweed farming, Belize: Cultural heritage and climate resilience in a Small Island Developing State. SST 8(1): 189–204. https://www.um.edu.mt/library/oar/handle/123456789/134794

Habeebullah SFK, Alagarsamy S (2023) Bioactive secondary metabolites from marine macroalgae. In: Algal metabolites. Apple Academic Press

Hagan M, Anyangwe N (2023) Vitamin content in seaweeds: a systematic review on water-soluble and fat-soluble vitamins for adult daily intake. Funct Food Sci 3(12):305. https://doi.org/10.31989/ffs.v3i12.1273

Haque MM, Mahmud MN (2025) Potential role of aquaculture in advancing sustainable development goals (SDGs) in Bangladesh. Aquac Res 2025(1):6035730. https://doi.org/10.1155/are/6035730

Hofmann LC, Strauss S, Shpigel M, Guttman L, Stengel DB, Rebours C, Gjorgovska N, Turan G, Balina K, Zammit G, Adams JMM, Ahsan U, Bartolo AG, Bolton JJ, Domingues R, Dürrani Ö, Eroldogan OT, Freitas A, Golberg A et al (2025) The green seaweed *Ulva*: tomorrow's "wheat of the sea" in foods, feeds, nutrition, and biomaterials. Crit Rev Food Sci Nutr 65(19):3728–3763. https://doi.org/10.1080/10408398.2024.2370489

Hedberg N, von Schreeb K, Charisiadou S, Jiddawi NS, Tedengren M, Nordlund LM (2018) Habitat preference for seaweed farming – A case study from Zanzibar, Tanzania. Ocean Coast Manag 154, 186–195. https://doi.org/10.1016/j.ocecoaman.2018.01.016

Islam MS, Ahmed MK, Habibullah-Al-Mamun M, Masunaga S (2015) Assessment of trace metals in foodstuffs grown around the vicinity of industries in Bangladesh. J Food Compos Anal 42:8–15

Ismail MM, El Zokm GM, Miranda Lopez JM (2023) Nutritional, bioactive compounds content, and antioxidant activity of brown seaweeds from the Red Sea. Front Nutr 10. https://doi.org/10.3389/fnut.2023.1210934

Jayakody MM, Vanniarachchy MPG, Wijesekara I (2022) Seaweed derived alginate, agar, and carrageenan based edible coatings and films for the food industry: a review. J Food Meas Charact 16(2):1195–1227. https://doi.org/10.1007/s11694-021-01277-y

Kamleshbhai B, Bambhaniya I, Vaghela D, Bhadarka M, Ram M (2024) Seaweed: nature's dual role as soil conditioner and water purifier. Trends Agric Sci 3:58–62. https://doi.org/10.5281/zenodo.10670892

Kasato Y, Mwansisya A, Islam S, Juma NS, Massawe A, Shuaibu A, Ying X (2025) Seaweed in Africa: a mini review of industry, benefits and safety concerns. Int J Res Stud Biosci 11(1):1–8. https://doi.org/10.20431/2349-0365.1101001

Khan N, Sudhakar K, Mamat R (2024) Macroalgae farming for sustainable future: navigating opportunities and driving innovation. Heliyon 10(7). https://doi.org/10.1016/j.heliyon.2024.e28208

Khatri K, Rathore MS, Agrawal S, Jha B (2019) Sugar contents and oligosaccharide mass profiling of selected red seaweeds to assess the possible utilization of biomasses for third-generation biofuel production. Biomass Bioenergy 130:105392. https://doi.org/10.1016/j.biombioe.2019.105392

Kuznetsova TA, Andryukov BG, Makarenkova ID, Zaporozhets TS, Besednova NN, Fedyanina LN, Kryzhanovsky SP, Shchelkanov MY (2021) The potency of seaweed sulfated polysaccharides for the correction of hemostasis disorders in COVID-19. Molecules 26(9):2618. https://doi.org/10.3390/molecules26092618

Larson S, Stoeckl N, Fachry ME, Dalvi Mustafa M, Lapong I, Purnomo AH, Rimmer MA, Paul NA (2021) Women's well-being and household benefits from seaweed farming in Indonesia. Aquaculture 530:735711. https://doi.org/10.1016/j.aquaculture.2020.735711

Lee Y, Park HJ, Jo M, Ha KS, Mok JS (2025) Analysis and risk assessment of total iodine content in edible seaweeds in South Korea. Foods 14(16):2865. https://doi.org/10.3390/foods14162865

Liu Y, Zhou M, Diao Q, Ma T, Tu Y (2025) Seaweed as a feed additive to mitigate enteric methane emissions in ruminants: opportunities and challenges. J Integr Agric 24(4):1327–1341. https://doi.org/10.1016/j.jia.2024.09.036

Lodhia R (2023) Tanzania seaweed guide: opportunities for increased productivity, traceability, and sustainability

Lomartire S, Gonçalves AMM (2023) Algal phycocolloids: bioactivities and pharmaceutical applications. Mar Drugs 21(7):384. https://doi.org/10.3390/md21070384

Lumenyela RA, Magasha O, Dimosso P (2023) Adoption of seaweed improved farming techniques among farmers in Zanzibar: An Application of Adoption and Diffusion Outcome Prediction Tool. Soc Sci Humanit Open 7(1), 100472. https://doi.org/10.1016/j.ssaho.2023.100472

Mendes M, Hillinger A, Ihle K, Cascais M, Andrade P, Cotas J, Pacheco D, Marques J, Pereira L, Gonçalves A (2024) Seaweed aquaculture importance in sustainable economy in an era of climate change. In: Oceanography - relationships of the oceans with the continents, their biodiversity and the atmosphere. https://doi.org/10.5772/intechopen.114366

Mensah EO, Kanwugu ON, Panda PK, Adadi P (2023) Marine fucoidans: structural, extraction, biological activities and their applications in the food industry. Food Hydrocoll 142:108784. https://doi.org/10.1016/j.foodhyd.2023.108784

Mirera DO, Kimathi A, Ngarari MM, Magondu EW, Wainaina M, Ototo A (2020) Societal and environmental impacts of seaweed farming in relation to rural development: the case of Kibuyuni village, south coast, Kenya. Ocean Coastal Manage 194:105253. https://doi.org/10.1016/j.ocecoaman.2020.105253

Msuya F (2020) Seaweed resources of Tanzania: status, potential species, challenges and development potentials. Bot Mar 1. https://doi.org/10.1515/bot-2019-0056

Msuya FE, Bolton J, Pascal F, Narrain K, Nyonje B, Cottier-Cook EJ (2022) Seaweed farming in Africa: current status and future potential. J Appl Phycol 34(2):985–1005. https://doi.org/10.1007/s10811-021-02676-w

Mupambwa H, Hausiku-Ikechukwu M (2018) Seaweed amended rice straw substrate and its influence on health related nutrients, trace elements, growth and yield of edible white elm mushroom (Hypsizygus ulmarius). Int J Agric Biol 20:2763–2769. https://doi.org/10.17957/IJAB/15.0831

Mutjida PS (2024) A geochemical baseline study of giant kelp (macrocystis pyrifera) at the kelp blue farm to assess carbon sequestration potential. University of Namibia. http://hdl.handle.net/11070/3877

Matoju I, Le Masson V, Montalescot V, Ndawala MA, Msuya FE (2022) A resilience lens to explore seaweed farmers' responses to the impacts of climate change in Tanzania. Appl. Phycol 3 (1): 132–148. https://doi.org/10.1080/26388081.2022.2091951
Mtemvu JB, Adebayo ST (2025) An evaluation of hazard analysis and personal protective equipment in mitigating maritime accidents in Tanzania. Ayden Int J Basic Appl Sci 13(3): 23–40. https://doi.org/10.5281/zenodo.17046078
Munguti J, Muthoka M, Kyule D, Nzioka A, Ogello E, Magondu EW, Mukami M, Wainaina M (2025) Current status and production trends of commercially cultured mariculture species along the Kenyan coast: Challenges, opportunities and future prospects. Aquacult Fish Fish 5(5): e70118. https://doi.org/10.1002/aff2.70118
Nadeeshani H, Hassouna A, Lu J (2022) Proteins extracted from seaweed Undaria pinnatifida and their potential uses as foods and nutraceuticals. Crit Rev Food Sci Nutr 62(22):6187–6203. https://doi.org/10.1080/10408398.2021.1898334
Nakhate P, van der Meer Y (2024) Global initiatives, future challenges, and trends for the wider use of seaweed. In: Ozogul F, Trif M, Rusu A (eds) Seaweeds and seaweed-derived compounds: meeting the growing need for healthy biologically active compounds. Springer International Publishing, pp 521–544. https://doi.org/10.1007/978-3-031-65529-6_19
Naseri A, Marinho GS, Holdt SL, Bartela JM, Jacobsen C (2020) Enzyme-assisted extraction and characterization of protein from red seaweed *Palmaria palmata*. Algal Res 47:101849. https://doi.org/10.1016/j.algal.2020.101849
Ngajilo D, Adams S, Baatjies R, Jeebhay MF (2025) Seaweed farming in Tanzania: A study of working practices and geospatial analysis of occupational hazards and adverse health outcomes. J Agromed 30(2):237–249. https://doi.org/10.1080/1059924X.2024.2407986
Olsson J, Toth G, Albers E (2020) Biochemical composition of red, green and brown seaweeds on the Swedish west coast. J Appl Phycol 32. https://doi.org/10.1007/s10811-020-02145-w
Pereira L (2018) Preface—therapeutic and nutritional uses of algae, pp v–xxxi. https://doi.org/10.1201/9781315152844
Permani R, Muflikh YN, Sjahruddin F (2024) Mapping the complex web of policies for seaweed industry development in Indonesia: what is the role of a national roadmap? Ocean Coastal Manage 259:107464. https://doi.org/10.1016/j.ocecoaman.2024.107464
Polat S, Trif M, Rusu A, Šimat V, Čagalj M, Alak G, Meral R, Özogul Y, Polat A, Özogul F (2023) Recent advances in industrial applications of seaweeds. Crit Rev Food Sci Nutr 63(21):4979–5008. https://doi.org/10.1080/10408398.2021.2010646
Pradhan B, Bhuyan PP, Ki J-S (2023) Immunomodulatory, antioxidant, anticancer, and pharmacokinetic activity of Ulvan, a seaweed-derived sulfated polysaccharide: an updated comprehensive review. Mar Drugs 21(5):300. https://doi.org/10.3390/md21050300
Peñalver R, Lorenzo JM, Ros G, Amarowicz R, Pateiro M, Nieto G (2020) Seaweeds as a Functional Ingredient for a Healthy Diet. Mar Drugs, 18(6): 301. https://doi.org/10.3390/md18060301
Ramu M, Chavadi M (2020) Biotechnology of microalgae: a green approach towards exploitation of Omega-3 fatty acids. https://doi.org/10.1201/9781003019510-12
Reynolds CK, Humphries DJ, Künzel S, Rodehutscord M, Lignou S, Fagan CC, Methven L, Norton G, Alzahrani A, Feldmann J, Desnica N, Gunnlaugsdottir H, Pétursdóttir ÁH (2025) Effects of feeding dairy cows seaweed mixtures on feed intake, methane emission, milk production, and milk mineral and fatty acid composition. J Dairy Sci 108(9):9487–9500. https://doi.org/10.3168/jds.2025-26577
Rimmer MA, Larson S, Lapong I, Purnomo AH, Pong-Masak PR, Swanepoel L, Paul NA (2021) Seaweed aquaculture in Indonesia contributes to social and economic aspects of livelihoods and community wellbeing. Sustainability 13(19):10946. https://doi.org/10.3390/su131910946
Rocha CP, Pacheco D, Cotas J, Marques JC, Pereira L, Gonçalves AMM (2021) Seaweeds as valuable sources of essential fatty acids for human nutrition. Int J Environ Res Public Health 18(9):4968. https://doi.org/10.3390/ijerph18094968
Salehi B, Sharifi-Rad J, Seca AML, Pinto DCGA, Michalak I, Trincone A, Mishra AP, Nigam M, Zam W, Martins N (2019) Current trends on seaweeds: looking at chemical composition, phy-

topharmacology, and cosmetic applications. Molecules 24(22):4182. https://doi.org/10.3390/molecules24224182
Shafie MH, Kamal ML, Zulkiflee FF, Hasan S, Uyup NH, Abdullah S, Mohamed Hussin NA, Tan YC, Zafarina Z (2022) Application of Carrageenan extract from red seaweed (*Rhodophyta*) in cosmetic products: a review. J Indian Chem Soc 99(9):100613. https://doi.org/10.1016/j.jics.2022.100613
Souza A (2024) Phytoplankton dynamics in a cultivated kelp farm off the coast of Namibia. Master Thesis. https://studenttheses.uu.nl/handle/20.500.12932/47504
Spillias S, Valin H, Batka M, Sperling F, Havlík P, Leclère D, Cottrell R, O'Brien K, McDonald-Madden E (2023) Reducing global land-use pressures with seaweed farming. Nat Sustainability 6:1–11. https://doi.org/10.1038/s41893-022-01043-y
Subbiah V, Xie C, Dunshea FR, Barrow CJ, Suleria HAR (2023) The quest for phenolic compounds from seaweed: nutrition, biological activities and applications. Food Rev Intl 39(8):5786–5813. https://doi.org/10.1080/87559129.2022.2094406
Sultana F, Wahab MA, Nahiduzzaman M, Mohiuddin M, Iqbal MZ, Shakil A, Mamun A-A, Khan MSR, Wong L, Asaduzzaman M (2023) Seaweed farming for food and nutritional security, climate change mitigation and adaptation, and women empowerment: a review. Aqua Fish 8(5):463–480. https://doi.org/10.1016/j.aaf.2022.09.001
Summers N, Fragoso GM, Johnsen G (2023) Photophysiologically active green, red, and brown macroalgae living in the Arctic polar night. Sci Rep 13(1):17971. https://doi.org/10.1038/s41598-023-44026-5
Spagnuolo D, Genovese G (2024) Macroalgal diseases: Exploring biology, pathogenesis, and management strategies. Phycology, 4(3), 450–464. https://doi.org/10.3390/phycology4030026
Takaichi S (2025) Distribution, biosynthesis, and function of carotenoids in oxygenic phototrophic algae. Mar Drugs 23(2):62. https://doi.org/10.3390/md23020062
Thiviya P, Gamage A, Gama-Arachchige NS, Merah O, Madhujith T (2022) Seaweeds as a source of functional proteins. Phycology 2(2):216–243. https://doi.org/10.3390/phycology2020012
Webb P, Somers NK, Thilsted SH (2023) Seaweed's contribution to food security in low- and middle-income countries: benefits from production, processing and trade. Glob Food Sec 37:100686. https://doi.org/10.1016/j.gfs.2023.100686
Windarto S, Herawati VE, Wijaya YJ, Indriati DA, Rachmasari Y, Elfitasari T (2025) Exploring the physicochemical properties and nutritional value of abundant seaweed species along the Jepara Coast, Indonesia. Thalassas: Int J Marine Sci 41(2):82. https://doi.org/10.1007/s41208-025-00846-y
World Bank (2023) Global seaweed: new and emerging markets report. World Bank Publications, Washington, DC
Xie C, Lee ZJ, Ye S, Barrow CJ, Dunshea FR, Suleria HAR (2024) A review on seaweeds and seaweed-derived polysaccharides: nutrition, chemistry, bioactivities, and applications. Food Rev Intl 40(5):1312–1347. https://doi.org/10.1080/87559129.2023.2212055
Yadav P (2024) Harvesting the future: the growing industry of SeaweedFfarming and its sustainable contributions. Scieng Publications, pp 22–27
Yarish C, Brummett R, Hansen S, Bjerregaard R, Valderrama D, Sims N, Radulovich R, Diana J, Capron M, Forster J, Goudey C, Hopkins K, Rust M, McKinnie C (2016) Seaweed aquaculture for food security, income generation and environmental health in tropical developing countries
Zhukova NV, Yakovleva IM (2021) Low light acclimation strategy of the brown macroalga Undaria pinnatifida: significance of lipid and fatty acid remodeling for photosynthetic competence. J Phycol 57(6):1792–1804. https://doi.org/10.1111/jpy.13209

Chapter 11
Harnessing the Food and Raw Material Potential of *Salicornia* Species in Africa

Martha K. Hausiku-Ikechukwu and Hupenyu Allan Mupambwa

Abstract *Salicornia* spp., a genus of halophytic plants adapted to saline environments, present a multifaceted opportunity to address food security and environmental challenges in Africa. This chapter describes the nutritional, therapeutic, industrial, and ecological applications of this plant, highlighting its value as a mineral-rich vegetable and dietary salt substitute enriched with bioactive phytochemicals, such as polyphenols and flavonoids. The medicinal properties of *Salicornia* include antimicrobial, antioxidant, anti-inflammatory, cardiovascular, and metabolic effects, with emerging relevance for pharmaceutical and nutraceutical development. Industrially, its high biomass and oil-rich seeds offer the potential for the production of biofuels, biodegradable plastics, cosmetics, and fodder. Environmentally, *Salicornia* contributes to soil desalination, phytoremediation, carbon sequestration, biofiltration, and wastewater treatment, reinforcing its role in sustainable agroecosystems. Although extensive studies from Europe, the Middle East, and Asia have elucidated these attributes, empirical research within the African context remains sparse and constrained by taxonomic ambiguities, inadequate agronomic protocols, underdeveloped value chains, research funding deficits, and policy gaps. This chapter underscores that the direct extrapolation of foreign data is insufficient; unlocking *Salicornia*'s potential in Africa necessitates germplasm collection, breeding of locally adapted varieties, comprehensive nutritional profiling, and region-specific clinical validation of health benefits. Ultimately, this work advocates for coordinated, continent-wide research, policy support, and market development to harness *Salicornia* spp. as a climate-resilient crop that contributes to food security, economic development, and environmental sustainability in Africa.

Keywords *Salicornia* · Halophytes · Food Security · Climate Resilience · Saline Agriculture · Nutraceuticals · Phytoremediation · Biofuel · Value Chains

M. K. Hausiku-Ikechukwu (✉) · H. A. Mupambwa
Sam Nujoma Marine and Coastal Resources Research Centre, Sam Nujoma Campus, University of Namibia, Henties Bay, Namibia
e-mail: mikechukwu@unam.na

H. A. Mupambwa et al. (eds.), *Harnessing Coastal and Marine Resources for Improved Food Security in Africa*, Sustainability Sciences in Asia and Africa,
https://doi.org/10.1007/978-981-95-9725-3_11

11.1 Introduction

Food security in Africa is increasingly threatened by the nexus of climate change, land degradation, and population growth. Conventional crops often fail in saline and arid environments, which are expanding across the continent. This reality necessitates a strategic pivot towards resilient, non-traditional crops. Halophytic (salt-tolerant) plants are gaining prominence as sustainable food and feed resources, with *Salicornia* spp., commonly known as sea asparagus or pickleweed, emerging as a particularly promising genus (Ekanayake et al. 2023). For Africa, *Salicornia* represents not merely a novel crop but also a potential paradigm shift in utilising marginal lands for sustainable production.

Globally, *Salicornia* is recognised for its nutritional value and is rich in essential minerals, vitamins, and bioactive compounds (Cardoso et al. 2022). Its applications extend beyond food to animal feed, biofuels, and phytoremediation (Chaturvedi et al. 2021; Fredsgaard et al. 2023). However, a critical analysis of the literature revealed a significant geographical bias. The vast majority of empirical studies on agronomy, nutritional profiling, and commercial cultivation originate in Europe, the Middle East, and Asia. This chapter, therefore, has a dual purpose: to synthesise global knowledge of *Salicornia* and to critically examine its specific potential and challenges within the African context, explicitly highlighting where research and data are most urgently needed.

11.2 Botanical and Agronomic Characteristics: Foundations for African Cultivation

Salicornia spp. are succulent halophytic annual herbs within the Amaranthaceae family that are anatomically adapted for survival in saline habitats, such as salt marshes and inland salt pans. Their distinctive morphological features include highly reduced, jointed, and photosynthetic stems that appear leafless, with true leaves existing on tiny, fleshy scales (Narasimha Rao and Pragada 2013). This simplified structure, coupled with significant phenotypic plasticity, has resulted in considerable taxonomic complexity, with an estimated 25–30 species frequently subject to misidentification (Tuğ and Başköse 2023). This persistent taxonomic confusion presents a primary barrier in Africa. Without accurate species delineation, breeding programs and agronomic recommendations are built upon an unstable foundation, hindering the development of optimised cultivation systems.

Ecologically, *Salicornia* species are hygrohalophytes that thrive across a spectrum of saline environments, from coastal marshes to inland salt lakes, spanning boreal to subtropical regions, including parts of southern Africa (Orzoł et al. 2025). Their tolerance to hypersalinity and periodic immersion is facilitated by key physiological adaptations, such as salt excretion and succulent water storage in specialised tissues (Mendis et al. 2025). The genus comprises both diploid and tetraploid lineages, with evidence suggesting that tetraploids often exhibit a competitive advantage in the most saline zones of a marsh (Chatrenoor and Akhani 2021). This inherent resilience makes *Salicornia* a promising candidate for the phytoremediation and rehabilitation of Africa's extensively degraded and salinized lands. However, successful cultivation is contingent on species-specific thresholds of moisture, salinity, and temperature (Giménez et al. 2024; Katel et al. 2023).

Agronomically, *Salicornia* is characterised as a low-input crop that is well suited to marginal environments. Plants require minimal fertilisation due to their adaptation to nutrient-poor saline soils, although they demonstrate positive responses to organic amendments and specific micronutrients (Giménez et al. 2024). The high salinity conditions that support *Salicornia* growth naturally suppress weed competition, whereas pest pressures remain generally limited, with occasional damage from lepidopteran larvae representing a primary concern (Davy et al. 2001). A defining agronomic advantage is the plant's ability to complete its life cycle using saline or brackish water irrigation, constituting a significant advantage for water-scarce regions across Africa. Harvest management is determined by the intended use, with young succulent stems collected for culinary markets, whereas mature plants are harvested for oil-rich seeds in designated varieties (Ventura et al. 2011).

Although pilot projects in Tunisia and other regions demonstrate fundamental feasibility (Katel et al. 2023), the central implication for Africa is that the universal cultivation protocols developed in other continents will be inadequate. The performance of species from the Middle East cannot be directly extrapolated to the Sahel or Southern African contexts. Scaling production, therefore, necessitates addressing critical continent-specific research priorities, beginning with comprehensive germplasm collection to identify and characterise indigenous African species and ecotypes possessing desirable adaptive traits. Concurrently, irrigation management should be optimised for the specific hydrogeological conditions of major African basins, and participatory breeding approaches should engage local farmers in developing varieties aligned with regional market preferences and farming systems. Advancing genetic improvement remains paramount, although it is still in the early developmental stages globally. Taxonomic complexity and limited genomic resources complicate traditional breeding, necessitating greater reliance on molecular tools for proper germplasm characterisation and marker-assisted selection (Collard and Mackill 2008; Jamdade et al. 2022). Therefore, strategic investments in foundational genomic infrastructure, comprehensive germplasm banks, and collaborative breeding networks are essential prerequisites for unlocking the genetic potential of *Salicornia* for African food security and environmental applications.

Salicornia spp. represents a unique bio-resource with transformative potential to address interconnected challenges of food security, economic development, and environmental sustainability across Africa. The plant's diverse applications form a synergistic framework of benefits, as illustrated in Fig. 11.1, which categorises its contributions into four interconnected domains directly relevant to African development priorities: Nutritional & Culinary Applications, Industrial & Economic Applications, Therapeutic & Pharmaceutical Applications, and Environmental Benefits & Ecosystem Services. This integrated value proposition positions

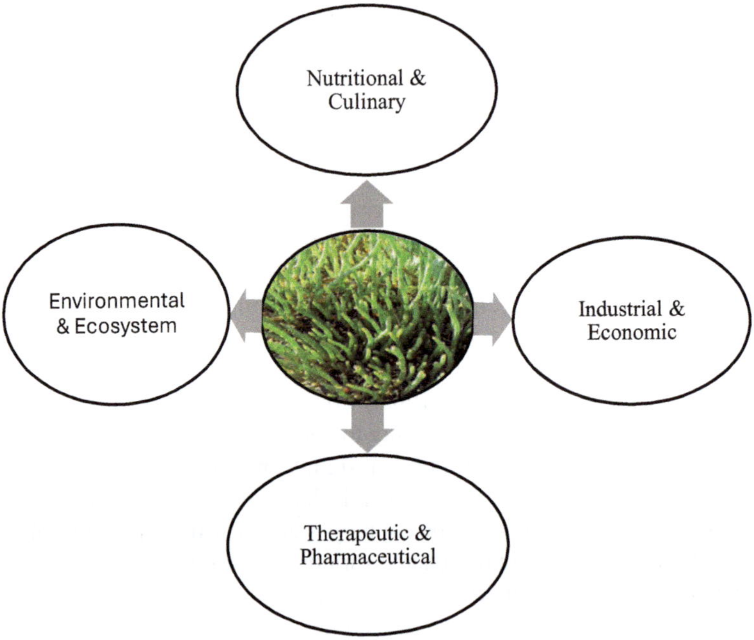

Fig. 11.1 Potential applications of *Salicornia* in Africa

Salicornia not merely as an alternative crop, but also as a foundational component of climate-resilient bio-economies across the continent.

11.3 Nutritional, Culinary, and Therapeutic Value: Bridging Potential and Application

Salicornia spp. presents a compelling profile as a functional food and source of therapeutic compounds, offering multifaceted benefits that are highly relevant to African nutritional and public health challenges. The plant's nutritional composition, detailed in Table 11.1, reveals a low-calorie food source rich in dietary fibre, essential amino acids, and a remarkable density of critical minerals (Harboub et al. 2025; Kamal et al. 2024). Its high concentrations of iron and zinc are particularly significant, directly addressing the widespread deficiencies of these micronutrients that contribute to anaemia and impaired immune function across the continent, especially among vulnerable groups. However, a critical analysis of the literature reveals substantial variation in reported mineral values—iron content can range from 2 to 4 mg/100 g and iodine from 5 to 50 μg/100 g—highlighting that composition is heavily influenced by species, geography, and edaphic conditions (Alfheeaid et al. 2022; Harboub et al. 2025; Kamal et al. 2024). This inconsistency underscores that promoting *Salicornia* as a uniform nutritional solution is premature; a strategic

Table 11.1 Nutritional Composition of *Salicornia* spp. (per 100 g fresh weight)

Nutrient component	Composition	Significance for African nutrition
Energy & macronutrients	Energy: 22–30 kcal; Carbohydrates: 4–5 g; Dietary Fiber: ~3 g; Protein: 2–3 g; Fat: <0.5 g	Provides low-calorie nutrition with satiating fibre and plant-based protein.
Key minerals	Sodium: 900–1500 mg; Potassium: 150–300 mg; Iron: 2–4 mg; Zinc: 1–2 mg; Calcium: 50–100 mg; Iodine: 5–50 μg	Targets widespread mineral deficiencies (iron, zinc); high sodium offers salt substitution potential.
Vitamins	Vitamin A (as carotenoids): 100–300 μg; Vitamin C: 20–50 mg; B Vitamins (B1, B2, Folate): 0.1–0.5 mg	Addresses vitamin A deficiency (VAD); supports immune function and metabolic health.

Sources: Harboub et al. (2025), Kamal et al. (2024), Alfheeaid et al. (2022)

Table 11.2 Culinary applications of *Salicornia* spp.

Application	Preparation method	Potential African integration
Fresh/cooked vegetable	Sautéed, steamed, or stir-fried as a side dish.	Seasonal green for coastal communities; niche vegetable for urban markets and agro-tourism.
Seafood garnish/ accompaniment	Used fresh to provide a saline contrast to fish and shellfish.	Direct application in fishing communities along Africa's extensive coastline.
Flavouring agent & salt substitute	Dried and powdered as "green salt" or used in soups/stews.	Sustainable, local alternative to imported salt; enhances mineral intake in staple dishes.

Sources: Cardoso et al. (2022), Patel (2016)

research imperative for Africa should include the active selection and cultivation of genotypes that not only thrive in local saline environments but also demonstrate reliable accumulation of targeted micronutrients.

The culinary applications of *Salicornia*, summarised in Table 11.2, leverage its characteristic salty taste and crisp texture, thereby providing versatile opportunities for integration into African food systems. Its inherent salinity positions it as a natural, nutrient-dense salt alternative or flavour enhancer, which could help reduce reliance on imported processed seasonings while simultaneously boosting mineral intake. Practical integration strategies include its use as a fresh or cooked vegetable in coastal communities, a garnish for traditional seafood dishes, and processing into a mineral-rich "green salt" powder (Harboub et al. 2025; Kim et al. 2021; Patel 2016). High sodium content (Alfheeaid et al. 2022; Kamal et al. 2024), while beneficial as a salt substitute, necessitates the development of public health guidance regarding its appropriate use, particularly for population groups with hypertension or specific sodium-sensitive health conditions.

Beyond its direct nutritional value, *Salicornia* possesses a rich repertoire of bioactive compounds, including polyphenols, flavonoids, and saponins, that underpin

Table 11.3 Bioactive compounds and documented therapeutic effects of *Salicornia* spp.

Bioactive compound class	Documented health effects	Research status & considerations
Polyphenols & flavonoids	Potent antioxidant and anti-inflammatory activities; hepatoprotective effects.	Strong in vitro evidence; limited human clinical trials.
Saponins	Anti-inflammatory, cholesterol-lowering, and potential antidiabetic properties.	Bioavailability and effective dosing in humans require validation.
Complex phytochemical spectrum	Antimicrobial, cardioprotective, and wound-healing potential.	Mostly preliminary studies; mechanisms of action not fully elucidated.

Sources: Alfheeaid et al. (2022), Limongelli et al. (2022), Fredsgaard et al. (2023)

its documented in vitro pharmacological activities. These compounds, detailed in Table 11.3, exhibit antioxidant, anti-inflammatory, antidiabetic, and hepatoprotective properties in laboratory studies (Fredsgaard et al. 2023; Limongelli et al. 2022). However, there is a significant translational gap between these promising in vitro results and practical human health applications. Most pharmacological evidence is derived from concentrated extracts rather than whole-plant consumption, and critical questions regarding bioavailability, dosing, and efficacy in human subjects remain largely unanswered.

For Africa, this evidence gap presents both a caution against overstatement and a significant opportunity. The high cost of imported pharmaceuticals and the growing burden of non-communicable diseases make local sources of evidence-based nutraceuticals highly attractive. Therefore, research priority should shift toward validating health claims through well-designed clinical trials using African-sourced *Salicornia*. Initial research should strategically focus on its potential role in managing widespread conditions, such as type 2 diabetes and hypertension, thereby moving from traditional use and laboratory promise toward evidence-based, locally relevant health solutions. The translation of these bioactive constituents into practical applications, ranging from nutraceuticals and functional foods to cosmeceuticals, will require parallel investment in optimising extraction protocols, standardising preparations, and understanding mechanisms of action to ensure consistent safety and efficacy, as illustrated in Fig. 11.2.

Therefore, the integration of *Salicornia* into African food and health systems requires a dual-track approach: first, the strategic development of its nutritional potential through targeted breeding and culinary integration, and second, the systematic investigation of its therapeutic properties through rigorous, locally relevant clinical research. This comprehensive strategy would maximise the potential of the plant to simultaneously contribute to improved nutrition, economic development through value-added products, and accessible health solutions tailored to African needs.

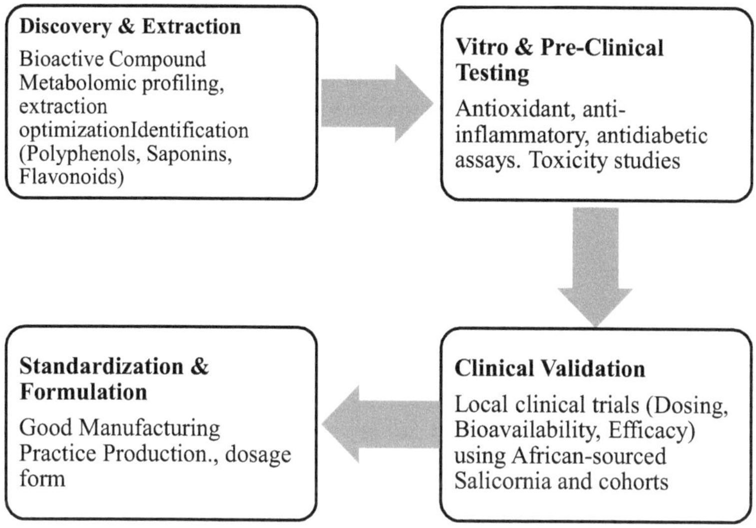

Fig. 11.2 Translational research pathway for therapeutic applications of *Salicornia*

Table 11.4 Pharmaceutical Development Pathways for *Salicornia* spp.

Application area	Bioactive components	development status & considerations
Nutraceuticals & functional foods	Polyphenols, saponins, minerals, fibre	Ready for product development; requires standardisation and clinical validation for specific health claims
Cosmeceuticals & dermatological	Flavonoids, polyphenols, minerals	Formulation development underway; efficacy testing needed for commercial products
Therapeutic candidates	Flavonoids, saponins, phenolic compounds	Preclinical investigation stage; requires mechanism studies and safety profiling
Biopharmaceutical leads	Extracted metabolites	Early discovery phase; needs compound identification and target validation

Sources: Limongelli et al. (2022), Nájar et al. (2023), Al (2025)

11.4 Pharmaceutical and Industrial Applications: From Bioactive Compounds to Economic Value

The diverse bioactivity of *Salicornia* components provides a robust foundation for their translation into pharmaceutical and industrial products, representing significant economic opportunities for African development. Its pharmacological potential spans multiple application areas, as detailed in Table 11.4, ranging from preventive health supplements to targeted therapeutic formulations. In the nutraceutical sector, standardised extracts rich in polyphenols and saponins show promise for developing supplements aimed at antioxidant and anti-inflammatory support (Limongelli et al. 2022; Nájar et al. 2023), while the plant's unique mineral profile supports its use in functional foods as a natural salt substitute and ingredient in antidiabetic dietary plans (Al 2025; Chrigui et al. 2023). Dermatological applications are equally promising, with flavonoids and polyphenols being investigated as cosmeceuticals targeting anti-aging, UV protection, and skin moisturising (Limongelli et al. 2022; Zhao et al. 2014).

However, the transition from laboratory findings to clinically validated products requires addressing significant research gaps (Fig. 11.3). Although in vitro studies have demonstrated compelling anticancer activity against specific cell lines and antimicrobial properties against various pathogens (Al 2025; Limongelli et al. 2022), these effects are primarily derived from concentrated extracts rather than whole-plant preparations. The bioavailability of these compounds in humans and their effective therapeutic dosages remain poorly characterised. For African applications, the priority should be to validate these pharmacological properties through clinical trials using locally sourced *Salicornia* material, with an initial focus on conditions of high regional prevalence, such as metabolic disorders and infectious diseases.

Beyond pharmaceutical applications, *Salicornia* offers substantial industrial potential through a cascading utilisation model that aligns with the principles of circular bioeconomy. Seed oil serves as a non-competing feedstock for biodiesel and sustainable aviation fuel (AlYammahi et al. 2024), whereas residual seed cake provides a protein-rich supplement for animal feed (Cárdenas-Pérez et al. 2021), although determining safe inclusion rates for African livestock breeds represents a critical research need. The high biomass yield supports applications in biocomposites and biodegradable materials (Fredsgaard et al. 2023), reducing the dependence

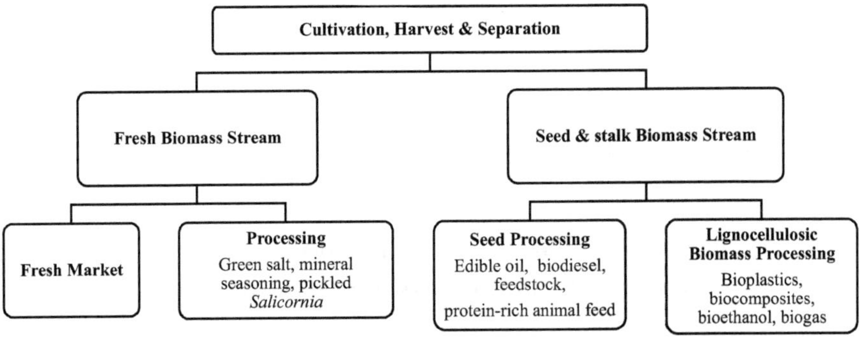

Fig. 11.3 Industrial value chain development of *Salicornia* spp. Economic modelling for *Salicornia* cultivation in Africa requires careful localisation rather than direct transfer of data from other regions. While international markets offer premium prices for fresh shoots (USD 10–20/kg) and specialised products (Ekanayake et al. 2023), developing viable African value chains demands the accurate determination of local production costs, achievable yields under African growing conditions, and domestic market demand patterns. Strategic development should prioritise establishing processing infrastructure for multiple product streams, creating mechanisms for equitable benefit-sharing with smallholder farmers, and building partnerships between research institutions and private sector entities to drive innovation along the entire value chain

on petroleum-based alternatives. Perhaps most significantly for African environmental challenges, *Salicornia*'s capacity for phytoremediation enables the rehabilitation of saline and degraded soils, transforming unproductive lands into agricultural assets while simultaneously generating harvestable biomass.

The integration of pharmaceutical and industrial applications creates a powerful economic model in which high-value health products subsidise broader cultivation efforts, whereas industrial applications ensure complete biomass utilisation. This integrated approach maximises economic returns while supporting environmental restoration and positioning *Salicornia* as a cornerstone species for sustainable development strategies across Africa's marginal lands.

11.5 Pharmaceutical and Industrial Applications: Translating Bioactivity into Economic Value

The diverse bioactivity profile of *Salicornia* spp. establishes a compelling foundation for its development across the pharmaceutical and industrial sectors, representing significant economic opportunities for African nations (Table 11.5). Its pharmacological potential spans multiple therapeutic categories, from preventive health formulations to targeted therapeutic applications. Standardised extracts rich in polyphenols and saponins show particular promise for developing nutraceutical supplements targeting oxidative stress and inflammatory conditions, while the plant's distinctive mineral composition supports its application in functional foods as both a salt substitute and an anti-diabetic dietary component. The convergence of flavonoids and polyphenols with essential minerals further enables the dermatological applications of cosmeceuticals focused on anti-aging, photoprotection, and skin barrier enhancement.

However, the translational pathway from laboratory evidence to clinically validated products present significant methodological challenges. Current pharmacological data are primarily derived from in vitro studies using concentrated extracts (Al 2025; Limongelli et al. 2022), creating uncertainty regarding bioavailability,

Table 11.5 Pharmaceutical Development Pathways for *Salicornia* spp.

Application area	Bioactive components	Development status & considerations
Nutraceuticals & functional foods	Polyphenols, saponins, minerals, fibre	Requires standardisation protocols and clinical validation for health claims
Cosmeceuticals & dermatological	Flavonoids, polyphenols, minerals	Needs formulation optimisation and efficacy testing for commercial viability
Therapeutic candidates	Flavonoids, saponins, phenolic compounds	Requires mechanism of action studies and comprehensive safety profiling
Biopharmaceutical leads	Extracted metabolites	Demands compound identification and target validation studies

Sources: Limongelli et al. (2022), Nájar et al. (2023), Al (2025)

therapeutic dosing, and efficacy in humans. For African applications, research priorities should include clinical validation using locally sourced materials, with an initial focus on conditions that demonstrate a high regional disease burden, particularly metabolic disorders and infectious diseases. The establishment of standardised extraction protocols and safety profiles is an essential prerequisite for advancing these applications to the commercial stage.

Beyond pharmaceutical applications, *Salicornia* demonstrates substantial potential within industrial sectors through an integrated cascading utilisation model that aligns with circular economy principles. Seed oil provides a non-competitive feedstock for biodiesel and sustainable aviation fuel production, whereas residual seed meal offers protein-rich supplementation for animal feed formulations, although determining optimal inclusion rates for indigenous African livestock breeds remains a critical research requirement. The substantial biomass production supports downstream applications in bio-composites and biodegradable polymers, reducing the dependency on conventional petroleum-based materials. Most significantly within the African context, *Salicornia*'s demonstrated phytoremediation capabilities enable simultaneous rehabilitation of degraded saline soils and generation of valuable biomass, effectively transforming environmental liabilities into productive assets.

Economic modelling of *Salicornia* cultivation in Africa necessitates context-specific adaptation rather than direct extrapolation from international data. While global markets demonstrate premium pricing for fresh shoots (USD 10–20/kg) and specialised derivatives, establishing viable African value chains requires accurate quantification of localised production parameters, including region-specific yield potentials, operational cost structures, and domestic market absorption capacity. Strategic development should emphasise creating a processing infrastructure capable of handling multiple product streams, implementing equitable benefit-sharing mechanisms for smallholder participants, and fostering innovation ecosystems through research-private sector partnerships.

The synergistic integration of pharmaceutical and industrial applications establishes an economically resilient framework in which high-value health products support cultivation expansion, whereas industrial applications ensure comprehensive biomass utilisation. This integrated approach optimises economic returns while advancing environmental restoration objectives, positioning *Salicornia* as a foundational component of sustainable development strategies for Africa's marginal lands. However, the successful realisation of this potential depends on establishing robust translational research pipelines and developing context-appropriate business models that address both economic and ecological considerations simultaneously.

11.6 Environmental Benefits, Ecosystem Services, and Adoption Challenges

Salicornia spp. offers critical environmental services that extend far beyond agricultural production, offering nature-based solutions to Africa's most pressing ecological challenge (Table 11.6). Through specialised physiological adaptations, including

Table 11.6 Environmental benefits and ecosystem services of *Salicornia* spp.

Environmental benefit	Mechanism & impact	Relevance to African context
Soil rehabilitation	Salt excretion and ion compartmentalisation enable growth in hypersaline soils, reclaiming degraded lands	Addresses soil salinisation affecting approximately 80 million hectares of African farmland
Carbon sequestration	High biomass production captures atmospheric carbon in both plant tissue and soil organic matter	Contributes to climate change mitigation while improving soil fertility
Coastal protection	Dense root systems stabilise sediments and buffer wave energy, reducing shoreline erosion	Protects vulnerable coastal communities and infrastructure from sea-level rise and storm surges
Water purification	Nutrient uptake from wastewater and contaminated water sources (85% nitrogen, 73% phosphorus removal)	Low-cost biofiltration solution for agricultural runoff and municipal wastewater
Biodiversity support	Provides habitat for 40+ bird species, aquatic fauna, and halophytic plant communities	Enhances ecological resilience in degraded coastal and inland saline ecosystems

Sources: Katel et al. (2023), Ventura et al. (2011), Galparsoro et al. (2014)

salt excretion and ion compartmentalisation, these halophytes demonstrate a remarkable capacity for phytoremediation of saline and degraded soils, effectively rehabilitating lands otherwise unsuitable for conventional agriculture (Katel et al. 2023). Their ability to accumulate heavy metals and remove pollutants, such as selenium, from contaminated soils and water systems further enhances their remediation potential (Katel et al. 2023; Ventura et al. 2011). These characteristics, combined with rapid biomass production that enriches soil organic matter and improves water retention, make *Salicornia* a multifunctional tool for ecosystem restoration.

The ecosystem services provided by *salicornia*-dominated habitats contribute significantly to ecological stability and climate resilience. These saline ecosystems function as vital habitats for diverse fauna, including waterfowl, shorebirds, and commercially important fish species, while simultaneously supporting nutrient cycling and controlling eutrophication through efficient nutrient uptake (Galparsoro et al. 2014; Katel et al. 2023). For African nations, particularly vulnerable to climate change impacts, these services represent invaluable natural capital that justifies public investment in *Salicornia*-based restoration initiatives.

11.7 Challenges to the Adoption of *Salicornia* spp.

The potential of *Salicornia* to contribute to food security, create economic opportunities, and rehabilitate saline-degraded soils in Africa is significant. However, its development is trapped in a self-reinforcing cycle of underutilisation, shaped by the

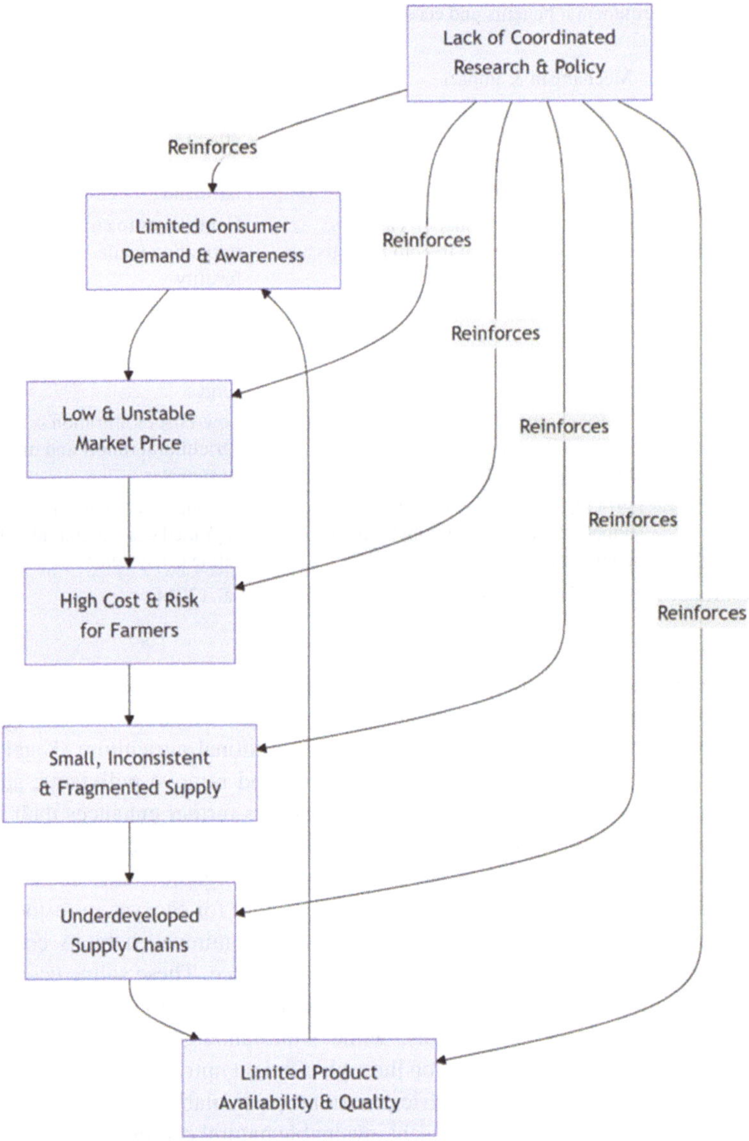

Fig. 11.4 Challenges hindering the exploitation of *Salicornia* spp. for improved food security in Africa

continent's unique challenges (Fig. 11.4). The cycle begins with limited consumer awareness and low perceived value. Outside of a few localised communities, particularly in North and West Africa, Salicornia is largely unknown. Urban and emerging middle-class consumers, who drive market trends, are unfamiliar with its culinary use and nutritional benefits (Lago et al. 2021). This absence of market "pull" makes retailers and restaurants hesitant to dedicate valuable shelf space or menu items to an unknown product, which directly suppresses its market price and stability. Without a receptive market, Salicornia cannot achieve the economies of scale necessary to transition from a niche novelty to a mainstream commodity, resulting in prices that are too low and volatile to attract large-scale investment (Lombardi et al. 2022; Lyra et al. 2021). This directly results in unstructured markets and non-existent value chains (Chaturvedi et al. 2021; Ranjbar et al. 2022). Unlike established commodities, there are no formal trading platforms, price benchmarks, or organised off-take agreements for Salicornia (Bazihizina et al. 2024; Hasnain et al. 2023a).

The absence of a structured market translates into low and unpredictable incomes for harvesters, which perpetuates its status as a subsistence resource rather than a viable cash crop. This high commercial risk discourages investment, leading to high perceived agronomic risk for smallholder farmers (Rwema et al. 2025). African farmers, often operating with minimal capital, are risk-averse (Simutowe et al. 2024). Without proven, context-specific cultivation techniques for African saline soils and water, and no guarantee of a reliable market, farmers are hesitant to allocate precious land and labour to Salicornia instead of familiar subsistence crops (Hasnain et al. 2023b). This is compounded by a lack of access to quality seeds or planting materials tailored to local conditions (Garza-Torres et al. 2020). Consequently, the supply remains fragmented, informal, and based on wild harvest. Reliance on wild populations leads to unsustainable harvesting pressures, inconsistent quality, and unpredictable availability (Khan et al. 2016). This fragmented supply cannot meet the volume and consistency requirements for entry into larger retail markets or for use as a raw material by food processors. This, in turn, prevents the development of critical post-harvest infrastructure and processing capabilities. The lack of investment in cold chains, transportation, cleaning, and grading facilities means that the highly perishable fresh product has a short shelf-life and poor market presentation. Furthermore, there is a severe lack of small- to medium-scale processing facilities to create value-added products (e.g., dried spices, pickles, or oil) that could reduce post-harvest losses and reach wider markets (Patel 2016).

The result is minimal product availability and poor quality in formal markets. Even if consumer interest were to grow, formal retail outlets would be unable to source Salicornia reliably. The product that does reach distant markets is often of poor quality, reinforcing the low perception of its value and closing the negative feedback loop of low demand. Underpinning this entire cycle is a critical cross-cutting barrier: a profound gap in context-specific research, policy, and institutional support. Research on Salicornia in Africa is scarce and often not translational for smallholder contexts. There is an urgent need for research into domestication, agronomic practices, and low-tech processing suitable for African conditions (Hasnain et al. 2023b). Moreover, agricultural extension services are not equipped with knowledge on halophyte cultivation, and government policies rarely recognise or support the development of saline agriculture or niche crops, focusing instead on staple foods (Khan et al. 2016; Kule et al. 2025). This institutional vacuum leaves individual farmers, entrepreneurs, and communities without the technical or financial support needed to break the cycle, locking a promising resource into a state of underutilisation.

Breaking this cycle of underutilisation requires a coordinated, multi-pronged strategy that addresses both technical and socioeconomic constraints simultaneously. A strategic framework should provide a roadmap for building a supportive ecosystem for *Salicornia* development in Africa, emphasising the need for parallel progress in research, policy, and market development to realise the full environmental and economic potential of this promising halophyte.

11.8 Strategic Framework for Advancing *Salicornia* spp. in Africa

Overcoming the multifaceted barriers to *Salicornia* adoption requires a coordinated multistakeholder approach that integrates scientific innovation, strategic policy, and market development (Table 11.7). Since the challenges are interconnected, simultaneous interventions across the research, infrastructure, finance, and policy domains are necessary to break the cycle of underutilisation and establish a viable bioeconomy around this halophyte.

A strategic framework for advancing *Salicornia* in Africa should prioritise targeted investments along the entire value chain (Table 11.7). Critical research priorities include optimising saline irrigation management, improving seed quality, and developing pest control strategies adapted to African conditions (Singh et al. 2014). Genetic improvement of *Salicornia* is advancing rapidly through genomic tools that identify genes associated with salt tolerance, yield, and nutritional quality. Breeding programs should leverage both conventional and biotechnological approaches, including genomic tools and CRISPR, to enhance yield, stress tolerance, and nutritional content while ensuring the development of locally adapted varieties (Siridewa

Table 11.7 Integrated strategy for *Salicornia* development in Africa

Strategic pillar	Key interventions	Implementation framework
Research & innovation	Germplasm collection; precision irrigation; genomic breeding; participatory adaptation trials	Establish regional research consortia; develop technology transfer protocols
Policy & governance	Integration into climate-smart policies; financial incentives; food safety standards; environmental safeguards	Multi-ministerial task forces; community engagement mechanisms; monitoring systems
Market systems	Value chain development; farmer cooperatives; consumer awareness; export certification	Public-private partnerships; market linkage programs; branding initiatives
Capacity building	Farmer training; entrepreneurial skills; extension services; south-south collaboration	Digital extension platforms; demonstration farms; peer learning networks

Sources: Katel et al. (2023), Ihou and Mansingh (2025), Negacz et al. (2021), Ulian et al. (2020)

et al. 2025; Tarolli et al. 2024). Marker-assisted selection, hybridisation, and emerging genome-editing technologies have enabled precise trait modifications (Siridewa et al. 2025). Parallel investments in physical infrastructure, including specialised irrigation systems, processing hubs, and cold storage facilities, are essential for reducing postharvest losses and maintaining product quality (Ihou and Mansingh 2025). However, fully realising these advances in Africa requires expanded genomic infrastructure, comprehensive germplasm collection, and specialised training programs to deliver resilient, high-yielding varieties to farmers.

Policy reforms should create an enabling environment that recognises *Salicornia*'s climate-smart attributes and potential for land rehabilitation (Katel et al. 2023). Effective instruments include subsidies for saline agricultural inputs, tax incentives for processing investments, and research grants that encourage private sector participation. The establishment of robust food safety and quality standards is crucial for local and global market competitiveness (Fortin 2023), while environmental safeguards should address the potential risks of soil erosion, water depletion, and biodiversity loss (Navarro-Torre et al. 2023). Critically, policy development should employ participatory approaches that ensure equitable benefit-sharing and respect for traditional knowledge (Ulian et al. 2020).

Recent agronomic innovations have offered promising pathways for enhancing productivity and sustainability. The utilisation of saline or brackish water irrigation, complemented by precision methods such as drip systems, maximises water-use efficiency while conserving freshwater resources (Zhang et al. 2023). Soil health can be improved through organic amendments and biofertilizers, while optimised planting densities and intercropping with complementary halophytes can increase the overall system productivity. Improved post-harvest techniques, including solar drying and value-added processing, extend shelf life and enhance market value (Wang et al. 2023).

The resilience of *Salicornia* to harsh environments makes it a strategic crop for climate-smart agriculture. Its integration into diversified farming systems reduces crop failure risk and rehabilitates degraded soils through agroecological practices, while simultaneously contributing to carbon sequestration and phytoremediation (Ranjbar et al. 2022). Scaling up cultivation requires coordinated efforts involving infrastructure investment, inclusive value chain development, farmer cooperatives, and policy frameworks. Public-private partnerships can drive innovation and risk reduction, while consumer awareness campaigns and product standards build market confidence as summarised in Table 11.8.

Ultimately, embedding *Salicornia* development into national agricultural and environmental strategies ensures long-term sustainability and impact. By learning from global halophyte success and focusing on context-specific innovations in agronomy, markets, and policy, Africa can establish *Salicornia* as a catalyst for food security, economic growth, and climate resilience. Future efforts should prioritise inclusive participation, SDG alignment, and knowledge co-creation to ensure lasting benefits across the continent.

Table 11.8 Stakeholder engagement framework across research, policy, and practice for building a supportive ecosystem for realising the transformative potential of *Salicornia* spp. in Africa

Stakeholder group	Key actors	Primary role and responsibilities
Research & academia	National Agricultural Research Institutes (e.g., ISRA in Senegal), Universities, CGIAR Centres (e.g., IWMI)	Conduct context-specific R&D on germplasm & agronomy; Lead M&E; build local capacity and training programs.
Policy & government	Ministries of Agriculture & Environment, Local Governments, Investment Promotion Agencies	Develop supportive policies & saline land leases; Fund R&D; Integrate halophytes into national food security & climate adaptation strategies.
Private sector f finance	Agri-Processors, Impact Investors, Local Banks, Farmer Cooperatives	Invest in processing infrastructure & value chains; Provide financing to farmers; Develop markets and distribution channels.
Practice & community	Smallholder Farmers, Local Communities, NGOs & Cooperatives	Co-develop farming practices; Provide land and labour; Own and operate community-based enterprises.

11.9 Conclusion

Salicornia spp. represents a paradigm-shifting opportunity to address Africa's interconnected challenges of food insecurity, environmental degradation, and economic vulnerability. These halophytic species offer a sustainable agricultural alternative for the continent's extensive saline and marginal lands, transforming environmental constraints into productive assets. The plant's unique biological attributes enable a multifaceted value proposition: as a nutrient-dense food source rich in essential minerals and bioactive compounds; as a feedstock for diverse industrial applications, including biofuels and biodegradable materials; and as a therapeutic resource with demonstrated pharmacological properties. Crucially, *Salicornia* cultivation delivers significant ecosystem services through phytoremediation of degraded soils, carbon sequestration, and coastal protection, thereby contributing directly to climate change adaptation and environmental restoration.

Despite this compelling potential, *Salicornia* remains substantially underutilised across Africa, constrained by persistent barriers, including limited research investment, underdeveloped value chains, and inadequate policy support. The absence of localised agronomic data, coupled with taxonomic complexities and insufficient processing infrastructure, has hindered the transition from experimental cultivation to commercial production. Furthermore, the translational gap between laboratory findings and practical applications, particularly in nutritional and therapeutic domains, requires targeted investigation using African-sourced materials and conditions.

Realising the full potential of *Salicornia* demands a fundamental shift from dependency on global data to the generation of context-specific African evidence.

Strategic priorities should include establishing comprehensive germplasm collections and breeding programs for locally adapted varieties, validating health claims through clinical trials focused on African population needs, developing integrated biorefinery models that maximise value from biomass, and implementing policies that recognise *Salicornia*'s role in climate-resilient agriculture. This necessitates coordinated investment across the research-development-application continuum, with particular emphasis on building scientific capacity, fostering public-private partnerships, and creating inclusive market systems that benefit smallholder farmers and rural communities.

The successful integration of *Salicornia* into African agricultural and economic systems will require sustained commitment to collaborative, transdisciplinary approaches that bridge scientific innovation, policy reform, and community engagement. By embracing this multifunctional crop within broader sustainable development strategies, African nations can simultaneously advance progress toward food sovereignty, environmental security, and economic diversification. Ultimately, harnessing the potential of *Salicornia* spp. represents more than an agricultural innovation—it constitutes a strategic pathway toward building more resilient and self-reliant food systems across the continent, transforming ecological challenges into sustainable opportunities for growth and development.

References

Al S (2025) دراسة التركيب الكيميائي لنبات Salicornia europaea L المنتشر في المنطقة الشاطئية المقابلة لكورنيش جبلة في الساحل السوري. *Iraqi* J Aquac 22(1):203–213

Alfheeaid HA, Raheem D, Ahmed F, Alhodieb FS, Alsharari ZD, Alhaji JH, BinMowyna MN, Saraiva A, Raposo A (2022) Salicornia bigelovii, S. brachiata and S. herbacea: their nutritional characteristics and an evaluation of their potential as salt substitutes. Foods 11(21):3402

AlYammahi J, Chelaifa H, Hasan A, Darwish AS, Lemaoui T, Hernandez HH, Rios-Galvan A (2024) *Salicornia* seed oil: a high-yielding and sustainable halophytic feedstock for biodiesel and energy in underutilized hypersaline coastal deserts. Energy Convers Manag 318:118914. https://doi.org/10.1016/j.enconman.2024.118914

Bazihizina N, Papenbrock J, Aronsson H, Hamed K, Elmaz O, Dafku Z, Custódio L, Rodrigues MJ, Atzori G, Negacz K (2024) The sustainable use of halophytes in salt-affected land: state-of-the-art and next steps in a saltier world. Plants 13:2322. https://doi.org/10.3390/plants13162322

Cárdenas-Pérez S, Piernik A, Chanona-Pérez JJ, Grigore MN, Perea-Flores MJ (2021) An overview of the emerging trends of the Salicornia L. genus as a sustainable crop. Environ Exp Bot 191:104606

Cardoso M, Silva H, Patinha C, Costa N, Nunes S, Cunha Â (2022) From the saltpan to the plate: an evaluation of the use of the edible halophyte Salicornia ramosissima in catering. Ann Appl Biol 180(1):99–108. https://doi.org/10.1111/aab.12714

Chatrenoor T, Akhani H (2021) An integrated morpho-molecular study of Salicornia (Amaranthaceae-Chenopodiaceae) in Iran proves Irano-Turanian region the major center of diversity of annual glasswort species. Taxon 70(5):989–1019. https://doi.org/10.1002/tax.12538

Chaturvedi T, Christiansen AH, Gołębiewska I, Thomsen MH (2021) Salicornia species: current status and future potential. In: Future of sustainable agriculture in saline environments. CRC Press, pp 461–482. https://www.taylorfrancis.com/chapters/oaedit/10.1201/9781003112327-31/salicornia-species-tanmay-chaturvedi-aslak-christiansen-iwona-go%C5%82%C4%99biewska-mette-thomsen

Chrigui S, Hadj Taieb S, Jemai H, Mbarek S, Benlarbi M, Feki M, Haouas Z, Zemmel A, Chaouacha-Chekir RB, Boudhrioua N (2023) Anti-obesity and anti-Dyslipidemic effects of Salicornia arabica decocted extract in Tunisian Psammomys obesus fed a high-calorie diet. Foods 12(6):1185

Collard BCY, Mackill DJ (2008) Marker-assisted selection: an approach for precision plant breeding in the twenty-first century. Philos Trans R Soc B 363(1491):557–572. https://doi.org/10.1098/rstb.2007.2170

Davy A, Bishop GF, Costa C (2001) Salicornia L. (Salicornia pusilla J. Woods, S. Ramosissima J. Woods, S. europaea L., S. obscura P.W. Ball & Tutin, S. Nitens P.W. Ball & Tutin, S. fragilis P.W. Ball & Tutin and S. dolichostachya Moss). J Ecol 89:681–707. https://doi.org/10.1046/j.0022-0477.2001.00607.x

Ekanayake S, Egodawatta C, Attanayake RN, Perera D (2023) From salt pan to saucepan: *Salicornia* , a halophytic vegetable with an array of potential health benefits. Food Front 4(2):641–676. https://doi.org/10.1002/fft2.214

Fortin ND (2023) Global governance of food safety: the role of the FAO, WHO, and codex Alimentarius in regulatory harmonization. In: Research handbook on international food law. Edward Elgar Publishing, pp 227–242. https://www.elgaronline.com/edcollchap/book/9781800374676/book-part-9781800374676-24.xml

Fredsgaard M, Kaniki SEK, Antonopoulou I, Chaturvedi T, Thomsen MH (2023) Phenolic compounds in Salicornia spp. and their potential therapeutic effects on H1N1, HBV, HCV, and HIV: a review. Molecules 28(14):5312. https://doi.org/10.3390/molecules28145312

Galparsoro I, Borja A, Uyarra MC (2014) Mapping ecosystem services provided by benthic habitats in the European North Atlantic Ocean. Front Mar Sci 1. https://doi.org/10.3389/fmars.2014.00023

Garza-Torres R, Troyo-Dieguez E, Nieto-Garibay A, Lucero G, Magallón-Barajas F, García-Galindo E, Fimbres Y, Amador B (2020) Environmental and management considerations for adopting the halophyte Salicornia bigelovii Torr. As a sustainable seawater-irrigated crop. Sustainability 12:707. https://doi.org/10.3390/su12020707

Giménez A, Gallegos-Cedillo VM, Benaissa RR, Egea-Gilabert C, Signore A, Ochoa J, Gruda NS, Arnao MB, Fernández JA (2024) Enhancing the cultivation of Salicornia fruticosa with agro-industrial compost leachates in a cascade cropping system: evaluating the impact of melatonin application. Front Plant Sci 15:1441884. https://doi.org/10.3389/fpls.2024.1441884

Harboub N, Mighri H, Bennour N, Dbara M, Pereira C, Chouikhi N, Custódio L, Abdellaoui R, Akrout A (2025) Nutritional profile, chemical composition and health promoting properties of *Salicornia emerici* Duval-Jouve and *Sarcocornia alpini* (Lag.) Rivas Mart. From southern Tunisia. Biocatal Agricult Biotechnol 64:103502. https://doi.org/10.1016/j.bcab.2025.103502

Hasnain M, Abideen Z, Ali F, Hasanuzzaman M, El-Keblawy A (2023a) Potential of halophytes as sustainable fodder production by using saline resources: a review of current knowledge and future directions. Plants 12(11):2150. https://doi.org/10.3390/plants12112150

Hasnain M, Abideen Z, Ali F, Hasanuzzaman M, El-Keblawy A (2023b) Potential of halophytes as sustainable fodder production by using saline resources: a review of current knowledge and future directions. Plants 12:2150. https://doi.org/10.3390/plants12112150

Ihou AFY, Mansingh JP (2025) Pathways to farmers' entrepreneurship: the role of entrepreneurial mindset. Front Sustain Food Syst 9. https://doi.org/10.3389/fsufs.2025.1584522

Jamdade R, Al-Shaer K, Al-Sallani M, Al-Harthi E, Mahmoud T, Gairola S, Shabana HA (2022) Multilocus marker-based delimitation of Salicornia persica and its population discrimination assisted by supervised machine learning approach. PLoS One 17(7):e0270463. https://doi.org/10.1371/journal.pone.0270463

Kamal M, Lele W, Shuzhen T, Jiandi L, Rongyan Q, Yanfeng L, Wenqi W, Xiangyu C, Cheng Y (2024) Influence of dietary Salicornia europaea L. extract supplementation on feed efficiency of Altay sheep by modifying their gastrointestinal bacteria communities. Front Microbiol 15. https://doi.org/10.3389/fmicb.2024.1377314

Katel S, Yadav SPS, Turyasingura B, Mehta A (2023) Salicornia as a salt-tolerant crop: potential for addressing climate change challenges and sustainable agriculture development. Turk J Food Agric Sci 5(2):55–67. https://doi.org/10.53663/turjfas.1280239

Khan M, Ozturk M, Gul B, Ahmed M (2016) Halophytes for food security in dry lands

Kim S, Lee E-Y, Hillman PF, Ko J, Yang I, Nam S-J (2021) Chemical structure and biological activities of secondary metabolites from Salicornia europaea L. Molecules 26(8):2252

Kule E, Obia A, Agole D, Okello D, Odongo W (2025) Farmer perceptions and their implications for adoption of sustainable agricultural intensification practices. Discov Sustain 6:210. https://doi.org/10.1007/s43621-025-00929-z

Lago J, Pinto D, Moreira M, Grosso C, Fernandes V, Delerue-Matos C, Rodrigues F (2021) Salicornia ramosissima bioactive composition and safety: eco-friendly extractions approach (microwave-assisted extraction vs. conventional maceration). Appl Sci 11:4744. https://doi.org/10.3390/app11114744

Limongelli F, Crupi P, Clodoveo ML, Corbo F, Muraglia M (2022) Overview of the polyphenols in Salicornia: from recovery to health-promoting effect. Molecules 27(22):7954. https://doi.org/10.3390/molecules27227954

Lombardi T, Bertacchi A, Pistelli L, Pardossi A, Pecchia S, Toffanin A, Sanmartin C (2022) Biological and agronomic traits of the Main halophytes widespread in the Mediterranean region as potential new vegetable crops. Horticulturae 8(3):195. https://doi.org/10.3390/horticulturae8030195

Lyra, D.-A., Lampakis, E., Muhairi, M. A., Bin Tarsh, F. M., Hamyd Dawoud, M. A., Khawaldeh, B. A., Moukayed, M., Plewa, J., Cobre, L., Al Masjedi, O. S., Al Marzouqi, K. M., Ahmadzai, H., Al Tamimi, M. K., & Abou Dahr, W. A. (2021). From desert farm to fork. In K. Negacz, P. Vellinga, E. Barrett-Lennard, R. Choukr-Allah, & T. Elzenga, Future of sustainable agriculture in saline environments (1st edn, pp. 181–200). CRC Press. https://doi.org/10.1201/9781003112327-11

Mendis CL, Padmathilake RE, Attanayake RN, Perera D (2025) Learning from Salicornia: physiological, biochemical, and molecular mechanisms of salinity tolerance. Int J Mol Sci 26(13):5936. https://doi.org/10.3390/ijms26135936

Nájar AM, Romero-Bernal M, del Río C, Montaner J (2023) A review on polyphenols in Salicornia ramosissima with special emphasis on their beneficial effects on brain ischemia. Nutrients 15(3):793. https://doi.org/10.3390/nu15030793

Narasimha Rao G, Pragada P (2013) Morphological and anatomical features of Salicornia brachiata. Roxb 30:887–891

Navarro-Torre S, Garcia-Caparrós P, Nogales A, Abreu MM, Santos E, Cortinhas AL, Caperta AD (2023) Sustainable agricultural management of saline soils in arid and semi-arid Mediterranean regions through halophytes, microbial and soil-based technologies. Environ Exp Bot 212:105397. https://doi.org/10.1016/j.envexpbot.2023.105397

Negacz K, Vellinga P, Barrett-Lennard E, Choukr-Allah R, Elzenga T (2021) Future of sustainable agriculture in saline environments, 1st edn. CRC Press. https://doi.org/10.1201/9781003112327

Orzoł A, Głowacka K, Pätsch R, Piernik A, Gallegos-Cerda SD, Cárdenas-Pérez S (2025) The local environment influences salt tolerance differently in four Salicornia europaea L. inland populations. Sci Rep 15(1):13128. https://doi.org/10.1038/s41598-025-97394-5

Patel S (2016) Salicornia: evaluating the halophytic extremophile as a food and a pharmaceutical candidate. 3 Biotech 6(1):104. https://doi.org/10.1007/s13205-016-0418-6

Ranjbar G, Dehghani F, Rahimian M (2022) Endemic species of Salicornia with economic value in the flora of Iran. Iran Nat 7(1):45–50. https://doi.org/10.22092/irn.2022.356592.1416

Rwema M, Safari B, Sylla MB, Roininen L, Laine M (2025) Understanding farmers' knowledge, perceptions, and adaptation strategies to climate change in eastern Rwanda. Sustainability 17(15):6721. https://doi.org/10.3390/su17156721

Simutowe E, Ngoma H, Manyanga M, Silva J, Baudron F, Nyagumbo I, Kalala K, Habeenzu M, Thierfelder C (2024) Risk aversion, impatience, and adoption of conservation agriculture practices among smallholders in Zambia. Heliyon 10:e26460. https://doi.org/10.1016/j.heliyon.2024.e26460

Singh D, Buhmann AK, Flowers TJ, Seal CE, Papenbrock J (2014) Salicornia as a crop plant in temperate regions: selection of genetically characterized ecotypes and optimization of their cultivation conditions. AoB Plants 6. https://doi.org/10.1093/aobpla/plu071

Siridewa K, De Silva W, Ratnayake RMCS, Wijesundara S, Perera D, Attanayake RN (2025) Species identification and pollination biology of an economically important true halophyte, *Salicornia brachiata* Roxb. Aquat Bot 196:103827. https://doi.org/10.1016/j.aquabot.2024.103827

Tarolli P, Luo J, Park E, Barcaccia G, Masin R (2024) Soil salinization in agriculture: mitigation and adaptation strategies combining nature-based solutions and bioengineering. iScience 27(2):108830. https://doi.org/10.1016/j.isci.2024.108830

Tuğ G, Başköse İ (2023) Illustration of succulent taxa from Amaranthaceae: problems and proposed solutions. Herbarium Turcicum *0*(1):10–13. https://doi.org/10.26650/HT.2023.1239982

Ulian T, Diazgranados M, Pironon S, Padulosi S, Liu U, Davies L, Howes M-JR, Borrell JS, Ondo I, Pérez-Escobar OA, Sharrock S, Ryan P, Hunter D, Lee MA, Barstow C, Łuczaj Ł, Pieroni A, Cámara-Leret R, Noorani A et al (2020) Unlocking plant resources to support food security and promote sustainable agriculture. Plants People Planet 2(5):421–445. https://doi.org/10.1002/ppp3.10145

Ventura Y, Wuddineh WA, Shpigel M, Samocha TM, Klim BC, Cohen S, Shemer Z, Santos R, Sagi M (2011) Effects of day length on flowering and yield production of *Salicornia* and *Sarcocornia* species. Sci Hortic 130(3):510–516. https://doi.org/10.1016/j.scienta.2011.08.008

Wang N, Zhao Z, Zhang X, Liu S, Zhang K, Hu M (2023) Plant growth, salt removal capacity, and forage nutritive value of the annual euhalophyte Suaeda salsa irrigated with saline water. Front Plant Sci 13:1040520. https://doi.org/10.3389/fpls.2022.1040520

Zhang C, Gong R, Zhong H, Dai C, Zhang R, Dong J, Li Y, Liu S, Hu J (2023) Integrated multi-locus genome-wide association studies and transcriptome analysis for seed yield and yield-related traits in Brassica napus. Front Plant Sci 14. https://www.frontiersin.org/journals/plant-science/articles/10.3389/fpls.2023.1153000

Zhao Y, Wang H, Liu T, Xin Z (2014) The individual lipid compositions produced by Cunninghamella sp. Salicorn 5, an endophytic oleaginous fungus from Salicornia bigelovii Torr. Eur Food Res Technol 238(4):621–633. https://doi.org/10.1007/s00217-013-2141-4

Chapter 12
Climate Change and the Prevalence of Parasitic Diseases in Coastal and Marine Fisheries: Risks to Food Security in Africa

Ekaette Godwin Edelduok and Ubong Uwem George

Abstract Climate change has a major influence on ecological and environmental changes, with a significant impact on coastal and marine fisheries. One of the less assessed but increasingly important effects of climate change is its impact on the prevalence and distribution of parasitic diseases in aquatic ecosystems, and its associated risks to food security. Utilizing a narrative literature review approach, this chapter examines how rising sea temperatures, ocean acidification, and salinity changes influence parasite–host dynamics, fish health, and fisheries productivity, with a focus on Africa. Relevant peer-reviewed studies published between 2013 and 2025 were sourced from PubMed, Scopus, Web of Science, and Google Scholar, complemented by gray literature from FAO and related organizations. Evidence indicates that climate change accelerates parasite development, alters their geographic distribution, and increases infection risks, leading to reduced fish growth, survival, and availability. These effects heighten food insecurity, especially for vulnerable coastal communities dependent on fish protein. This review identifies critical research gaps and advocates adaptive interventions, including strengthened monitoring systems, climate-resilient aquaculture, ecosystem-based fisheries management, livelihood diversification, and focused research on parasite–host ecology to safeguard fisheries and promote sustainable food systems across Africa.

Keywords Climate change · Marine parasites · Fish health · Coastal fisheries · Africa · Food security · Adaptation

E. G. Edelduok (✉)
Department of Zoology, Akwa Ibom State University, Mkpat Enin, Akwa Ibom, Nigeria
e-mail: ekaetteedelduok@aksu.edu.ng

U. U. George
Department of Fisheries and Aquaculture, Akwa Ibom State University, Mkpat Enin, Akwa Ibom, Nigeria

H. A. Mupambwa et al. (eds.), *Harnessing Coastal and Marine Resources for Improved Food Security in Africa*, Sustainability Sciences in Asia and Africa,
https://doi.org/10.1007/978-981-95-9725-3_12

12.1 Introduction

Climate change is an important environmental challenge currently experienced in the world, and it has significant implications for biodiversity, ecosystem functioning, and the well-being of humans. Significant among its myriad impacts is the effect of climate change on coastal and marine fisheries. This is of great importance since coastal and marine fisheries are essential in food systems and economic livelihoods. Globally, coastal and marine fisheries support millions of people, providing food and employment, especially in developing regions of the world. Unfortunately, climate change has influenced the marine ecosystem such that there has been an increase in the emergence and spread of parasitic diseases among fish populations. Apart from leading to a decline in fish stocks and compromising fish health, parasitic diseases also constitute significant threats to food safety and security. This chapter evaluates the intersection of climate change, parasitic diseases, and food security in the context of coastal and marine fisheries in Africa.

12.2 Climate Change and Aquatic Ecosystems

Climate change can be observed in marine environments through increases in sea surface temperatures, ocean acidification, altered salinity levels, and greater frequencies of extreme weather events. Over the years, there have been gradual and persistent rises in ocean temperatures (Morice et al. 2021).

The most recent decadal estimates, which span the five-year period from 2020 to 2024, indicate that rainfall in North and Southern Africa is decreasing and temperatures are continuing to rise, while the Sahel region is experiencing an increase in rainfall (Bichet and Diedhiou 2018; Ogunrinde et al. 2024). For example, the global climate data for January–December 2024 show that oceans were 1.27 °C higher than average, making it the warmest January on record since 1850. These changes disrupt ecological balances, negatively impacting fish distribution, spawning cycles, and migration patterns. Decreased dissolved oxygen levels are experienced in warmer waters, resulting in impaired fish metabolism and immune responses. Increased carbon dioxide absorption, which often culminates in ocean acidification, alters the availability of calcium carbonate, and this impairs shell-forming organisms and the structure of the food web. Moreover, nutrient runoff from increased rainfall can result in eutrophication and hypoxic zones, and this further stresses marine life.

12.3 Parasitic Diseases in Marine and Coastal Fisheries

Parasitic diseases, caused by a variety of parasites including protozoans, helminths, and crustaceans, often occur in marine and coastal fish. *Anisakis* (nematode), *Cryptocaryon* (protozoan), and *Neobenedenia* (trematode) are notable examples (Table 12.1). Infection with these parasites often results in the manifestation of

Table 12.1 Major marine and coastal fish parasites (Compiled by authors from published data)

Parasite group	Example species	Common host species	Common host sites	Transmission mode	Effects on fish	Characteristics	References
Protozoa (Ciliate)	*Cryptocaryon irritans*	Groupers, Pomfrets, Snappers	Skin, fins, gills	Direct—encysted tomont releases infective theronts	White nodules, appetite loss, mortality in severe cases	Oval, ciliated protozoan causing "marine white spot disease"; three life stages—tomont, theront, trophont.	Zhang et al. (2023)
	Brooklynella hostilis	Clownfish, Tangs	Gills, skin	Direct—contact or contaminated water	Heavy mucus, respiratory distress, sloughing of skin	Pear-shaped ciliate; feeds on epithelial cells; causes "clownfish disease."	Noga (2010)
Monogenean trematode	*Neobenedenia girellae*	Golden pompano, Groupers, Snappers)	Skin, fins, eyes	Direct—eggs hatch and oncomiracidia attach to new hosts	Skin lesions, hemorrhages, reduced growth, mortality in high infestations	Flat, transparent worm with haptor; broad host range; no intermediate host.	Brazenor et al. (2018)
Digenean trematode	*Heterophyes heterophyes*	Mullet, Tilapia)	Intestine (metacercariae encysted in muscles)	Indirect—snail → fish → bird/ mammal	Minor tissue damage in fish; zoonotic risk to humans	Small intestinal fluke; complex life cycle involving snails and fish.	Chai et al. (2021)
Cestode (Tapeworm)	*Trypanorhyncha* spp.	Mackerel, Sardines	Musculature, viscera	Indirect—crustacean → fish → shark	Muscle encystment, unappealing fillets; reduced market value	Scolex with retractable tentacles and hooks; larvae encyst in tissue.	Mahmoud et al. (2015)
	Bothriocephalus acheilognathi	Marine fish	Intestine	Indirect—copepod intermediate host	Nutrient depletion, intestinal blockage	Ribbon-like segmented worm with two bothria for attachment.	Scholz et al. (2019)

(continued)

Table 12.1 (continued)

Parasite group	Example species	Common host species	Common host sites	Transmission mode	Effects on fish	Characteristics	References
Nematode (Roundworm)	*Anisakis simplex*	Herring, Cod, Mackerel	Viscera, muscles	Indirect—crustacean → fish → marine mammal	Light infections in fish; zoonotic (anisakiasis) in humans	White, coiled nematode larvae; zoonotic significance; allergenic potential.	Cipriani et al. (2018)
	Contracaecum multipapillatum	Marine fish	Liver, body cavity	Indirect—crustacean → fish → bird/mammal	Organ damage, retarded growth, inflammation	Thick-bodied nematode; encysted larvae visible in organs.	Shamsi (2019)
Crustacean (Copepod)	*Caligus rogercresseyi*	Salmonids, Groupers, Mullets)	Skin, fins	Direct—copepodid larva attaches to host	Skin lesions, hemorrhage, secondary infections, stress	Flattened copepod with long antennae; causes major aquaculture losses.	Bravo et al. (2023)
	Caligus elongatus	Salmon and wild pelagic fish	Skin, fins	Direct—free-swimming larvae infect host	Skin erosion, anemia, poor feed conversion	Oval-shaped copepod; important ectoparasite in marine aquaculture.	Hemmingsen et al. (2020)
	Lernaea spp. ("Anchor Worm")	Asian seabass, Common carp	Skin, fins, gills, muscles	Direct—copepod larvae embed in the host	Ulcers, hemorrhage, secondary infections	Worm-like female burrows into tissue with anchor-like appendages.	Raja et al. (2023)

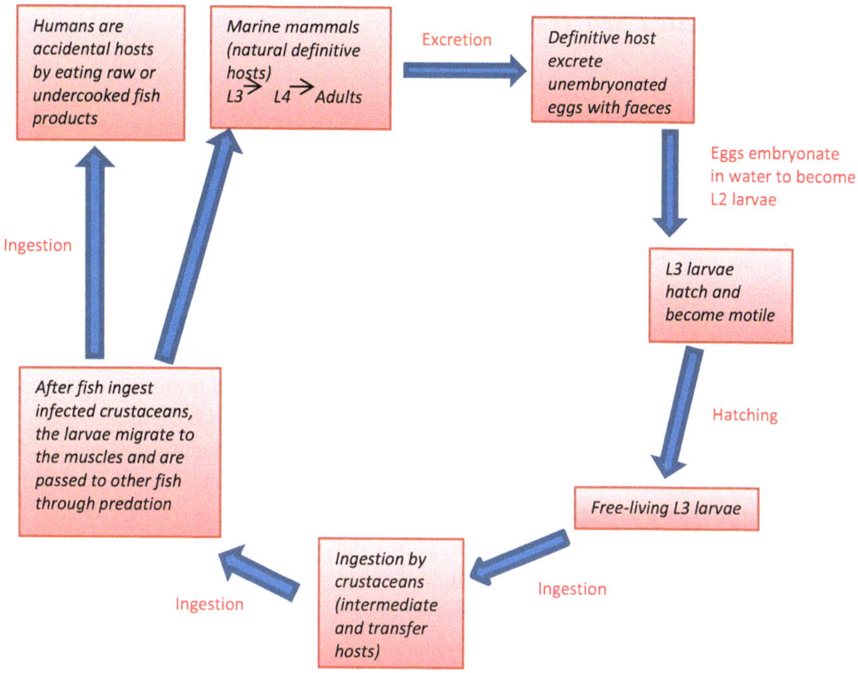

Fig. 12.1 Illustration by authors. (Based on CDC (2019))

pathological conditions such as granulomatous inflammation, muscle degeneration, and compromised organ function.

The life cycles of many parasites are complex, involving multiple hosts and environmental stages, and this makes their transmission sensitive to changes in the ecosystem. An example is the life cycle of *Anisakis* spp. (Fig. 12.1). Even though *Anisakis* species that infect humans can be found in many marine fish species, they are mostly common in predatory species such as cod, herring, and mackerel. Environmental factors such as salinity, water temperature, and pollution have been known to influence the survival and infectivity of most parasites. Therefore, the importance of parasitic diseases in marine and coastal fisheries cannot be overemphasized.

12.4 Relationship Between Climate Change and Parasitic Disease Prevalence

Climate change influences the epidemiology of parasitic diseases in coastal and marine ecosystems in various ways. Climate change often results in elevated sea temperatures, ocean acidification, changes in precipitation patterns and runoff, and extreme weather events such as storms and heatwaves.

12.4.1 Elevated Sea Temperatures

One of the impacts of climate change is that it results in rising sea temperatures, and this has a direct impact on the life cycles of parasites. Temperature is known to influence the life cycles of many marine parasites, such as the copepods, monogeneans, and some endoparasites. Parasite development and reproduction rates may increase because of warmer waters, leading to increases in the frequency and intensity of infections. For example, high temperatures may trigger the hatchability and maturation of eggs and cercariae of trematodes, resulting in increased prevalence of parasitic infections. Studies exploring the sensitivity of host–parasite parameters to changes in temperature have indicated that temperature changes and changes in the sensitivity of host survival to temperature can easily transcend non-viable parasites to optimal parasites (Santos and Ebert 2022; Yin et al. 2023; Okon et al. 2024). A decline in healthy fish stocks because of the occurrence of disease can reduce the availability of affordable protein, directly impacting the food security and nutritional status of coastal communities.

Additionally, increased water temperatures may result in thermal stress in fish, weakening the immune systems of fish, and subjecting them to increased susceptibility to parasitic infections (Shameena et al. 2021; Roychowdhury et al. 2024). This may increase the mortality of fish, leading to a decline in their availability and subsequently posing a risk to food security.

The geographical range of parasites and their intermediate hosts can be widened by rising temperatures, resulting in the introduction of infections to regions that were previously unaffected. On the other hand, parasites may migrate out of marine waters that become too warm. In both cases, the prevalence of parasitic diseases is increased.

12.4.2 Ocean Acidification

Ocean acidification is the gradual and persistent decrease in the pH of the Earth's oceans, caused mainly by the absorption of excess carbon dioxide (CO_2) from the atmosphere. It occurs because of climate change and has significant negative impacts on host defenses.

It has been noted that ocean acidification can cause stress in marine organisms, including fish (Chen and Mai 2024; Devergne et al. 2023). This could result in a weakening of their immune responses, altering their physiology, and making them more susceptible to parasitic infections.

Ocean acidification can influence the calcification of shell-forming organisms (Guo et al. 2023; Devergne et al. 2023), which often serve as intermediate hosts for various parasites (e.g., snails for trematodes). This could adversely affect the populations and health of these intermediate hosts and could alter the transmission dynamics of the parasites.

12.4.3 Changes in Precipitation Patterns and Runoff

One of the consequences of climate change is that it leads to altered precipitation patterns and runoff. While an increase in rainfall and river discharge can lower coastal salinity, prolonged droughts can elevate it. Since many parasites are sensitive to salinity changes, this could adversely affect them. For example, some marine parasites might be stressed by sudden drops in salinity (Byers 2021).

Changes in precipitation can lead to an increase in runoff from land, resulting in more nutrients and pollutants being carried into coastal waters. This can result in algal blooms, including harmful algal blooms, leading to fish being stressed, and creating an enabling environment for some opportunistic pathogens and parasites to thrive.

12.4.4 Extreme Weather Events

Climate change can result in extreme weather events, such as the occurrence of storms. Severe storms can disrupt the marine ecosystem, displacing fish populations, altering water currents, and potentially transmitting parasites to new habitats or concentrating them in others (Bacheler et al. 2019).

Heatwaves, another extreme weather event, can result in localized "dead zones" or can cause fish to migrate to smaller and cooler regions, thus elevating stress and stocking densities, which are optimal conditions for the transmission and outbreaks of parasites (McCartan et al. 2025; Tomamichel et al. 2025). This poses a direct challenge for the sustainability of fisheries, and an indirect risk to food security.

In all these, communities heavily dependent on fish as staple food may experience reduced availability or an increase in the prices of fish because of disease outbreaks, and this would directly undermine their food security.

Undeniably, climate change is changing the landscape of marine parasitic diseases in African coastal waters. The complex interplay of elevated temperatures, ocean acidification, and altered hydrological cycles provides an environment conducive to increased parasite prevalence of parasites, ranges expansion, and enhanced disease intensity.

12.5 Case Studies

The case studies presented below further highlight the relationship between climate change and the prevalence of parasitic diseases.

12.5.1 Case Study 1: Climate Change and the Prevalence of Parasitic Diseases in West African Coastal Waters

Several studies report high parasite loads in West African coastal fishes, and evidence suggests climate change could increase parasite impacts (Okon et al. 2024; Rowley et al. 2024; Sarre et al. 2024; Tossavi et al. 2025). These environmental shifts have resulted in higher parasite prevalence, weakening fish stocks, and threatening food security (Marcogliese 2025). For example, research in the Gulf of Guinea shows that parasite loads are increasing in commercially important species like croaker (Mahe et al. 2024). Sea surface temperature (SST) is positively correlated with increasing prevalence of gill flukes and nematodes, thus having negative implications on local food security (Kouadio et al. 2021). Another study on West African oyster fisheries found that the mangrove oyster (*Crassostrea tulipa*) is highly vulnerable to climate stressors such as temperature fluctuations, salinity changes, and sea-level rise (Mahu et al. 2022). A study on Ghana's marine artisanal fisheries found that increasing sea surface temperatures are negatively impacting shellfish production, making them more susceptible to parasitic infections (Atindana et al. 2025).

12.5.2 Case Study 2: Climate Change and the Prevalence of Parasitic Diseases in Eastern African Marine Ecosystems

Research in the Western Indian Ocean region has explored how climate change is altering host–parasite dynamics in estuarine and near-shore environments (Poulin and Mouritsen 2006; Byers 2020; Byers 2021). Warmer waters and ocean acidification have contributed to higher infection rates among marine species. Reports on climate change impacts in East Africa highlight how rising temperatures and changing precipitation patterns are affecting marine biodiversity and ecosystem health (IPCC 2022; UNECA 2023).

12.5.3 Case Study 3: Shellfish and Fish Diseases in African Fisheries

Reports have indicated that increased prevalence of disease outbreaks in commercially important species like shrimp, oysters, and other shellfish is exacerbated by climate change (Pratoomyot 2024; Rowley et al. 2024). A study on Ghana's marine artisanal fisheries indicated that elevated sea surface temperatures are adversely impacting shellfish production, making them more vulnerable to parasitic infections (Atindana et al. 2025). Reports from the Food and Agriculture Organization (FAO) reveal that there is an increase in the prevalence of fish diseases in African fisheries because of environmental changes due to climate change (Barange et al. 2018). These pose an increasing threat to food security, requiring urgent policy interventions.

These case studies highlight the critical need for enhanced disease monitoring, sustainable fisheries management, and climate adaptation strategies to protect African coastal communities and promote food security.

12.6 Impacts of Marine Parasite Infections on Fisheries and Food Security

Marine parasite infections have myriad impacts on African fisheries and food security, affecting aquaculture, wild fish populations, and ultimately, human livelihoods and access to healthy food. Parasitic infections in fish reduce growth rates, impair reproduction, and increase mortality, leading to declining fish stocks and economic losses in the fisheries sector. Infected fish may also be unmarketable due to visible lesions or concerns about food safety. Some parasites are zoonotic, capable of infecting humans through the consumption of raw or undercooked fish, posing public health risks.

12.6.1 Reduction in Fish Production and Decline in Fisheries

Severe parasitic infections can result in direct mortality in fish, especially in overcrowded habitats which enable the rapid spread and transmission of parasites (Schaal et al. 2022; Marcogliese 2025). Where the mortality rate is high, it leads to significant losses in fish stock.

In some instances, being infected with parasitic diseases may not result in the death of fish. However, parasites can adversely impact the health of these fish (Sures and Nachev 2022). They can damage the host's organs, weaken the fish, and compete for nutrients, thus reducing growth rates and lowering reproductive success. This reduces the overall biomass and yield of fisheries.

Parasitic infections result in compromised immune systems of the host (fish) (Holzer et al. 2021). This increases their susceptibility to secondary infections, creating a cascade of health issues that further impact negatively on fish populations.

Even though the impacts of parasitic infections are more pronounced in aquaculture, multiple recent surveys demonstrate that wild marine fish stocks are also significantly parasitized: 15% prevalence in Moroccan coastal species (Tanjaoui et al. 2025) and 92% in West African Sardinella (Tossavi et al. 2025). This can result in localized declines in fish availability and abundance, impacting negatively on traditional fishing communities.

12.6.2 Economic Losses for Fisheries

Direct financial losses may be experienced by fishers. This could be a result of reduced yields and fish mortality due to parasitic infections in marine fish.

Fish infected with parasites may have reduced marketability because of their poor physical appearance, lesions, or compromised flesh quality (Tanjaoui et al. 2025). These make them unappealing to consumers, and result in lower market prices or outright rejection.

Some countries observe strict import regulations with respect to fish health and disease (EC 2021; CFIA 2025). This can negatively impact export opportunities for African fisheries products affected by parasitic diseases.

Fishing is a major source of income and employment for most coastal and inland communities in Africa. Parasitic disease outbreaks can devastate these livelihoods, resulting in poverty and, in extreme cases, displacement.

12.6.3 Threats to Food Security and Nutrition

Protein is essential for good health, and fish is an important source of animal protein. Where there is a decline in fish production because of parasitic diseases, it will directly reduce the availability of this critical food source, increasing malnutrition and food insecurity.

A decline in fish supplies because of infection with parasitic diseases will result in an increase in prices. The end result is that fish will become less affordable for vulnerable populations, especially those with low incomes.

Parasitized fish raise safety concerns. Even though parasitized fish are not harmful to humans if properly cooked, some can transmit zoonotic diseases if the fish is eaten raw or undercooked (Ziarati et al. 2022; Oros et al. 2023). People often perceive parasitized fish as being unhealthy, and this can lead to consumer rejection, even if proper preparation reduces the risk of being infected.

12.7 Mitigation and Adaptation Strategies

To mitigate the growing threat of climate change to the increased prevalence of parasitic diseases in coastal and marine fisheries, and the subsequent risk to food security in Africa, the following recommendations are important:

12.7.1 Enhanced Monitoring and Surveillance

Comprehensive and integrated surveillance systems combining epidemiological, ecological, and climatic data should be adopted in the effective monitoring of parasitic diseases in fish populations (Riuzzi et al. 2024). Prompt detection and outbreak prediction can be enhanced by the use of molecular diagnostic tools and remote sensing technologies (Peters et al. 2018; Jaies et al. 2024). Identification of high-risk areas, by projecting future alterations in sea temperatures and host distributions, can be aided by climate models (Cárdenas-Sánchez et al. 2022; Sarre et al. 2024). Robust and coordinated regional monitoring programs for fish health and parasitic diseases in African coastal and marine waters should be established. This should include regular and routine sampling, increasing diagnostic capabilities, and sharing data among African countries.

12.7.2 Research and Capacity Building

If we are to better understand the specific parasites affecting African marine species, their life cycles, and their sensitivity to environmental changes, there is a need for investment in research. There should be an enhanced and increased local capacity in fish parasitology, pathology, and ecological study and modeling to predict the outbreaks of parasitic diseases.

12.7.3 Integrated Ecosystem Management

An integrated ecosystem approach to fisheries management that takes into account the cumulative impacts of climate change, pollution, and disease on marine resources should be adopted.

12.7.4 Awareness and Training

It is pertinent that awareness of the relationship between parasitic diseases, climate change, and fish health be raised among fishing communities. Training on best practices for fish handling, disease recognition, and basic preventive measures should be carried out regularly.

12.7.5 Regional and International Collaboration

Strong regional and international collaboration for sharing knowledge, resources, and best practices in marine disease management in the face of climate change should be fostered and encouraged.

12.8 Research Gaps and Challenges

12.8.1 Limited Research and Data

Africa is faced with limited data and research on fish diseases and parasitic infections compared to other regions. This constitutes a challenge in accurately assessing the full extent of the problem and developing efficient and effective control strategies.

12.8.2 Inadequate Diagnostic Capacity

Most African countries do not have the necessary diagnostic capacity. There is inadequate infrastructure, laboratory space, and trained personnel for the prompt and accurate diagnosis of parasitic diseases of fish. This slows down intervention and enhances the transmission and spread of diseases.

12.9 Future Research Needs

Even though there is an increasing awareness, significant knowledge gaps still exist in understanding parasite ecology under changing climate conditions. Consequently, future research should focus on long-term ecological studies, the development of climate-resilient fish breeds, and innovations in diagnostic technologies. Interdisciplinary collaboration among parasitologists, fishery experts, climate scientists, and public health professionals is important if holistic solutions are to be developed.

12.10 Conclusion

The intersection of climate change, parasitic diseases, and food security in coastal and marine fisheries is a critical challenge that demands urgent and coordinated action. As climate stressors increase, the risks associated with parasitic infections will similarly increase, impacting negatively on human nutrition, fish health, and economic stability. It is possible to mitigate these risks and protect the future of marine food systems in Africa by adopting a One Health approach and integrating science, policy, and community action. If the nexus of climate change and marine parasitic diseases is proactively addressed, African nations can work towards safeguarding their invaluable marine resources, strengthening food security, and building more resilient coastal communities in a changing climate.

References

Atindana SA, Ofori-Danson PK, Brucet S (2025) Modelling the effects of climate change on shellfish production in marine artisanal fisheries of Ghana. AAS Open Res 2:16. https://doi.org/10.12688/aasopenres.12956.2

Bacheler NM, Shertzer KW, Cheshire RT, MacMahan JH (2019) Tropical storms influence the movement behavior of a demersal oceanic fish species. Sci Rep 9(1):1481. https://doi.org/10.1038/s41598-018-37527-1

Barange M, Bahri T, Beveridge MCM, Cochrane KL, Funge-Smith S, Poulain F (2018) Impacts of climate change on fisheries and aquaculture: synthesis of current knowledge, adaptation and mitigation options (FAO Fisheries and Aquaculture Technical Paper No. 627). Food and Agriculture Organization of the United Nations. https://www.fao.org/3/i9705en/I9705EN.pdf

Bichet A, Diedhiou A (2018) West African Sahel has become wetter during the last 30 years, but dry spells are shorter and more frequent. Clim Res 75:155–162. https://doi.org/10.3354/cr01515

Bravo S, Silva MT, Contreras G (2023) The management of sea lice in Chile: a review. Rev Aquac 15(4):2301–2319

Brazenor AK, Bertozzi T, Miller TL, Whittington ID, Hutson KS (2018) DNA profiling reveals *Neobenedenia girellae* as the primary parasitic monogenean in global fisheries and aquaculture. Mol Phylogenet Evol 129:130–137. https://doi.org/10.1016/j.ympev.2018.05.012

Byers JE (2020) Effects of climate change on parasites and disease in estuarine and nearshore environments. PLoS Biol 18(11):e3000743. https://doi.org/10.1371/journal.pbio.3000743

Byers JE (2021) Marine parasites and disease in the era of global climate change. Annu Rev Mar Sci 13:397–420. https://doi.org/10.1146/annurev-marine-031920-100429

Cárdenas-Sánchez D, Sampayo AM, Rodríguez-Prieto M, Feged-Rivadeneira A (2022) A comparative framework to analyze convergence on Twitter electoral conversations. Sci Rep 12:19062. https://doi.org/10.1038/s41598-022-21861-6

CDC (2019) Anisakiasis. In: National Center for Emerging and Zoonotic Infectious Diseases (NCEZID). Division of Parasitic Diseases and Malaria. https://www.cdc.gov/dpdx/anisakiasis/index.html. Accessed 14 Jun 2025

CFIA (2025) Aquatic animal imports. Canadian Food Inspection Agency Government of Canada. https://inspection.canada.ca/en/animal-health/aquatic-animals/imports

Chai JY, Lee SH, Jeon HK (2021) Foodborne trematode infections and zoonotic risks. Foodborne Pathog Dis 18(5):328–336

Chen L, Mai RD (2024) Impacts of ocean acidification on marine ecosystems and mitigation strategies. Int J Marine Sci 14(3):231–244. https://doi.org/10.5376/ijms.2024.14.0027

Cipriani P, Mattiucci S, Santoro M (2018) *Anisakis* larvae in commercial fish and human health risk assessment. Food Control 84:284–292

Devergne J, Loizeau V, Lebigre C, Bado-Nilles A, Collet S, Mouchel O, Iaria U, Le Gall M-M, Madec L, Turiès C, Servili A (2023) Impacts of long-term exposure to ocean acidification and warming on three-spined stickleback (*Gasterosteus aculeatus*) growth and reproduction. Fishes 8(10):523. https://doi.org/10.3390/fishes8100523

EC (2021) Regulation (EU) 2016/429 of the European Parliament and of the Council on transmissible animal diseases and related implementing acts. European Commision Official Journal of the European Union. https://eur-lex.europa.eu/legal-content/EN/TXT/?uri=CELEX:32021R0404

Guo X, Huang M, Luo X, You W, Ke C (2023) Impact of ocean acidification on shells of the abalone species *Haliotis diversicolor* and *Haliotis discus hannai*. Mar Environ Res 192:106183. https://doi.org/10.1016/j.marenvres.2023.106183

Hemmingsen W, MacKenzie K, Heuch PA (2020) Sea lice (*Caligus elongatus*) as parasites of farmed salmonids: a review. Aquaculture:735160

Holzer AS, Piazzon MC, Barrett D, Bartholomew JL, Sitjà-Bobadilla A (2021) To react or not to react: the dilemma of fish immune systems facing myxozoan infections. Front Immunol 12:734238. https://doi.org/10.3389/fimmu.2021.734238

IPCC (2022) Climate change 2022: impacts, adaptation and vulnerability. Working Group II contribution to the Sixth Assessment Report of the Intergovernmental Panel on Climate Change (IPCC). (Chapter 9: Africa). Cambridge University Press. https://www.ipcc.ch/report/ar6/wg2/chapter/chapter-9/

Jaies I, Shah FA, Qadiri SSN, Qayoom I, Bhat BA, Dar SA, Bhat FA (2024) Immunological and molecular diagnostic techniques in fish health: present and future prospectus. Mol Biol Rep 51(1):551. https://doi.org/10.1007/s11033-024-09344-5

Kouadio JM, Ndiaye W, Jean-Baptiste Kassi A, Niang T, Djagoua EV (2021) Seasonal variability of sea surface temperature and chlorophyll concentration and its correlation with pelagic fish catch in Senegalese Exclusive Economic Zone (EEZ). Agric For Fish 10(5):176–182. https://doi.org/10.11648/j.aff.20211005.12

Mahe K, Ekperusi AO, Diouf K, Munschy C, Aminot Y, Andrialovanirina N, Dussuel A, Diop M (2024) Growth assessment of croakers (*Pseudotolithus* species) in West African waters. Front Fish Sci 2. https://doi.org/10.3389/frish.2024.1498784

Mahmoud NE, Alhindy MK, Fahmy MM (2015) Trypanorhynch cestode infecting Mediterranean sea fish, Egypt: *Callitetrarhynchus gracilis* Larvae (Pintner, 1931) as a bio-indicator of heavy metals pollution. Oceanography 3:133

Mahu E, Sanko S, Kamara A, Chuku EO, Effah E, Sohou Z, Zounon Y, Akinjogunla V, Akinnigbagbe RO, Diadhiou HD, Marchant R (2022) Climate resilience and adaptation in West African oyster fisheries: an expert-based assessment of the vulnerability of the oyster *Crassostrea tulipa* to climate change. Fishes 7(4):205. https://doi.org/10.3390/fishes7040205

Marcogliese DJ (2025) Climate change and parasitism in aquatic ecosystems. In: Smit NJ, Sures B (eds) Aquatic parasitology: ecological and environmental concepts and implications of marine and freshwater parasites. Springer, Cham. https://doi.org/10.1007/978-3-031-83903-0_22

McCartan N, O'Keeffe F, Zhang G, Luijckx P (2025) Impact of heatwave amplitude, duration, and timing on parasite fitness at different baseline temperatures. PLOS Clim 4(6):e0000632. https://doi.org/10.1371/journal.pclm.0000632

Morice CP, Kennedy JJ, Rayner NA, Winn JP, Hogan E, Killick RE, Dunn RJH, Osborn TJ, Jones PD, Simpson IR (2021) An updated assessment of near-surface temperature change from 1850: the HadCRUT5 data set. J Geophys Res Atmos 126(3):e2019JD032361

Noga EJ (2010) Fish disease: diagnosis and treatment, 2nd edn. Wiley-Blackwell, Hoboken

Ogunrinde AT, Adeyeri OE, Xian X, Yu H, Jing Q, Faloye OT (2024) Long-term spatiotemporal trends in precipitation, temperature, and evapotranspiration across arid Asia and Africa. Water 16(22):3161. https://doi.org/10.3390/w16223161

Okon EM, Oyesiji AAA, Okeleye ED, Kanonuhwa MM, Khalifa NE, Eissa EH, Mathew RT, Eissa ME, Alqahtani MA, Abdelnour SA (2024) The escalating threat of climate change-driven diseases in fish: evidence from a global perspective—a literature review. Environ Res 263(3):120184. https://doi.org/10.1016/j.envres.2024.120184

Oros M, Barčák D, Antolová D, Fecková M, Scholz T (2023) Zoonotic marine nematode infection of fish products in landlocked country, Slovakia. Emerg Infect Dis 29(12):2578–2580. https://doi.org/10.3201/eid2912.230674

Peters L, Spatharis S, Dario MA, Dwyer T, Roca IJT, Kintner A, Kanstad-Hanssen Ø, Llewellyn MS, Praebel K (2018) Environmental DNA: a new low-cost monitoring tool for pathogens in salmonid aquaculture. Front Microbiol 9:3009. https://doi.org/10.3389/fmicb.2018.03009

Poulin R, Mouritsen KN (2006) Climate change, parasitism and the structure of intertidal ecosystems. J Helminthol 80(2):183–191. https://doi.org/10.1079/JOH2006341

Pratoomyot G (2024) Climate change and the growing epidemic of marine pathogens. J Aquacult Res Dev 15:911. https://doi.org/10.35248/2155-9546.24.15.911

Raja RA, Patil PK, Avunje S, Kumaran M, Periyakaruppan A, Kondusamy A, De D, Jithendran KP, Alavandi SV, Vijayan KK (2023) Natural infestation of an anchor worm, *Lernaea* sp. in cage culture of Asian seabass (*Lates calcarifer*) juveniles and its control using an anti-parasitic drug, emamectin benzoate. J Parasit Dis 47(2):306–318. https://doi.org/10.1007/s12639-023-01571-0

Riuzzi G, Sbettega F, Tavornpanich S, Posen P, Contiero B, Mazzucato M, Segato S, Ferrè N (2024) Eliciting expert opinion on GIS use for surveillance and disease response in the aquatic animal health domain. Front Mar Sci 11:1429634. https://doi.org/10.3389/fmars.2024.1429634

Rowley AF, Baker-Austin C, Boerlage AS, Caillon C, Davies CE, Duperret L, Martin SAM, Mitta G, Pernet F, Pratoomyot J, Shields JD, Shinn AP, Songsungthong W, Srijuntongsiri G, Sritunyalucksana K, Vidal-Dupiol J, Webster TMU, Taengchaiyaphum S, Wongwaradechkul R, Coates CJ (2024) Diseases of marine fish and shellfish in an age of rapid climate change. iScience 27(9):110838. https://doi.org/10.1016/j.isci.2024.110838

Roychowdhury P, Aftabuddin M, Pati MK (2024) A review on the impact of thermal stress on fish biochemistry. Aquat Sci Eng 39(2):121–129. https://doi.org/10.26650/ASE20231341460

Santos J, Ebert D (2022) The effects of temperature and host-parasite interactions on parasite persistence in a planktonic crustacean. J Freshw Ecol 37:555–568. https://doi.org/10.1080/02705060.2022.2134219

Sarre A, Demarcq H, Keenlyside N, Krakstad JO, Ayoubi S, Jeyid AM, Faye S, Mbaye A, Sidibeh M, Brehmer P (2024) Climate change impacts on small pelagic fish distribution in Northwest Africa: trends, shifts, and risk for food security. Sci Rep 14(1):12684. https://doi.org/10.1038/s41598-024-61734-8

Schaal P, Cheaib B, Kaufmann J, Phillips K, Ryder L, McGinnity P, Llewellyn M (2022) Links between host genetics, metabolism, gut microbiome and amoebic gill disease (AGD) in Atlantic salmon. Anim Microbiome 4(1):53. https://doi.org/10.1186/s42523-022-00203-x

Scholz T, Kuchta R, Brabec J (2019) Cestodes of fish: a global perspective on biodiversity and impact. Parasitol Res 118(7):2107–2120

Shameena SS, Kumar S, Kumar K, Raman RP (2021) Role of temperature and co-infection in mediating the immune response of goldfish. Microb Pathog 156:104896. https://doi.org/10.1016/j.micpath.2021.104896

Shamsi S (2019) *Contracaecum* species in fish: emerging issues in aquaculture and food safety. Int J Parasitol Parasites Wildl 10:142–150

Sures B, Nachev M (2022) Effects of multiple stressors in fish: how parasites and contaminants interact. Parasitology 149(14):1822–1828. https://doi.org/10.1017/S0031182022001172

Tanjaoui Y, Dahani S, Bouchriti N, El Hariri O (2025) Occurrence of parasites in fishery species along the Moroccan coastline. Italian J Food Saf 5:13584. https://doi.org/10.4081/ijfs.2025.13584

Tomamichel MM, Lowe KI, Arnold KMH, Frischer ME, Irwin BJ, Osenberg CW, Hall RJ, Byers JE (2025) Warmer is deadlier: a meta-analysis reveals increasing temperatures accentuate disease effects on fisheries hosts. Ecol Lett 28(7):e70156. https://doi.org/10.1111/ele.70156

Tossavi ND, Gandé JGJ, Togla AI, Bouko B, Aboki-Onzo AA, Sinaré Y, Ibikounlé M (2025) Comparison of parasite prevalence in marine fish *Sardinella maderensis* (Teleostei: Clupeidae) according to different length and fish sex in the waters near the Coast of Benin, West Africa. J Adv Biol Biotechnol 28(1):984–993. https://doi.org/10.9734/jabb/2025/v28i11955

UNECA (2023) State of the climate in Africa 2023. United Nations Economic Commission for Africa & World Meteorological Organization. https://library.wmo.int/idurl/4/70827

Yin F, Yin J, Xie X, Jiang L (2023) Water temperature affects *Cryptocaryon irritans* development, cryptocaryoniasis occurrence and an auxiliary treatment decision-making webpage. Aquaculture 574:739694. https://doi.org/10.1016/j.aquaculture.2023.739694

Ziarati M, Zorriehzahra MJ, Hassantabar F, Mehrabi Z, Dhawan M, Sharun K, Emran TB, Dhama K, Chaicumpa W, Shamsi S (2022) Zoonotic diseases of fish and their prevention and control. Vet Q 42(1):95–118. https://doi.org/10.1080/01652176.2022.2080298

Zhang Y, Li M, Zhao X (2023) Outbreak of *Cryptocaryon irritans* infection in silver pomfret (*Pampus argenteus*) cultured in China. Dis Aquat Org 154:37–48

Chapter 13
Fish Exports and Economic Growth Nexus: Evidence from Namibia

Vistorina Shithigona and Canicio Dzingirai

Abstract The fisheries and aquaculture sector is crucial to Namibia in relation to income generation, trade, food security, and protein provisioning. Considering this paramount importance, the primary objective of this chapter is to investigate the relationship between fish exports and inclusive growth in Namibia, focusing mainly on two secondary objectives using nonlinear autoregressive distribution (NARDL) and vector-error correction (VECM) analysis techniques covering the period 2008Q1–2020Q4. First, on the asymmetric impact of fish exports on inclusive growth, the results from the ARDL show that in both the short run and long run, a decrease in fish export revenue poses greater and nuanced welfare costs to Namibians than the benefits brought about by a similar magnitude increase in fish exports. Second, on the direction of the causal relationship between fish exports and inclusive growth, the evidence from VECM confirms the presence of bidirectional causality between fish exports and growth in the long run and unidirectional causality from both positive and negative fish export shocks to economic growth in the short run. The policy implication from this chapter is for relevant authorities to promote measures that foster and safeguard fish production through sustainable harvesting by adherence to fishing quotas and supporting small-scale inland fish producers. This avoids recurrence of the *tragedy-of-the-commons* problem and dangers to the fisheries and aquaculture sector posed by events like the "fishrot" scandal.

V. Shithigona
Social Security Commission, Windhoek, Namibia

C. Dzingirai (✉)
Department of Economics, University of Namibia, Windhoek, Namibia
e-mail: cdzingirai@unam.na

H. A. Mupambwa et al. (eds.), *Harnessing Coastal and Marine Resources for Improved Food Security in Africa*, Sustainability Sciences in Asia and Africa,
https://doi.org/10.1007/978-981-95-9725-3_13

13.1 Introduction

The fisheries and aquaculture sector plays a significant role in global and national economies. In 2022, global fisheries and aquaculture production reached 223.2 million tons, valued at approximately US$452 billion, with fishery exports totaling around 65 million tons and generating US$178 billion in export value, 60% of which came from developing countries (Food and Agriculture Organization [FAO] 2024). In Namibia, the fisheries sector contributes over N$9 billion annually, accounting for 4.1% of GDP, and employs more than 15,000 people, making it the third largest employer after mining and agriculture (International Labor Organization [ILO] 2022; Moyo 2022). Thus, the fishing sector holds a pivotal position within Namibia's economy, recognized as the third largest contributor to the nation's Gross Domestic Product (GDP) and a significant source of employment, ranking third in Africa and 30th globally in terms of annual marine landings. In 2022, in Africa, at third position, Namibia's share of aquatic animal exports was 9.3% of Africa's total share and 0.3% of Africa's world contribution of 4.4% (FAO 2025). In 2023, fish products made up 14% of Namibia's total exports, ranking as the second largest export category after commodities, and more than 90% of the national fish output is exported.

The export-led growth hypothesis has been widely studied, though findings remain mixed. Some scholars argue that exports drive economic growth (Faisal et al. 2017; Sunde 2017), while on the contrary, other views are that they have minimal impact (Abbas 2012; Mahmood and Munir 2018) and a negative impact on the local community (Institute for Public Policy Research [IPPR] 2014). Fish exports can foster food security, economic growth, and welfare by integrating economies into global markets (United Nations Conference on Trade and Development [UNCTAD] 2017). For Namibia, several studies emphasize the importance of exports for economic growth, suggesting that enhancing export performance could drive development (Amavilah 2003; Jordaan and Eita 2007; Simasiku and Sheefeni 2017; Mosikari and Eita 2020).

Despite this potential, Namibia's fisheries sector has been shrouded with scandals like the so-called "*Fishrot scandal*" and has seen limited structural transformation since independence and remains heavily reliant on the export of unprocessed fish products (Chiripanhura and Teweldemedhin 2016; Namibia Statistics Agency (NSA) 2022; IPPR 2014). These challenges potentially cause socioeconomic problems in Namibia like food insecurity, unemployment, and inequality, just to mention a few topical ones. Calls have been made for further research into the sector's socioeconomic effects using econometric models specifically to gain better insights into its economic impact. While the government has introduced export-promoting initiatives like the Export Processing Zone (EPZ), the overall structure of Namibia's economy has remained largely unchanged (National Planning Commission [NPC] 2016).

The Ministry of Fisheries and Marine Resources has raised serious concerns over the depletion of fish stocks, especially in Namibia's inland rivers, lakes, and dams due to unsustainable fishing methods, weak law enforcement, and lack of ownership. Weak enforcement of the customary land tenure systems, sections 22 and 29 of the Inland Fisheries Resources Act of 2003, and opaque ownership over these open-access fisheries are the main drivers of this observed *tragedy of the commons* problem. Climate change also brings in another layer of threat to Namibia's fisheries and aquaculture sector, which results in socioeconomic challenges, especially to small-scale, recreational, and rock lobster fisherman, for example, of small pelagic fisherman due to the collapse of pilchard (Engelhard et al. 2024). Besides the ongoing concerns about declining fish stocks for several species since independence, other problems include the pervasive issue of bycatch (unintentional capture of non-target species), growing environmental impacts stemming from other industrial activities like seismic surveys related to oil and gas exploration, and the threats posed by proposed phosphate mining (Chiripanhura and Teweldemedhin 2016).

Despite the pivotal role of the fishery and aquaculture sector in food security, nutrition, employment, income generation, and trade in Namibia, interestingly, to date, no econometric study has specifically focused on examining the relationship between fish exports and economic growth in Namibia. Though Mosikari and Eita (2020) looked at the asymmetric impact of manufactured, diamonds, food, and live animal exports on Namibia's economic growth, they did not specifically focus on the fish sector exports. Therefore, no study to the best of our knowledge has used a nonlinear ARDL (NARDL) to examine the fish exports—growth nexus in Namibia. This lack of empirical evidence highlights the need for rigorous investigation to inform policy decisions. As Natale et al. (2015) argue, evidence-based policymaking is crucial for gaining insights and maximizing the socioeconomic benefits of key sectors like fisheries. It is for that reason that this chapter examines the asymmetric effects of fish exports on economic growth in Namibia using NARDL.

13.2 Related Theoretical and Empirical Literature

International trade theories provide the foundation for understanding how nations can benefit from engaging in global commerce. These theories offer varying perspectives on the mechanisms through which trade enhances economic efficiency, resource allocation, and national income. The relevant theories considered are discussed below.

13.2.1 Absolute Advantage Theory

Adam Smith's theory of absolute advantage argues that countries should specialize in the production of goods they can produce more efficiently, either at a lower cost or faster than others and trade for goods they produce less efficiently (Smith 1977). Smith opposed protectionist policies, suggesting they often lead to retaliation and inefficiencies in global trade. For example, if Namibia can produce fish and fish products more efficiently than its trading partners, it should focus on producing and exporting these, while importing goods that are more efficiently produced elsewhere. This specialization improves productivity and innovation, raising living standards, which Smith viewed as the true measure of a nation's wealth, not the accumulation of gold and silver.

13.2.2 Comparative Advantage Theory

On the other hand, Ricardo (1817) extended Smith's work by introducing the theory of comparative advantage, which emphasizes opportunity costs. Ricardo (1817) argued that even if a country is more efficient in producing all goods, it should still specialize in those with the lowest opportunity cost and import the rest. For instance, if Namibia produces fish at a lower opportunity cost than other goods, it should concentrate on fish production and trade the surplus for other products. This mutual specialization ensures that both countries benefit from trade by maximizing resource use and overall production.

13.2.2.1 Coase Theorem

The Coase Theorem posits that, under ideal conditions of well-defined property rights and negligible transaction costs, private parties can negotiate to resolve externalities and manage to provide public goods and services more efficiently than the government, irrespective of the initial allocation of those rights (Coase 1960). This contradicts Pigou's (1932) view that such market failures like over-fishing necessitated government intervention, typically through taxes or subsidies, which will enforce social efficiency and eradicate the market failure problem. Therefore, these two approaches aimed to align private costs with social costs, thereby guiding resource allocation toward a socially optimal outcome from different angles, namely, private and public routes. There are six strong assumptions that make the Coase theorem applicable, namely, clearly defined property rights, low transaction costs, few affected parties, rational behavior, perfect information, and no wealth effects. Though some of the assumptions are applicable to the Namibia fishing sector like existence of open-access fisheries and defined property rights through its quota system, traditional customary land tenure systems and community-managed fisheries or fishery reserves created under sections 22 and 29 of the Inland Fisheries Resources Act of 2003, the nontransferability of these quotas and the presence of high transaction costs fundamentally limit the direct applicability of a pure Coasean bargaining approach. The main challenge is on the enforcement of the quota system, the traditional customary land tenure system, and sections 22 and 29 of the Inland Fisheries Resources Act of 2003. This led to current challenges of bycatch, overfishing, and scandals like "*fishrot*," which are an epitome of "*tragedy of the commons*" problem currently bedeviling Namibia's fish sector.

13.2.3 The Export-Led Growth Hypothesis

The export-led growth hypothesis (ELGH) posits that increasing exports is associated with higher national income. The ELGH advocates that in addition to increasing overall growth by increasing labor and capital, countries can also expand their level of exports (Medina-Smith 2000). Since exports are a component of GDP as indicated by the Keynesian national income identity, theoretically, the spillovers from international trade could considerably stimulate national output.

The export-led growth hypothesis has been examined by several researchers around the world (Michaely 1977; Balassa 1978, 1985; Ram 1985, 1987; Bhagwati 1988; Alam 1991; Salvatore and Hatcher 1992; Onafowora and Owoye 1998; Tivatyi et al. 2022). They concluded that export expansion promotes economic growth. Ali et al. (2018) added to the evidence in support of the export-led growth hypothesis in Somalia using data from 1970 to 1991. In addition, supporting the ELGH and import-led-growth hypothesis (ILGH), Bakari and Mabrouki (2017) find evidence confirming that exports and imports Grangercause economic growth in Panama.

13.2.4 The Growth-Led Exports Hypothesis

The growth-led export hypothesis is the opposite of the exports-led growth hypothesis. Proponents of the growth-led exports hypothesis hold contrary to the export-led hypothesis that it is rather an increase in output that stimulates exports (Bhagwati 1988; Abbas 2012; Shihab et al. 2014; Tivatyi et al. 2022).

In his study, Bhagwati (1988) postulates that the growth-led exports hypothesis reinforces the neoclassical trade theory where the author believes that economic development should focus on developing both demand and supply sides of the economy. Among others, Mahmood and Munir (2018) found an insignificant positive long run relationship between agricultural exports and economic growth in Pakistan and evidence in support of the growth exports hypothesis.

Also, Shihab et al. (2014) provided further evidence of the growth-led exports hypothesis in Jordan. The data covered the period from 2000 to 2012. The results present evidence of unidirectional causality between export and economic growth in Jordan and the direction of causality is strictly from economic growth to exports. In the same vein, Shihab et al. (2014) showed evidence of unidirectional Granger causality from economic growth to exports. This means that policy makers should increase production to increase the level of exports.

13.2.5 Relevant Empirical Evidence

Kalaitzi and Chamberlain (2020) conducted a study to test the export-led-growth (ELG) hypothesis in five Gulf Cooperation Council (GCC) countries, Bahrain, Kuwait, Oman, Saudi Arabia, and the United Arab Emirates, using annual data from 1975 to 2016. They employed the Johansen cointegration test to assess long-run relationships and applied multivariate Granger causality tests within a vector autoregressive (VAR) model, modified Wald test, to examine causal directions in both the short and long run. The findings revealed that the ELG hypothesis holds in the short run for the United Arab Emirates, suggesting that increases in exports can stimulate short-term economic growth. Additionally, the study identified a bidirectional causal relationship between exports and economic growth in Kuwait, implying mutual reinforcement between the two variables. In the long run, the ELG hypothesis was validated in the case of Bahrain, where exports were found to drive economic growth. Conversely, in Kuwait and Saudi Arabia, the direction of causality runs from economic growth to exports, indicating that it is economic expansion that spurs trade activity in these countries rather than the other way around.

Similarly, Öztürk and İlbeyli (2020) assessed the validity of the export-led growth hypothesis in Turkey for the period 2002Q1–2018Q4. Their study applied the Toda-Yamamoto causality test. The results showed evidence in support of the growth-led exports in the manufacturing sector and the presence of export-led growth in the agriculture, forestry, and fishing sectors. However, the results found no evidence of causality between the variables under study in the service sector. The causal relationship between the total exports and GDP proved to be bidirectional. Lastly, it was advised that the government should implement policies that stimulate sustainable and inclusive smart agriculture, forestry, and livestock sectors to promote economic growth in the country.

Consistent with the export-led-growth hypothesis, Sunde (2017) investigated the nature of the relationship between economic growth, exports, and foreign direct investments in South Africa using annual data from 1970 to 2014. The study employed a linear Autoregressive Distributed Lag (ARDL) model and VECM Granger causality approach. In addition, co-integration results showed evidence of long-term relationships among the variables under study, unidirectional causality from foreign direct investments to exports, and bi-directional causality between exports and economic growth.

In another related study, Mahmood and Munir (2018) found an insignificant positive long-run relationship between agricultural exports and economic growth in Pakistan. The evidence supports the growth exports hypothesis. Their study considered data on agricultural export (AX), Consumer Price Index (CPI), Gross Domestic Product Growth (GDPG), Gross Capital Formation (GCF), and Labor Force Participation Rate (LFPR) from 1970 to 2014. The Granger causality test revealed evidence of causality from GCF to AX, from AX to LFPR, and from GDPG to AX. In contrast, Faisal et al. (2017) employed the linear ARDL bounds testing approach for cointegration to examine the nature of the relationship between economic growth, exports, and imports in Saudi Arabia. Their study employed annual data covering the period 1968–2014. The evidence confirms that exports pose a strong positive impact on economic growth as well as a high speed of adjustment toward the long-run equilibrium. Moreover, the Granger causality results show evidence of a unidirectional causality from exports to economic growth, validating the export-led-growth hypothesis. The study further found evidence of unidirectional causality from imports to exports.

Moreover, Simasiku and Sheefeni (2017) showed that agricultural exports in Namibia do not influence economic growth, while non-agricultural exports significantly impact growth. Their study used quarterly time series data from 1990 to 2014 using the Johansen cointegration approach. The results further show a long-run relationship between agricultural exports and economic growth in the country, which supports the export-led hypothesis.

In Asia, Sothan (2016) examined the impact of FDI and exports on the economic growth of 21 Asian countries for the period between 1980 and 2013 using panel cointegration and Granger causality analysis. The results confirmed the presence of a long-term equilibrium relationship between the variables under study. Furthermore, the panel Granger causality analysis found evidence of the existence of a strong bidirectional causal relationship between variables. It was therefore concluded that the export-led growth and the growth-led exports hypothesis hold for Asian countries.

13.3 Data and Methodology

13.3.1 Variables and Data Sources

Table 13.1 provides a description of variables. Due to complete secondary time series data availability and recent data constraints, especially for fish export revenue, the chapter covered the first quarter of 2008 to the last quarter of 2020. The data were adjusted for inflation.

13.3.2 Empirical Model

The following model was adopted by this study:

$$\text{GDPPC} = f\left(\text{FX}^{+}, \text{FX}^{-}, M2\right). \tag{13.1}$$

The elasticity model can be presented as

$$\ln \text{GDPPC}_t = \beta_0 + \beta_1 \ln \text{FX}_t^{+}, \beta_2 \ln \text{FX}_t^{-} + \beta_3 \ln \text{M2}_t + \varepsilon_t, \tag{13.2}$$

Table 13.1 Description of variables

Variable	Description	Source	Expected sign
Gross Domestic Product (GDPPC)	The Gross Domestic Product per Capita is the dollar value of all final goods and services produced within the borders of a country divided by the nation's population	Namibia Statistics Agency	Dependent variable
Broad Supply (M2)	M2 is a measure of money supply that includes currency in the hands of the public, checking deposits, money market, unit trusts, and forms of near money that can be easily changed to cash	Bank of Namibia	Positive
Fish Exports (FX)	This refers to the sale of fish products by a country to other countries. Fish export revenue is measured in billion N$	Namibian Statistics Agency	Positive

where ln $GDPPC_t$ is symbolized as per capita real GDPPC (N$), ln M_t is real money supply data (N$), $\ln \mathrm{FX}_t^+$ symbolizes positive changes in fish exports, and $\ln \mathrm{FX}_t^-$ symbolizes negative changes in fish exports. β_0 is a constant and β_1, β_2, and β_3 are the coefficients of the variables.

Following Shin et al. (2014), fish exports were decomposed into positive and negative shocks as reported in Eqs. 13.3 and 13.4.

$$\ln \mathrm{FX}_t^+ = \sum_{i=1}^{t} \Delta \ln \mathrm{FX}_t^+ = \sum_{i=1}^{t} \max\left(\Delta \mathrm{Ln}\, \mathrm{FX}_t, 0\right) \tag{13.3}$$

$$\mathrm{Ln}\, \mathrm{FX}_t^- = \sum_{i=1}^{t} \Delta \mathrm{Ln}\, \mathrm{FX}_t^- = \sum_{i=1}^{t} \min\left(\Delta \mathrm{Ln}\, \mathrm{FX}_t, 0\right) \tag{13.4}$$

Incorporating Eqs. 13.3 and 13.4 into Eq. 13.2 gives a fully NARDL form Eq. 13.5:

$$\begin{aligned}\Delta \ln \mathrm{GDPPC}_t &= \alpha_0 + \sum_{i=1}^{q1} \varphi_{0i} \Delta \ln \mathrm{GDPPC}_{t-i} + \sum_{i=0}^{q2} \varphi_{1i}^+ \Delta \ln \mathrm{FX}_{t-i}^+ + \sum_{i=0}^{q3} \varphi_{2i}^- \Delta \ln \mathrm{FX}_{t-i}^- \\ &+ \sum_{i=0}^{q4} \omega_{3i} \ln M2_{t-i} + \gamma_0 \mathrm{GDPPC}_{t-1} + \gamma_1^+ \ln \mathrm{FX}_{t-1}^+ + \gamma_2^- \ln \mathrm{FX}_{t-1}^- + \gamma_3 \ln M2_{t-1} + \varepsilon_t\end{aligned} \tag{13.5}$$

Accordingly, the short-run asymmetric effects of fish exports are given by φ_{2i}^+ for positive shocks and φ_{2i}^- for negative shocks in eq. 13.5. At the same time, the long-run asymmetric effects are given by γ_1^+ for positive changes and by γ_1^- for negative changes.

13.4 Results and Discussion

13.4.1 Descriptive Statistics

Table 13.2 presents the descriptive statistics of the variables considered by the study. The mean for per capita GDP (GDPPC) is N$5130.71; for positive changes in LNFX_POS, it is 2.89, for negative changes; it is −2.34; and for LNM2, it is 11.17. The skewness of all variables is close to zero, which shows that they are clustered around their means, an indication of normality, as also confirmed by insignificant Jarque–Bera statistics.

13.4.2 Unit Root Results

Unit root test results in Table 13.3 revealed that M2 and GDPPC are integrated of first order and while FX stationary in levels. The combination of mixed order of integration of the independent variable series, dependent variable being I(1) and absence of I(2) variables justifies NARDL as the most suitable technique to investigate the long-run and short-run relationship between fish export and Namibia's economic growth.

13.4.3 Cointegration Test Results

The bound cointegration test (Pesaran et al. 2001) results in Table 13.4 confirm the presence of cointegration, because the computed F-statistic of 6.465 is greater than the lower and upper bounds at 5%. This complements the unit root test in Table 13.3, further cementing the suitability of NARDL model.

Table 13.2 Descriptive statistics

	LNGDPPC	LNFX_POS	LNFX_NEG	LNM2
Mean	8.543	2.894	−2.339	11.171
Median	8.467	2.671	−2.048	11.217
Maximum	8.768	5.115	0.000	11.743
Minimum	7.798	0.149	−4.369	10.241
Std. Dev.	0.337	1.568	1.307	0.377
Skewness	−0.232	−0.058	0.061	−0.552
Kurtosis	1.425	1.718	1.817	2.894
Jarque–Bera	5.724	3.522	3.003	2.614
Probability	0.057	0.172	0.223	0.271
Sum	527.353	147.618	−119.278	569.679
Sum Sq. Dev.	5.683	122.857	85.438	7.089
Observations	**52**	**52**	**52**	**52**

Source: Authors' computation using EViews 10

Table 13.3 Unit root results

	ADF test		PP test	
Variable	Levels	First differences	Levels	First differences
lnGDPPC	−0.691	−7.863***	−0.524	−10.472***
lnM2	−0.761	−5.015***	−2.691	−7.206***
lnFX	−5.563***	–	−6.954***	–

Note: ***$p < 0.01$, **$p < 0.05$, *$p < 0.1$. indicate significance at 1%, 5%, and 10%, respectively
Source: Authors' computation using EViews10

Table 13.4 Cointegration results

Dependent: GDPPC	F-statistics	Level	Lower bound	Upper bound	Outcome
	6.465***	10%	2.723	3.775	Cointegrated
		5%	3.232	4.357	
		1%	4.291	5.618	

Note: ***, **, * denote significance at 1%, 5%, and 10% level, respectively
Source: Authors' computation using EViews 10

13.4.4 VECM Granger Causality Results

Table 13.5 presents the VECM Granger causality results. The results show that, in the long run, there is evidence of bidirectional causality among fishing exports and growth. In the short run, causality results show that there is unidirectional causality running from both lnFX_NEG and lnFX_POS to lnGDPPC. Therefore, to stimulate economic growth, the Namibian government, through the Ministry of Fisheries and Marine Resources, must take measures meant to boost fish production through sustainable harvesting and supporting small-scale inland fish producers. This contradicts bidirectional (unidirectional) causality evidence found by Sothan (2016) and Mahmood and Munir (2018) between fish exports and economic growth and supports the export-led-growth hypothesis found by Faisal et al. (2017) and Simasiku and Sheefeni (2017).

13.4.5 Long-Run and Short-Run NARDL Results

The long- and short-run NARDL results are shown in Table 13.6.

Both short- and long-run results showed that both lnFX_NEG and FX_POS have significant impacts on growth, with negative (positive) fish export shocks posing an elastic (inelastic) impact on Namibia's economic growth. This asymmetric evidence shows that a decrease in fish export revenue poses a greater welfare cost to Namibia than benefits brought about by a similar increase in fish exports. These results are consistent with the findings by Jaunky (2011) and Jawaid et al. (2019), who presented evidence of strong long-run relationships between fish exports and economic growth. The speed of adjustment toward the long-run equilibrium of economic growth was found to be 11.2%.

Table 13.5 Granger causality results

	SR causality				LR causality
Dependent variable	Δln GDPPC	Δln FX_NEG	Δln FX_POS	Δln M2	ECT_{t-1}
ΔlnGDPPC	–	−0.338** (0.015)	0.125** (0.048)	−0.122 (0.071)	−0.027** (0.039)
ΔlnFX_NEG	0.046 (0.538)	–	0.302 (0.129)	0.049 (0.831)	0.156*** (<0.001)
ΔlnFX_POS	−0.322 (0.311)	0.401 (0.215)	–	0.358 (0.193)	0.112*** (<0.001)
ΔlnM2	0.017 (0.921)	−0.197** (0.032)	−0.013 (0.812)	–	−0.023 (0.237)

Note: ***, **, * denote significance at 1%, 5%, and 10% level, respectively
Source: Authors' computation using EViews10

Table 13.6 Long- and short-run NARDL results

Variable	Coefficient	t-Statistic	Prob.
Long-run model			
lnFX_POS	0.213***	3.024	0.0040
lnFX_NEG	−1.125***	4.013	<0.001
lnM2	−0.045	−0.318	0.7517
Short-run model			
ΔlnGDP (−1)	0.208	1.291	0.2048
ΔlnGDP (−4)	0.254*	1.873	0.069
ΔlnFX_POS	0.103**	2.214	0.033
ΔlnFX_POS (−1)	0.066	−1.495	0.144
ΔlnFX_POS (−2)	0.159***	3.670	<0.001
ΔlnFX_NEG (−1)	−1.108***	−3.305	0.002
ECT (−1)	**−0.112*** **	**−6.249**	**<0.001**

Note: ****p* 0.01, ***p* 0.05, **p* 0.1. indicate significance at 1%, 5%, and 10%, respectivel
Source: Authors' computation using EViews 10

13.4.6 Long-Run and Short-Run Asymmetry Test

The presence of long-run asymmetry was tested by evaluating the hypothesis below: $H_o : \frac{-\Upsilon_{1i}^{+}}{\Upsilon_{0i}} = \frac{-\Upsilon_{2i}^{-}}{\Upsilon_{0i}}$ (this means that the two impacts are the same, i.e., symmetric impact)

$H_A : \frac{-\Upsilon_{1i}^{+}}{\Upsilon_{0i}} \neq \frac{-\Upsilon_{2i}^{+}}{\Upsilon_{0i}}$ (this means that there is a long-run asymmetry impact)

While the presence of short-run asymmetry was tested by

$H_o : \sum_{i=0}^{q2} \varphi_{1i}^{+} = \sum_{i=0}^{q2} \varphi_{2i}^{-}$ (presence of short-run symmetric impact)

Table 13.7 Long and short-run asymmetry

F-statistics		*P*-value
Long-run	6.108**	0.019
Short-run	5.013**	0.033

Note: ***, **, and * indicate the level of significance at 1%, 5%, and 10%, respectively
Source: Authors' computation using EViews 10

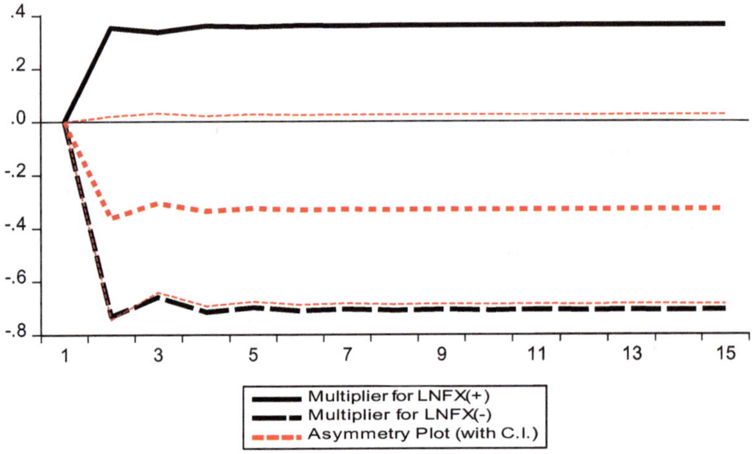

Fig. 13.1 Asymmetric cumulative dynamic multiplier. (Source: Authors' computation using EViews 10)

$$H_o : \sum_{i=0}^{q2} \varphi_{1i}^{+} \neq \sum_{i=0}^{q2} \varphi_{2i}^{-} \text{ (presence of short-run asymmetric impact)}$$

The results for long and short-run asymmetry in Table 13.7 revealed evidence of long- and short-run asymmetry relationship between lnGDPPC and positive and negative changes in lnFX. Figure 13.1 also confirms the presence of asymmetric effects of fish exports, where negative shocks have more nuanced and pronounced effects than positive shocks. This observation confirms the novelty of this study over previous extant studies that assumed and applied linear methodologies without testing for potential asymmetric relationships, especially when examining the export-growth nexus using Namibia as a case study. It also validates our evidence of the existence of asymmetric impact of positive and negative fish export revenue shocks on Namibia's economic performance.

Table 13.8 Diagnostics test results

Type of test	Test statistics	Probability
Breusch–Pagan–Godfrey	1.002641	0.4938
Breusch–Godfrey serial correlation LM	0.489692	0.6193
Ramsey Reset test	1.334176	0.2575
Jarque–Bera test	5.5581	0.9045

Source: Authors' computation using Eviews9

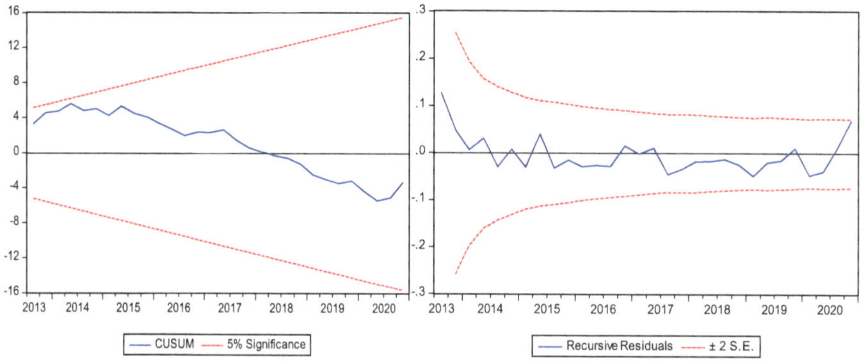

Fig. 13.2 CUSUM stability test. (Source: Authors' computation using EViews 10)

13.4.7 Post-estimation Diagnostic Tests

Table 13.8 presents the model diagnostics test results for the adopted model, which is a NARDL (4, 2, 1, 0).

The results of Breusch–Pagan–Godfrey and Breusch–Godfrey serial correlation LM tests show that there is no heteroskedasticity and autocorrelation since the corresponding probability values are more than 5%. Equally so, the probability value of the Ramsey RESET and Jarque–Bera tests of 0.1373 and 0.9045 NARDL (4, 2, 1, 0) is correctly specified and residuals normally distributed. The model is therefore stable and the null hypothesis that the model is stable is not rejected. This is further confirmed by the cumulative sum (CUSUM) and the CUSUM of squares (CUSUMQ) of recursive residuals presented in Fig. 13.2, which confirm model stability.

13.5 Conclusion and Policy Recommendations

Provoked by the importance of fisheries and aquaculture sector in Namibia in relation to food security and protein provisioning, income generation, and trade, the main aim of this study is to investigate the asymmetric nexus between fish exports and economic growth using Namibia as a laboratory. To achieve the main objective, the study employed a nonlinear Autoregressive Distributed Lag (NARDL) model and vector error correction (VEC) causality technique. Granger causality results support the export-led-hypothesis, where both positive and negative exports shocks Grange-cause economic growth in an asymmetric way. In both the short and long run, negative export revenue shocks have a more pronounced asymmetrical impact on economic growth than that of positive shocks of the same change magnitude. The policy implication arising from this study evidence is that, to stimulate economic growth, the Namibian government through the Ministry of Fisheries and Marine Resources must take measures to boost fish production through sustainable harvesting and supporting small scale inland fish producers. Proper enforcement and adherence to fishing quotas must be observed and hefty penalties to offenders should be implemented without fear or favor. It is also suggested that fish exports in Namibia can be increased by improving infrastructure and port facilities as well as improving transportation networks. Furthermore, to guarantee sustainable fish harvesting, the Ministry of Agriculture, Fisheries, Water and Land Reform should strengthen the Total Allowable Catch (TAC) fishing quota system, and unsure such fishing quotas are allocated in a more transparent manner which is free from corruption. In light of climate change, climate vulnerabilities mitigation measures must be implemented. The problem of bycatch needs to be legislated and guidelines should be clearly and succinctly elaborated to avoid reduction in fish stocks of some species like pilchards.

Declining fish stocks in inland rivers, lakes, and dams call for the Ministry of Fisheries and Marine Resources to engage communities to manage fishery resources within their traditional customary land tenure systems. These community-managed fisheries or fishery reserves created under sections 22 and 29 of the Inland Fisheries Resources Act of 2003 made Namibia a leading nation in Africa when it comes to empowering small-scale fishers in climate-smart governance of tenure and equitable access to valuable inland aquatic resources. To enhance food security and improve the nutrition of community members as well as socioeconomic development, the "*tragedy of the commons*" problem especially on open-access fishing areas need to be nipped in the bud through proper ownership legislation as well as training and equipping traditional leaders on best aquatic conservation under customary law. To address the challenges of Inland Fisheries, reduce unemployment, and inequality, the MFMR should also speed up the implementation of a National Plan of Action for Small-Scale Fisheries (NPOA-SSF) such that it will complement the Voluntary Guidelines on Securing Sustainable Small-scale Fisheries and Guidelines on Food Security and Poverty Eradication.

13.5.1 Study Limitations and Further Research Suggestions

A key limitation of this study is the unavailability of fish export data from 1990 to 2007, which restricted the analysis from capturing the relationship between fish exports and economic growth in the immediate post-independence period. Though the methodology used (NARDL) is powerful for both small and large samples, as a cautionary caveat, the findings should be interpreted with caution, acknowledging this temporal gap. Also, gross capital formation was omitted as control variable since the model was over-parameterized when we include it together with other control variables. But this potential omission bias did not seriously compromise the quality of the study since the model was found to be correctly specified based on relevant diagnostic check. Future research is encouraged to explore the relationship between fish exports and economic growth using alternative methodologies like threshold models or extended datasets that may capture earlier trends. Moreover, in light of *Fishrot* scandal, this study highlights the need for further investigation into effective regulatory mechanisms like institutional quality for Namibia's fishing sector, particularly to curb and prevent corruption. Ensuring transparency and accountability in the sector could enhance its contribution to sustainable economic development. Lastly, researchers are encouraged to broaden the scope of inquiry by examining the interaction between fish exports and other macroeconomic variables such as exchange rates, government revenue, and public debt, and Gross capital formation. Such studies would provide a more comprehensive understanding of the sector's role in Namibia's economic performance.

References

Abbas S (2012) Causality between exports and economic growth: investigating suitable trade policy for Pakistan. Eurasian J Bus Econ 1(5):91–98
Alam SM (1991) Trade orientation and macroeconomic performance in LDCs: an empirical study. Econ Dev Cult Chang 42(1):839–848
Ali AA, Ali AYS, Dalmar MS (2018) The impact of imports and exports performance on the economic growth of Somalia. Int J Econ Financ 10(1):110–119
Amavilah VH (2003) Exports and economic growth in Namibia, 1968–1992. University of Arizona Working Paper, Tucson, AZ
Bakari S, Mabrouki M (2017) Impact of exports and imports on economic growth: new evidence from Panama. J Smart Econ Growth 2(1):67–79
Balassa B (1978) Export incentives and export performance in developing countries: a comparative analysis. Rev World Econ 114(1):24–61
Balassa B (1985) Exports, policy choices, and economic growth in developing countries after the 1973 oil shock. J Dev Econ 18(1):23–35
Bhagwati JN (1988) Export-promoting trade strategy: issues and evidence. World Bank Res Obs 1(1):27–57
Chiripanhura B, Teweldemedhin M (2016) An analysis of the fishing industry in Namibia: the structure, performance, challenges, and prospects for growth and diversification, vol 1(2). International Food Policy Research Institute, Washington, DC, pp 66–70
Coase RH (1960) The problem of social cost. J Law Econ 3:1–44
Engelhard GH, Bova C, Gusha MNC et al (2024) Climate risk assessment of the fisheries in Namibia. Rev Fish Biol Fish 34:1231–1260
Faisal F, Türsoy T, Reşatoğlu NG (2017) Is export-led growth hypothesis exist in Saudi Arabia? Evidence from an ARDL bounds testing approach. Asian J Econ Model 5(1):110–117
FAO (2025) Fishery and aquaculture statistics yearbook 2022. Rome. Available at: FAO Yearbook of Fishery and Aquaculture Statistics 2022
Food and Agriculture Organization (FAO) (2024) The state of world fisheries and aquaculture 2024: blue transformation in action. FAO, Rome. https://www.fao.org/3/cc7749en/cc7749en.pdf

Institute for Public Policy Research (IPPR) (2014) Right now, I cannot survive: the legacy effects of the Fishrot corruption scandal on fisheries workers' lives. Available at: Fishrot HRI study report 2.indd
International Labor Organisation [ILO] (2022) Deep-dive research into decent work challenges and opportunities in Namibia's fishing supply chain. Available at: wcms_863628.pdf
Jaunky V (2011) Fish exports and economic growth: the case of SIDS. Coast Manag 39(4):377–395
Jawaid ST, Siddiqui MH, Atiq Z, Azhar U (2019) Fish exports and economic growth: the Pakistan's experience. Glob Bus Rev 20(2):279–296
Jordaan A, Eita J (2007) Export and economic growth in Namibia: a granger causality analysis. South Afr J Econ 75(3):540–547
Kalaitzi AS, Chamberlain TW (2020) Exports and economic growth: some evidence from the GCC. Int Adv Econ Res 26(2):203–205
Mahmood K, Munir S (2018) Agricultural exports and economic growth in Pakistan: an econometric reassessment. Qual Quant 52(4):1561–1574
Medina-Smith EJ (2000) Is the export-led growth hypothesis valid for developing countries? A case study of Costa Rica. Pol Issues Int Trade Commod 7(1):17–21
Michaely M (1977) Exports and growth: an empirical investigation. J Dev Econ 4(1):49–53
Mosikari TJ, Eita JH (2020) Modelling asymmetric relationship between exports and growth in a developing economy: evidence from Namibia. S Afr J Econ Manag Sci 23(1):a2905
Moyo T (2022) Fisheries sector set for rebound. Retrieved from: https://www.namibian.com.na/public/uploads/documents/626bdb6d0bbd2/fishing-low-res.pdf
Namibia Statistics Agency (2022) Gross domestic product (GDP) third quarter 2022. Windhoek, Namibia. Retrieved from: https://d3rp5jatom3eyn.cloudfront.net/cms/assets/documents/Third_Quarter_GDP_20222.pdf
Natale F, Borrello A, Motova A (2015) Analysis of the determinants of international seafood trade using a gravity model. Mar Policy 60(1):98–106
National Planning Commission (2016) NDP5 consultations. Windhoek, Namibia
Onafowora OA, Owoye O (1998) Can trade liberalization stimulate economic growth in Africa? World Dev 26(3):497–506
Öztürk F, İlbeyli MM (2020) Analysis of export-led growth hypothesis on sectoral basis in Turkey. J Instit Econ Dev Soc Res 1(6):581–597
Pesaran MH, Shin Y, Smith RJ (2001) Bounds testing approaches to the analysis of level relationships. J Appl Econ 16:289–326
Pigou AC (1932) The economics of welfare. MacMillan & Co, London
Ram R (1985) Exports and economic growth: some additional evidence. Econ Dev Cult Chang 33:415–425
Ram R (1987) Exports and economic growth in developing countries: evidence from time-series and cross-section data. Econ Dev Cult Chang 36(1):51–72
Ricardo D (1817) On the principles of political economy and taxation (John Murray, London). In: Sraffa P (ed) The works and correspondence of David Ricardo, vol 1. Cambridge University Press, Cambridge. 1951
Salvatore D, Hatcher T (1992) Inward oriented and outward oriented trade strategies. J Dev Stud 27:7–25
Shihab RA, Soufan T, Abdul-Khaliq S (2014) The causal relationship between exports and economic growth in Jordan. Int J Bus Soc Sci 5(3):302–308
Shin Y, Yu BC, Greenwood-Nimmo M (2014) Modelling asymmetric cointegration and dynamic multipliers in a nonlinear ARDL framework. In: Sickels R, Horrace W (eds) Festschrift in honor of Peter Schmidt: econometric methods and applications. Springer, New York, pp 281–314
Simasiku C, Sheefeni JP (2017) Agricultural exports and economic growth in Namibia. Eur J Basic Appl Sci 4(1):20–22
Smith A (1977) An inquiry into the nature and causes of the Wealth of Nations, reprints. J. M. Dent and Sons, London

Sothan S (2016) Foreign direct investment, exports, and long-run economic growth in Asia: panel cointegration and causality analysis. Int J Econ Financ 8(1):26–37

Sunde T (2017) Foreign direct investment, exports, and economic growth: ADRL and causality analysis for South Africa. Res Int Bus Financ 41(3):434–444

Tivatyi K, Shou J, N'Souvi K (2022) Study on import and export-led economic growth: cases of Botswana, Namibia, South Africa, and Zimbabwe in southern Africa. Open J Bus Manag 10:670–700

United Nations Conference on Trade and Development (UNCTAD) (2017) Fishery exports and the economic development of least developed countries: Bangladesh, Cambodia, The Comoros, Mozambique, Myanmar and Uganda. UNCTAD working paper No. 2

Chapter 14
Valorization of Marine Fishery Waste for Sustainable Aquafeed Production in Africa

Kenneth Prudence Abasubong, Mfon Timothy Udo, Ndakalimwe Naftal Gabriel, Ubong Uwem George, Aniefiokmkpong Okokon Okon, Michael Chukwuemeka Umehai, Umoh Uwana Emmanson, and Eyo Idorenyin Edem

Abstract The manufacture of aquafeed is still a major obstacle to the long-term growth of aquaculture in Africa, mainly due to the high price and restricted availability of imported feed ingredients. At the same time, the continent's fishing and coastal industries produce large amounts of nutrient-rich marine and fisheries waste, such as fish heads, bones, viscera, shrimp shells, and crustacean exoskeletons, which are frequently thrown away incorrectly and largely underutilized, seriously degrading the environment. The possibility of turning these leftovers into sustainable, alternative raw materials for regional aquafeed production is examined in this research. Fish waste may now be transformed into premium, protein-rich feed ingredients thanks to advancements in processing technologies, including rendering, fermentation, and enzymatic hydrolysis. This can increase fish growth and

K. P. Abasubong (✉) · U. U. George · U. U. Emmanson · E. I. Edem
Department of Fisheries and Aquaculture, Faculty of Agriculture, Akwa Ibom State University (AKSU), PMB 1167, Obio Akpa Campus, Oruk Anam, Akwa Ibom State, Nigeria
e-mail: kennethabasubong@aksu.edu.ng

M. T. Udo
Department of Fisheries and Aquaculture, Faculty of Agriculture, Akwa Ibom State University (AKSU), PMB 1167, Obio Akpa Campus, Oruk Anam, Akwa Ibom State, Nigeria

Department of Fisheries and Environmental Management, University of Uyo, Uyo, Akwa Ibom State, Nigeria

N. N. Gabriel
Department of Fisheries and Ocean Sciences, Faculty of Agriculture, Engineering, and Natural Sciences, Sam Nujoma Campus, University of Namibia, Henties Bay, Namibia

A. O. Okon
Department of Animal and Environmental Biology, University of Uyo, Uyo, Akwa Ibom State, Nigeria

M. C. Umehai
Department of Agricultural Extension, Delta State University, DELSU, Abraka, Nigeria

H. A. Mupambwa et al. (eds.), *Harnessing Coastal and Marine Resources for Improved Food Security in Africa*, Sustainability Sciences in Asia and Africa,
https://doi.org/10.1007/978-981-95-9725-3_14

health, decrease production costs, and lessen dependency on pricey imports. African nations can boost coastal economies, reduce environmental effects, increase resource efficiency, and promote food and nutritional security by incorporating marine waste valorization into aquaculture systems. This chapter outlines the environmental, economic, and scientific justifications for implementing waste-to-feed practices. It also identifies the main obstacles and opportunities for expanding these programs throughout the continent, which will support the growth of aquaculture sustainably and the circular bioeconomy.

Keywords Africa · Fish waste · Circular bioeconomy · Food security · Ocean · Valorization

14.1 Introduction

Harnessing the ocean's potential for sustainable aquaculture in Africa requires addressing a critical bottleneck: aquafeed production. Despite progress in countries like Egypt, Nigeria, and Uganda, the aquaculture sector remains heavily reliant on expensive imported high-protein floating pellets, which are largely inaccessible to the smallholder farmers that dominate the industry (Ragasa et al. 2022; Subasinghe et al. 2021). This dependence not only inflates production costs but also exposes the sector to global market fluctuations, trade restrictions, and supply chain disruptions (Olufemi-Phillips et al. 2024). At the same time, local feed production struggles due to limited availability and inconsistent quality of raw materials, rising prices of conventional protein sources such as fishmeal and soybean meal, and knowledge gaps in feed formulation and fish nutrition (Ndebele-Murisa et al. 2024). As a result, many farmers resort to substandard or homemade feeds that fail to meet species-specific needs, leading to poor fish growth, heightened disease risk, and lower profitability, hindering aquaculture's role in food security and economic development across Africa.

In this context, Africa's coastal regions hold an untapped, high-value resource: vast amounts of marine and fishery waste, including fish heads, bones, viscera, shrimp shells, and crustacean exoskeletons. Much of this waste is discarded improperly into landfills or marine environments, contributing to severe coastal pollution, eutrophication, hypoxia, and damage to vital marine habitats, thereby threatening biodiversity and the livelihoods of coastal communities (Moruf et al. 2022; Abasubong et al. 2023; Enock et al. 2025). However, these by-products are rich in proteins, lipids, calcium, chitin, and other bioactive compounds, making them ideal alternative raw materials for aquafeeds (Serra et al. 2024; Abasubong et al. 2025). Recent advances in processing technologies such as rendering, fermentation, enzymatic hydrolysis, and bioextraction provide scalable and cost-effective pathways to convert marine waste into valuable feed ingredients (Bilal et al. 2024; Torres-León et al. 2021). Countries with established fisheries sectors like Morocco, Senegal, and Ghana are especially poised to lead the development of circular bioeconomy models that integrate marine waste valorization into aquafeed supply chains, thus addressing environmental issues while promoting inclusive economic growth within Africa's blue economy.

This review synthesizes key evidence supporting marine waste valorization as a strategic response to Africa's aquafeed challenges. It highlights the biochemical composition, processing methods, and nutritional benefits of fishery by-products, while also considering their environmental and socioeconomic implications. Valorizing marine waste aligns with global sustainability goals by enhancing resource efficiency, alleviating pressure on wild fish stocks, and reinforcing coastal economies (Yusoff et al. 2024). To fully realize this potential, targeted investments are needed in affordable processing infrastructure, technical capacity building for feed formulation, and knowledge exchange among scientists, industry players, and local communities. By consolidating emerging innovations and practical strategies, this chapter aims to support researchers, policymakers, and aquaculture stakeholders in transforming marine waste from an ecological burden into a sustainable feed resource, thereby securing long-term feed independence and strengthening Africa's aquatic food systems.

14.1.1 Methodology

This study adopts a qualitative, desk-based review approach to assess the potential for valorizing marine fishery waste into sustainable aquafeed in Africa. The methodology integrates a structured literature review, guided by a systematic search strategy, inclusion/exclusion criteria, and a thematic analysis framework to identify key opportunities, challenges, and best practices.

Search Strategy

A comprehensive literature search was conducted across academic databases, including Scopus, Web of Science, ScienceDirect, Google Scholar, and PubMed, as well as institutional and policy reports from FAO, World Bank, AU-IBAR, and UNEP. Keywords and Boolean operators used in various combinations included:

- "*Fishery waste*" OR "*fish waste*" OR "*by-products*" OR "*fish processing waste*"
- AND "*aquafeed*" OR "*fish feed*" OR "*sustainable feed*"
- AND "*valorization*" OR "*utilization*" OR "*conversion*"
- AND "*Africa*" OR names of specific African countries (e.g., Nigeria, Ghana, Kenya)

The search covered publications from 2000 to 2025, with emphasis on the most recent and regionally relevant literature from 2010 onwards.

Inclusion and Exclusion Criteria

Inclusion Criteria

- Studies focused on the utilization or processing of marine fishery waste
- Research addressing aquafeed production or formulation
- Studies specific to or applicable in the African context
- Reports or policy documents from reputable international organizations
- Peer-reviewed articles, book chapters, theses, and gray literature with clear methodology

Exclusion Criteria

- Studies unrelated to aquafeed or outside the marine fisheries sector
- Articles not available in English
- Publications lacking empirical or technical relevance to the topic

Studies focused exclusively on freshwater or nonfishery agricultural waste

14.1.2 *Valorization of Marine and Fishery Waste*

Harnessing the ocean's full potential extends beyond fishing and aquaculture; it includes the strategic valorization of marine and fishery waste, a scientifically grounded approach that transforms discarded materials into valuable resources (Fig. 14.1). This process embodies the principles of a circular blue economy, where waste streams are no longer an endpoint but a new beginning for product development across the food, health, energy, and environmental sectors (Ling et al. 2024). As the global demand for marine resources increases, valorization provides a sustainable means of maximizing the ocean's yield without increasing the extraction pressure on ecosystems.

Marine industries generate vast quantities of organic and inorganic waste, including fish heads, bones, skins, viscera, crustacean shells, bycatch, processing wastewater, and discarded fishing gear (Ferraro et al. 2010). Traditionally overlooked or discarded, these materials are now being recovered, recycled, and repurposed using advanced biochemical, thermochemical, mechanical, and chemical methods. This harnessing of marine waste leads to the production of high-value compounds, such as fishmeal and oil, marine collagen, gelatin, chitin, and bioactive lipids (Thirukumaran et al. 2022). These are used in diverse applications, including cosmetics, nutraceuticals, pharmaceuticals, agriculture, and renewable energy.

Valorization not only reduces marine pollution and landfill burdens but also supports sustainable ocean resource management. By turning waste into economic assets, profitability within the fisheries sector increases, technological innovation is encouraged, and the resilience of coastal communities is strengthened through job creation and industrial diversification (Ore-Areche et al. 2024). As we continue to unlock the potential of the ocean, valorization stands out as a critical strategy to ensure that every part of the catch, visible and invisible, is utilized responsibly and efficiently, securing both environmental and economic benefits for future generations.

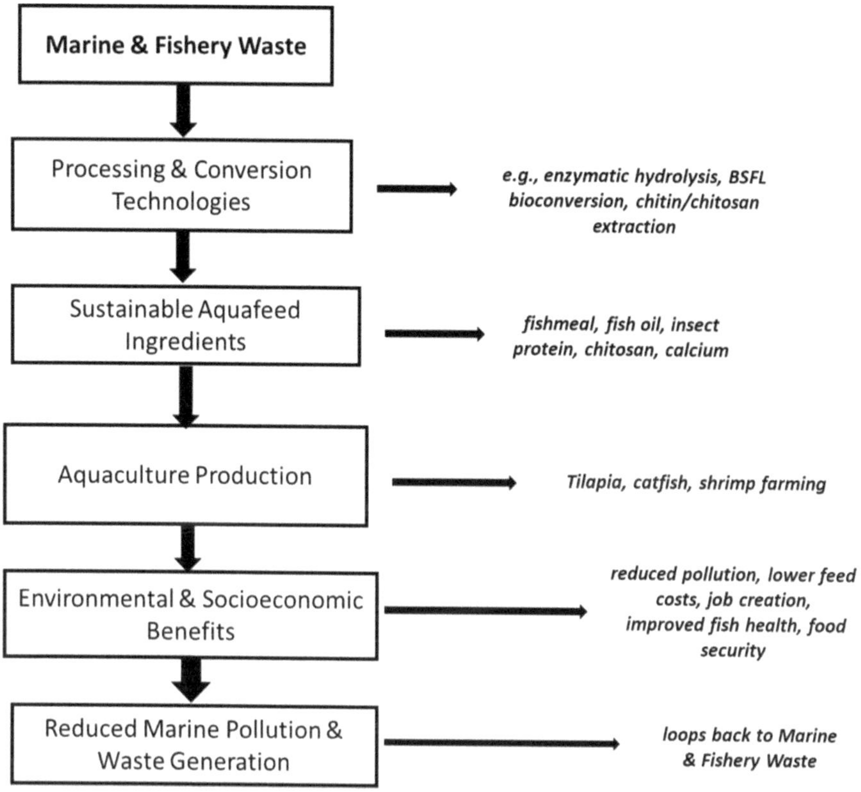

Fig. 14.1 Valorization of marine and fishery waste

14.2 Types and Composition of Marine and Fishery Waste

One of the most impactful applications of marine waste valorization is the extraction of innovative feed ingredients for aquaculture and livestock farming (Rudovica et al. 2021; Abasubong et al. 2025). With the rising demand for sustainable and cost-effective feeds across Africa, especially in coastal regions dependent on aquaculture, the reuse of marine waste offers a circular and environmentally sound alternative to conventional feed sources.

14.2.1 *Fishery By-products and Offal*

In many African coastal countries, large volumes of fishery by-products, such as heads, bones, skin, scales, and viscera, are discarded during fish processing despite being rich in proteins, lipids, essential amino acids, and minerals (Balange et al. 2023). These wastes can be transformed into valuable feed ingredients, such as fish protein hydrolysates (FPH), fishmeal, and fish oil, through mechanical pressing, cooking, or enzymatic hydrolysis (Siddik et al. 2021). Such by-products are ideal for formulating feeds for key aquaculture species, such as African catfish and tilapia, helping to reduce reliance on costly imported feed inputs.

Countries such as Nigeria, Egypt, and Morocco have seen small- and medium-scale enterprises adopt fish waste valorization to address feed shortages. For instance, Nigerian artisanal processors convert fish offal into feed-grade fishmeal, thereby reducing dependence on expensive soya and industrial fishmeal (Ayinla 2007). Egypt and Morocco have developed technologies to process fishery residues into high-protein feed components to support sustainable aquaculture growth (Ayinla 2007; Paruğ et al. 2024). This approach also mitigates coastal pollution by reducing organic waste discharge, which aligns with circular economy goals. Supported by initiatives from the EU, FAO, and World Bank, these efforts strengthen food security, promote local entrepreneurship, and drive job creation across Africa's blue economy.

14.2.2 Shellfish and Crustacean Waste

Shellfish and crustacean waste, especially shrimp, crab, and lobster shells, are underutilized resources in many African coastal regions. Often discarded at processing sites and markets, these materials are rich in chitin, chitosan, calcium, and other trace nutrients that are valuable to aquaculture (Chakravarty and Edwards 2022). Chitin and chitosan, extracted through chemical or enzymatic methods, serve as bioactive feed additives that improve digestibility, enhance immunity, and reduce the need for antibiotics in farmed fish (Kamilya and Khan 2020). Powdered shell waste also acts as a calcium supplement, promoting growth in tilapia and African catfish. Feeding trials in Ghana and Kenya have shown that low inclusion rates (2–5%) of shrimp shell powder improve weight gain, feed efficiency, and survival rate (Oucho 2017).

In Ghana, institutions such as the University of Cape Coast are working with local processors to collect and utilize shrimp shell waste in feed trials (Obirikorang et al. 2015), whereas in Kenya's Kilifi and Kwale counties, artisanal shrimp waste is being incorporated into smallholder aquafeeds (Kals et al. 2024). This practice supports circular aquaculture by reducing waste, recycling nutrients, and lowering feed costs, especially where conventional ingredients are scarce or expensive. Interest is growing in scaling up chitosan extraction for broader applications, including fish health, crop protection, and water treatment. Expanding these efforts will require investments in small-scale processing technologies and policy support from other shellfish-producing countries, such as Mozambique, Madagascar, Nigeria, and Tunisia.

14.2.3 *Seaweed and Algal Biomass*

Seaweed and algae harvested from aquaculture systems, natural blooms, or coastal cleanups can be dried and processed into feed additives that are rich in vitamins, minerals, and antioxidants (Salido et al. 2024). Species such as *Ulva*, *Gracilaria*, and *Sargassum*, which are common along the African coasts, have shown potential in promoting fish health and growth (Khan et al. 2024). In Tanzania and Madagascar, dried seaweed meal has been incorporated into tilapia and prawn feeds, with positive results in fish survival rates and immune function (Nur et al. 2020).

14.2.4 *Insect-Based Conversion from Organic Marine Waste*

Black Soldier Fly Larvae (BSFL) technology is gaining traction across Africa as an innovative solution for converting organic marine waste, such as fish viscera, scales, and skin, into high-protein insect meal for sustainable aquafeed production (Chaya et al. 2025). In coastal Kenya and Uganda, small-scale projects use BSFL to transform fish processing residues, which would otherwise pollute local environments, into nutrient-rich insect biomass (Soma et al. 2023). This insect meal serves as a low-cost, nutritious alternative to imported feeds, supporting species such as tilapia and African catfish while reducing feed expenses.

Beyond feed production, BSFL farming fosters a circular bioeconomy by minimizing organic waste, producing valuable by-products such as frass for organic fertilizers, and creating new livelihoods for local communities, especially for youth and women (Cattaneo et al. 2024). Supported by regional development initiatives, such as the African Union's Blue Economy Strategy and FAO programs, this approach is being piloted successfully in counties such as Kilifi and Mombasa in Kenya and the Lake Victoria Basin in Uganda (Otundo Richard 2024). With further technical support and policy incentives, BSFL-based valorization of marine waste holds great potential for scaling up in countries such as Tanzania, Senegal, Ghana, and Madagascar, enhancing aquafeed sustainability while tackling marine pollution.

14.3 Conversion Technologies: Composting, Bioconversion, and Pyrolysis

Conversion technologies are central to marine waste valorization, enabling the transformation of organic or mixed waste into valuable end products, such as animal or aquaculture feed. In African coastal contexts, four of the most promising and adaptable technologies are Mechanical Processing, composting, bioconversion, and pyrolysis, each offering unique advantages based on the waste type and desired outcomes.

14.3.1 Mechanical Processing and Enzymatic Hydrolysis

Mechanical processing and enzymatic hydrolysis are key technologies for converting marine and fishery waste into valuable feed ingredients for sustainable aquaculture (Khiari 2024). Mechanical processing involves physically breaking down fishery by-products, such as heads, bones, skin, and viscera, through grinding, pressing, and cooking to produce fishmeal and fish oil. This method efficiently recovers proteins and lipids but may not fully extract bioactive compounds or improve their digestibility (Benjakul et al. 2019).

In contrast, enzymatic hydrolysis uses specific enzymes to break down complex proteins into smaller peptides and amino acids, producing fish protein hydrolysates (FPH) with enhanced nutritional and functional properties. FPH is highly digestible and can improve feed palatability, nutrient absorption, and fish growth performance (Sapatinha et al. 2024). The combination of mechanical and enzymatic methods maximizes nutrient recovery from marine waste, reduces reliance on imported feed ingredients, and supports circular aquafeed production across African aquaculture systems.

14.3.2 Composting

Composting is an effective biological process that decomposes marine and fishery waste—such as fish offal, scales, shells, and other organic residues into nutrient-rich organic fertilizers (Balkhande 2020). Although composting does not directly produce aquafeed, it plays a crucial role in sustainable aquaculture by recycling valuable nutrients into agricultural soils, supporting integrated aquaculture-agriculture systems common in many African coastal regions (Hasimuna et al. 2023). This nutrient recycling improves soil fertility, enhances crop yields, and reduces reliance on chemical fertilizers, thereby contributing to overall ecosystem health. Moreover, composting reduces the volume of organic waste sent to landfills or discharged into coastal waters, thereby mitigating pollution, eutrophication, and habitat degradation (Hassan et al. 2023). In countries such as Ghana, Kenya, and Nigeria, small-scale composting initiatives are gaining traction, transforming fishery by-products into valuable inputs for local farming communities (Ononogbo et al. 2024). By diverting marine waste from the environment and turning it into a resource, composting supports circular bioeconomy principles and indirectly promotes sustainable aquafeed production through strengthened food systems in Africa.

14.3.3 Bioconversion

Bioconversion is an innovative and efficient method that transforms marine and fishery waste, such as fish offal, shrimp shells, and other organic by-products, into valuable protein-rich biomass for aquafeed production (Hassoun et al. 2022). A leading example is the use of Black Soldier Fly Larvae (BSFL), which consume high-moisture waste and convert it into nutrient-dense insect meal and oil suitable for fish feed. This process not only addresses the critical shortage and high cost of conventional fish feed ingredients in African aquaculture but also significantly reduces organic waste volume and associated environmental pollution (Macwan et al. 2023). Countries such as Kenya, Uganda, and Nigeria are pioneering BSFL-based bioconversion systems that provide affordable, locally sourced feed alternatives, enhance circular bioeconomy development, and support sustainable livelihoods in coastal communities. By closing the loop between waste and feed, bioconversion advances aquaculture resilience and contributes to food security in Africa.

14.3.4 Pyrolysis

Pyrolysis is a thermochemical process that heats marine and fishery waste, such as fish bones, shells, and offal, in the absence of oxygen to produce valuable by-products such as biochar, bio-oil, and syngas (Kumar and Panwar 2024). In the context of sustainable aquafeed production in Africa, pyrolysis offers an innovative way to valorize fishery waste by converting it into biochar—a stable carbon-rich material that can improve soil fertility and act as a carbon sink, thus supporting integrated aquaculture-agriculture systems (Reppas et al. 2025). Bio-oil derived from pyrolysis can be further refined for use as an energy source or feed additive, enhancing feed formulation options. This technology helps reduce organic waste volume, lowers environmental pollution, and promotes circular resource use (Ameh et al. 2024). Although still emerging in many African countries, pyrolysis holds great potential to complement other waste valorization methods, contributing to cost-effective and eco-friendly aquaculture development along the African coasts.

14.4 Innovative Feed Ingredient Extraction from Marine Waste

This approach focuses on harnessing the vast potential of the ocean by transforming discarded marine by-products (such as fish skins, shells, and bones) into valuable animal feed ingredients. By tapping into these underutilized resources, it reduces oceanic and coastal waste while extracting essential nutrients like proteins, omega-3 oils, and minerals. This innovation supports sustainable marine resource use, lowers dependence on traditional feed sources, and promotes a circular economy within aquaculture and livestock sectors, aligning with broader goals of ocean stewardship and environmental responsibility.

14.4.1 Fish Protein Hydrolysates and Fishmeal Production

One powerful example of this transformation is the production of fishmeal and fish protein hydrolysates (FPH). Fishmeal is produced by cooking, pressing, drying, and grinding discarded fish parts into a nutrient-rich powder used in aquafeeds, especially for carnivorous species such as tilapia and African catfish (Mısır 2022; Abasubong et al. 2025). FPH takes this a step further by using enzymatic hydrolysis to break down complex proteins into easily digestible peptides and amino acids, resulting in a feed ingredient that improves fish growth, enhances immunity, and increases feed efficiency (Nguyen et al. 2022).

In countries like Egypt, Nigeria, and Morocco, small- and medium-scale processors are turning the tide on marine waste by adopting these innovative approaches. By converting what would otherwise become ocean pollutants into high-value aquafeed, these enterprises not only improve fish farming sustainability but also help build resilient local supply chains (Kaleem and Bio Singou Sabi 2020).

In this context, harnessing the ocean means turning pollution into productivity. It's about leveraging marine by-products to reduce environmental damage, create economic value, and promote a circular blue economy—where nothing goes to waste, and every part of the catch contributes to ecological and social well-being (Fig. 14.2).

The production of fish protein hydrolysate (FPH) begins with the acquisition and mechanical processing of raw fish materials. The minced material is then subjected to either chemical (acidic or alkaline) or enzymatic hydrolysis, facilitated by the addition of water and specific reagents or enzymes. Hydrolysis is allowed to proceed until a predetermined degree of protein degradation is achieved, after which the reaction must be terminated (Shi and McHugh 2023). Depending on the hydrolysis method employed, this is accomplished through chemical neutralization or thermal inactivation. The resulting hydrolysate mixture is subsequently subjected to a solid-liquid separation step, in which insoluble residues are removed to isolate the aqueous protein fraction (Bedzo et al. 2024). If necessary, the liquid fraction undergoes thermal treatment (pasteurization) to reduce microbial activity. Unlike conventional dairy pasteurization, the conditions applied in FPH processing typically involve higher temperatures, reflecting the distinct physicochemical properties of fish-based substrates.

Once the primary hydrolysis and stabilization steps are complete, the focus shifts to refining and preserving the products. In cases where high product purity is essential, such as for applications in microbiological media, desalination may be applied to remove excess salts. The hydrolysate can then be concentrated to reduce

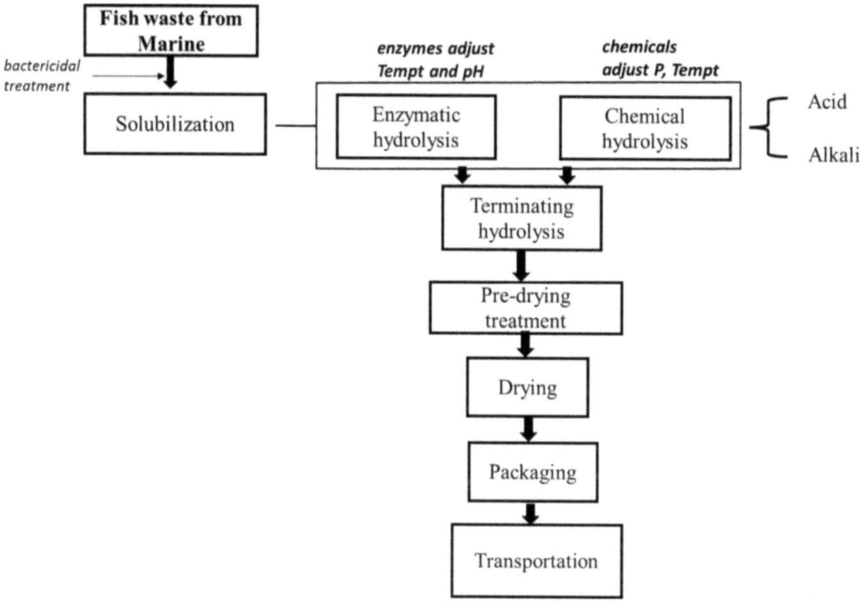

Fig. 14.2 Schematic overview of the fish protein hydrolysate (FPH) production process. (*Source*: Idowu et al. (2021) with a little modification)

its water content, thereby enhancing the efficiency of the subsequent drying process (Wu et al. 2024). However, this concentration step requires additional energy input and specialized equipment; therefore, its inclusion is determined based on specific production goals and economic considerations. The concentrated (or, in some cases, unconcentrated) hydrolysate is then dried to produce a stable powdered product. Following drying, the FPH is packaged and stored under controlled conditions, typically at or below 4 °C. Vacuum packaging is often employed to prevent lipid oxidation and preserve product quality during storage and transportation (He et al. 2013).

14.4.2 Fish Oil Extraction and Applications

Harnessing the ocean is not only about extracting seafood but also about reclaiming what is often discarded—polluted fish waste that too frequently ends up dumped into coastal waters. Viscera, heads, trimmings, and other by-products from fish processing are major contributors to marine pollution in many fishing communities. However, these waste materials are rich in marine oils, particularly polyunsaturated fatty acids (PUFAs) like eicosapentaenoic acid (EPA) and docosahexaenoic acid (DHA), which are essential for human and aquatic animal health (Qin et al. 2023; Pateiro et al. 2021).

By redirecting fish waste from ocean dumpsites into value-added production streams, we can transform a pollution problem into a sustainable economic opportunity. EPA and DHA, which originate from marine microorganisms and accumulate in fatty fish, are known for their role in improving cardiovascular, neurological, and developmental health in humans (Gow and Hibbeln 2014; DiNicolantonio and O'Keefe 2020) and enhancing growth, immunity, and stress resistance in aquaculture species (Tsoupras et al. 2022).

Technologies like urea complexation, molecular distillation, enzymatic hydrolysis, and supercritical CO_2 extraction allow us to isolate and purify these valuable omega-3s from fish waste with high efficiency and minimal degradation (Shahidi et al. 2019). Enzymatic hydrolysis using lipases such as *Candida rugosa* or Lipozyme IM preserves the bioactivity of EPA and DHA, while supercritical CO_2 extraction offers a solvent-free, oxidation-resistant method (Chandra et al. 2020; Lin et al. 2006). Urea complexation, especially when combined with hydrolysis, can achieve omega-3 concentrations of over 80%. Newer techniques, such as short-path distillation and polymeric membrane filtration (e.g., PVDF membranes), further improve purity and safety by removing contaminants (Magallanes et al. 2019).

In this way, harnessing the ocean means more than capturing seafood. It involves reclaiming polluted marine waste and converting it into high-value marine lipids. These oils not only support the growing demand for nutraceuticals but also serve as sustainable inputs in aquafeed. This circular approach reduces ocean pollution, adds value to the fisheries sector, and exemplifies how responsible innovation can turn environmental burdens into blue economy assets. FA profiles of different fish by-products are presented in Table 14.1 and Figs. 14.3 and 14.4.

14.4.3 Chitin and Chitosan from Crustacean Shells

In many coastal regions of Africa, the routine dumping of crustacean shell waste—mainly from shrimp, crab, and lobster—into the ocean has become a major source of marine pollution, threatening local ecosystems and posing sanitation risks to fishing communities (Topić Popović et al. 2023). However, this so-called "waste" is in fact a valuable resource. By harnessing the ocean's overlooked by-products, particularly shell waste, it is possible to shift from environmental degradation to circular innovation.

Crustacean shells are rich in chitin, a natural polysaccharide that can be converted into chitosan, a biodegradable, multifunctional polymer with a wide range of applications in aquaculture, agriculture, and industry. Comprising roughly 30–40% chitin, 30–50% calcium carbonate, and 20–30% proteins (Olugbojo et al. 2024a), these shells can be chemically processed through demineralization and deproteinization, followed by deacetylation to produce chitosan (Younes and Rinaudo 2015).

In the aquaculture sector, chitosan plays a transformative role: it enhances immune response, promotes gut health, and improves disease resistance in farmed species such as tilapia and catfish (Athanasopoulou et al. 2025). It acts as a

Table 14.1 FA profiles of different fish by-products

Species	Scientific name	Habitat	By product	FA amount	FA	Type FA and total content	References
Bonito	*Euthynnus pelamis*	Marine fish	Liver	PA	14.24–20.57%	∑ SFA = 26.2–32.07%	Watanabe et al. (1995)
				OA	8.93–13.19%	∑ MUFA = 11.76–17.68%	
			Heart	DHA	33.9%	∑ PUFA: n.a	
Common ling	*Molva molva*		Viscera	EPA, DHA	7.8–9.3%, 8.7–28.1%	∑ PUFA = 43.9%	Falch et al. (2006)
			Liver	EPA, DHA	4.8–6.6%, 5.4–9.4%	∑ PUFA = 30.8%	
			Trimming	EPA, DHA	6.2–9.3%, 6.9–28.1%	∑ PUFA = 40.2%	
Cod	*Gadus morhua*	Marine and brackish water fish	Liver	EPA, DHA	5.8–10.1%, 6.0–15.2%	∑ PUFA = 34.9%	Falch et al. (2006)
			Viscera	EPA, DHA	7.4–13.3% 8.5–22.0%	∑ PUFA = 38.1%	
			Trimming	EPA, DHA	6.7–11.5%, 8.9–23.3%	∑ PUFA = 40.4%	
Haddock	*Melanogrammus aegleinus*	Marine cold-water fish	Liver	EPA, DHA	9.0–13.5%, 5.6–14.1%	∑ PUFA = 37.8%	Falch et al. (2006)
			Viscera	EPA, DHA	7.7–15.7%, 10.3–20.2%	∑ PUFA = 50.7%	
			Trimming	EPA, DHA	5.4–15.3%, 10.4–25.1%	∑ PUFA = 39.0%	

(continued)

Table 14.1 (continued)

Species	Scientific name	Habitat	By product	FA amount	FA	Type FA and total content	References
Madeiran sardinella	*Sardinella maderensis*	Marine and brackish water fish	Skin	EPA	20.5%	$\sum$ PUFA = 33.6%	Njinkoué et al. (2002)
				PA	20.5%	$\sum$ SFA = 41.5%	
			Liver	PA	25.6%	$\sum$ SFA = 38.0%	
				OA	27.2%	$\sum$ MUFA = 44.2%	
				EPA	10.4%	$\sum$ PUFA = 22.5%	
Atlantic salmon	*Salmo salar*	Freshwater fish	Viscera	Myristic, PA, SA	7.1%, 18.8%, 4.8%	$\sum$ SFA = 30.8%	Sun et al. (2006)
				OA	18.9%	$\sum$ MUFA = 37.4%	
				EPA	8.5%	$\sum$ PUFA = 31.9%	
				DHA	6.6%		
				DPA	3.7%,		
				LA	4.4%		
Round sardinella	*Sardinella aurita*	Marine and brackish water fish	Skin	PA	20.5%	$\sum$ SFA = 40.3%	Sun et al. (2006)
				OA	15.5%	$\sum$ MUFA = 34.0%	
			Liver	PA	23.2%	$\sum$ SFA = 32.3%	
				OA	44.7%	$\sum$ MUFA = 54.0%	
Saithe	*Pollachius virens*	Marine fish	Liver	EPA, DHA	4.4–7.2%, 5.1–9.4%	$\sum$ PUFA = 33.3%	Falch et al. (2006)
			Viscera	EPA, DHA	5.0–10.2%, 5.9–21.9%	$\sum$ PUFA = 41.9%	
			Trimming	EPA, DHA	7.3–11.4%, 9.5–32.8%	$\sum$ PUFA = 47.8%	
Sardine	*Sardina pilchardus*	Marine and brackish water fish	Skin	EPA	22.28%	$\sum$ PUFA = 38.91%	Mkadem and Kaanane (2019)
			Meat	DHA	6.71%		
			Bones	ALA	2.36%		

Seabass	*Cephalopholis taeniops*	Marine fish	Skin	DHA	6.9%	∑ PUFA = 17.5%	Watanabe et al. (1995)
				PA	28.4%	∑ SFA = 46.4%	
				OA	12.5%	∑ MUFA = 33.6%	
			Liver	PA	33.9%	∑ SFA = 58.6%	
				EPA	8.16–15.07%,	∑ PUFA = 51.56–61.11%	
				DHA	30.80–32.91%	∑ SFA = 26.2–32.07%	
Tusk	*Brosme brosme*		Liver	EPA, DHA	6.7–7.8%, 9.0–10.9%	∑ PUFA = 26.6%	Falch et al. (2006)
			Viscera	EPA, DHA	5.6–8.5%, 7.2–18.9%	∑ PUFA = 37.2%	
			Trimming	EPA, DHA	3.6–7.4%, 8.3–16.5%	∑ PUFA = 31.8%	
Yellowfin tuna	*Hummus albacares*		Head	EPA, DHA	5.2%, 21.8%	∑ PUFA = 38.2%	
				PA, SA	21.4%, 6.9%	∑ SFA = 32.3%	
				OA	15%	∑ MUFA = 26.2%	
				EPA, DHA	5.1%, 22.0%	∑ PUFA = 38.2%	
			Eyeball	PA, SA	21.1%, 6.7%	∑ SFA = 31.6%	
				OA	16%	∑ MUFA = 27.4%	

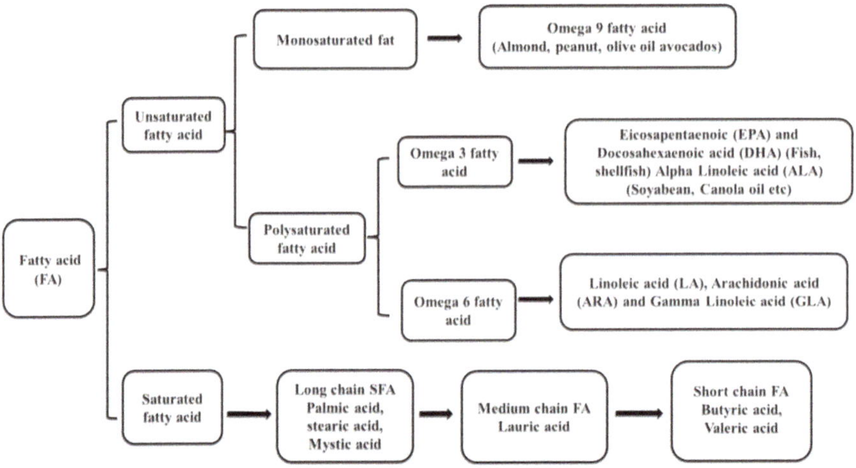

Fig. 14.3 Classification of fatty acids and their primary food sources

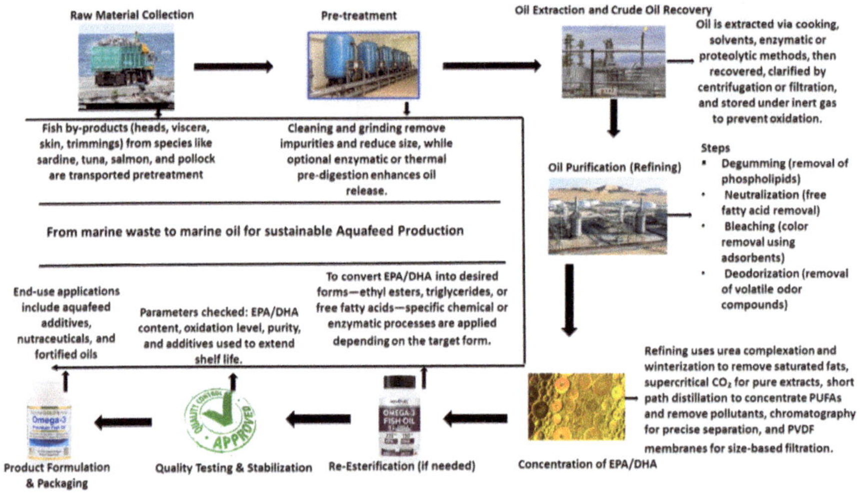

Fig. 14.4 Illustration of the transformation pathway of fish waste from coastal environments into the final product—omega-3 fatty acids. This figure presents a step-by-step overview of the valorization process, beginning with the collection of fish waste (such as heads, viscera, and trimmings) from coastal and fish processing facilities. It then traces the intermediate stages—such as sorting, rendering, oil extraction, and purification—leading to the production of high-quality omega-3 fatty acids, which are used in nutraceuticals, pharmaceuticals, functional foods, and aquafeed production

biostimulant, boosting growth and antioxidant activity (El-Naggar et al. 2021), while its antimicrobial and antifungal properties offer a sustainable alternative to antibiotics. Furthermore, chitosan contributes to wastewater treatment in aquaculture systems by flocculating suspended solids and improving water quality (Nechita 2017). In countries like Nigeria, Senegal, Morocco, and Mozambique, where large volumes of shell waste remain underutilized, harnessing this resource aligns with zero-waste and circular blue economy strategies. Valorizing shell waste reduces coastal pollution, creates economic opportunities for local processors, and replaces costly imported additives with locally produced, eco-friendly alternatives (Olugbojo et al. 2024b).

While challenges such as limited infrastructure, processing costs, and chemical safety persist (Wani et al. 2023), emerging biotechnological approaches, like microbial fermentation for chitin extraction, and expanding public-private partnerships offer scalable, sustainable solutions. By reimagining shell waste not as pollution but as a resource, we can harness the ocean in a way that supports environmental restoration, economic empowerment, and resilient aquaculture.

14.4.4 Emerging Bioactive Compounds and Functional Additives from Marine and Fishery Waste

Marine and fishery waste is increasingly recognized as a valuable source of bioactive compounds and functional additives that enhance the nutritional quality, health benefits, and sustainability of aquafeeds. As aquaculture expands across Africa, particularly in areas facing resource constraints and rising feed costs, the use of underutilized marine by-products is gaining momentum (Adeleke et al. 2020). Bioactive compounds such as peptides, polyunsaturated fatty acids (PUFAs), antioxidants, polysaccharides, and enzymes can be extracted from fish heads, skins, viscera, scales, and shellfish waste, like shrimp and crab shells. These compounds improve immune function, growth performance, stress tolerance, and disease resistance in species such as tilapia and catfish (Tufail et al. 2025). For example, fish protein hydrolysates, obtained through enzymatic processing, provide bioactive peptides that support gut health and act as natural immunostimulants (Nemati et al. 2024).

Functional additives like collagen, gelatin, and marine antioxidants are also drawing attention. Collagen from fish skin and scales supports bone and tissue development and enhances feed digestibility, while chitosan from crustacean shells offers antimicrobial properties and serves as a feed preservative (Irastorza et al. 2021). Fermented fish waste enriched with probiotics is emerging as another promising ingredient, helping to improve nutrient absorption and gut health. In Africa, the use of these compounds is still developing, with research institutions and local enterprises piloting applications. Though challenges such as limited technology, high costs, and standardization remain, integrating bioactive and functional ingredients from marine waste offers strong potential to boost fish health, reduce feed costs, and support a circular, sustainable aquaculture industry.

14.4.4.1 Innovative Aquafeed Formulations Using Marine Waste-Derived Ingredients

The aquaculture sector has increasingly turned to marine waste-derived ingredients as a solution to the mounting challenges of feed sustainability, economic viability, and environmental conservation. Marine waste—primarily sourced from fish processing by-products such as heads, bones, viscera, skin, scales, and shellfish exoskeletons represents a vast reservoir of untapped nutritional resources. For instance, it is estimated that nearly 30–50% of a whole fish is discarded during processing, yet these remnants are rich in high-quality proteins, lipids, essential amino acids, vitamins, and minerals (Ling et al. 2024). Through innovative bioprocessing technologies like enzymatic hydrolysis and microbial fermentation, these materials can be transformed into fish protein hydrolysates (FPH), fish silage, chitin, chitosan, and marine-derived oils. These feed additives and ingredients can effectively replace a portion of traditional fishmeal and fish oil in aquafeeds, reducing pressure on wild-capture fisheries and promoting circular economy principles within the seafood supply chain.

A notable example is the use of fish protein hydrolysates derived from salmon processing waste in salmonid aquafeeds. These hydrolysates, generated through the enzymatic breakdown of muscle proteins and collagen, have demonstrated not only high digestibility and palatability but also bioactive properties that support immune system modulation and gut health (Siddik et al. 2021). In Norway, aquafeed companies have successfully incorporated salmon-derived FPH into diets for Atlantic salmon (*Salmo salar*), improving feed conversion ratios and enhancing fish growth while reducing the environmental impact of feed production (Cantillo et al. 2025). Similarly, shrimp shell waste—rich in chitin and astaxanthin—has been used to formulate functional feeds for crustaceans and finfish (Ashikin et al. 2022). Chitin and its derivative chitosan have immunostimulatory and antimicrobial properties, which are particularly beneficial in disease-prone intensive aquaculture systems (Abu-Elala et al. 2024). For example, in India and Southeast Asia, chitosan-enriched feeds have been employed to improve disease resistance in *Penaeus vannamei* (whiteleg shrimp), leading to lower mortality rates during viral and bacterial outbreaks.

Another compelling case is the utilization of tuna and mackerel offal in the production of fish silage for tilapia and catfish farming. Through acid or microbial fermentation, these off-cuts are converted into a stable, protein-rich slurry that can be used directly or further processed into meal (Hasan 2001). In countries such as Ghana and the Philippines, community-level initiatives have developed small-scale fish silage production units, enabling local farmers to create affordable, nutrient-dense feed while minimizing waste (Rurangwa et al. 2015). Moreover, shellfish waste, including crab and shrimp shells, has been a valuable source of marine minerals like calcium and phosphorus, crucial for skeletal development in young fish (Suprijanto and Widowati 2024). Incorporating finely milled shell powder into larval diets has been shown to enhance survival and growth in hatchery-reared marine finfish species like grouper and sea bream (Hamre et al. 2013).

Beyond nutritional enhancements, marine waste-derived ingredients offer functional benefits that align with the growing trend of precision aquaculture. Bioactive peptides obtained from marine by-products have been shown to possess antioxidant, antimicrobial, and antihypertensive properties, offering potential as natural feed additives that reduce the need for synthetic chemicals or antibiotics. For instance, dietary inclusion of hydrolyzed fish skin peptides has been linked to increased antioxidant enzyme activity in rainbow trout (*Oncorhynchus mykiss*), leading to improved stress tolerance and reduced oxidative damage under high-density farming conditions (Akbarian et al. 2022; Giri and Ohshima 2012). Additionally, lipid fractions extracted from marine offal, particularly those containing long-chain omega-3 fatty acids (EPA and DHA), serve as effective substitutes for traditional fish oil, which is becoming increasingly scarce and expensive. In Chile and Peru, sardine and anchovy processing waste has been repurposed into concentrated fish oils for use in salmonid and seabass diets, helping maintain high levels of essential fatty acids without exerting further strain on pelagic fish populations (Pateiro et al. 2021; Tacon and Metian 2009).

14.4.4.2 Challenges of Valorization of Marine and Fishery Waste for Sustainable Aquafeed Production in Africa

Harnessing the ocean's full potential means not only exploiting its living resources but also reclaiming the polluted fish waste that clogs coastlines and contaminates marine environments. However, in many African countries, the conversion of this waste into sustainable aquafeed is hampered by a complex set of systemic and structural challenges (Amponsah 2025). Among the most critical barriers is the lack of adequate processing infrastructure. Many coastal and inland areas suffer from insufficient cold storage, drying equipment, and pelletizers, which results in rapid spoilage and loss of valuable nutrients before fish waste can be repurposed. This issue is exacerbated by the absence of standardized quality and safety protocols for handling fish waste, allowing contamination by pathogens, heavy metals, and chemical residues to undermine feed safety, public trust, and marketability.

The valorization of fish waste is further complicated by fragmented supply chains. Fish waste is often scattered across numerous small-scale fishers and processors, while poor road networks, inefficient transport systems, and a lack of cold chain infrastructure increase collection and distribution costs (Wong et al. 2016). Consequently, significant amounts of waste are illegally dumped back into the ocean, posing serious environmental and public health risks. Moreover, there is a shortage of technical expertise in advanced bioconversion technologies such as enzymatic hydrolysis, microbial fermentation, and insect-based protein farming (e.g., black soldier fly larvae), largely due to limited access to training and extension services (Siddiqui et al. 2024b). Financial limitations also restrict investments in necessary equipment and infrastructure, as small-scale entrepreneurs often lack access to affordable credit or subsidies. Cultural stigmas around using fish waste in feed, rooted in concerns over hygiene and quality, further hinder both production and consumer acceptance (Wong et al. 2016).

On the policy and market fronts, significant gaps persist. Most African countries lack comprehensive regulatory frameworks that address feed safety, environmental compliance, and waste management, creating legal uncertainty that discourages private-sector investment and limits access to export markets (Debrah et al. 2022). Weak market access and low consumer awareness also impede commercial viability, as producers struggle to build reliable marketing channels and cultivate buyer confidence (Kaleem and Bio Singou Sabi 2021). To overcome these intertwined challenges and turn fish waste pollution into a valuable resource, a coordinated strategy is vital. This includes investing in infrastructure and cold chain systems, expanding technical training and support, providing financial incentives and credit access, establishing clear regulatory frameworks, and launching public awareness campaigns. Only through such comprehensive efforts can Africa unlock the economic, nutritional, and environmental potential of ocean-based waste, transforming what pollutes into what sustains.

14.4.4.3 The Way Forward: Harnessing the Potential

Africa's oceans hold vast potential, not only through traditional fishing, but also by rethinking how we use what's typically discarded. Parts of marine harvests, such as heads, bones, skins, and shells, are often thrown away, leading to pollution and the loss of valuable resources (Bi et al. 2023). However, with the right strategies, these by-products can be transformed into income-generating products that support aquaculture, empower communities, and drive a circular blue economy. Realizing this opportunity requires a coordinated, cross-sector approach that includes investments in infrastructure, technology, and skills development. Many coastal areas currently lack the cold storage, transport systems, and processing units needed to preserve nutrients and prevent spoilage (Siddiqui et al. 2024c). Establishing decentralized cold chains and equipping processors with tools for drying, fermenting, or pelletizing can significantly reduce waste while increasing the supply of raw materials for products such as fishmeal and protein hydrolysates.

Just as important as infrastructure is the development of human capacity. Smallholders and processors often lack access to modern techniques like enzymatic hydrolysis or insect-based bioconversion, which are essential for converting fish waste into high-value products (Siddiqui et al. 2024a). Providing targeted training, extension services, and peer learning opportunities can enable local actors to adopt and scale these innovations. At the policy level, governments must introduce and enforce clear regulations that ensure product safety, traceability, and environmental protection. These standards are critical to protecting consumers, attracting investment, and maintaining industry credibility. Public awareness efforts are also needed to overcome cultural resistance to waste-derived products and to highlight their nutritional, economic, and environmental benefits.

Regional collaboration will be key to scaling these efforts. Harmonizing standards, sharing technologies, and developing cross-border value chains can accelerate innovation and amplify impact across the continent. The valorization of marine by-products fits squarely within Africa's broader blue economy agenda, which aims to sustainably harness ocean resources for inclusive growth and environmental stewardship. To make this vision a reality, governments, NGOs, research institutions, and the private sector must work together to invest in processing infrastructure, support research into scalable low-cost technologies like solar drying and fermentation, and implement policies that encourage responsible waste utilization. With the right mix of coordination, education, policy, and innovation, Africa can turn marine waste from an environmental challenge into a powerful driver of economic opportunity and sustainable development.

14.5 Conclusion

Africa's marine ecosystems offer far more than just fish for consumption—they hold untapped potential in the form of marine waste by-products. Properly harnessed, these by-products can reduce environmental pollution, create jobs, support aquaculture growth, and add value to fishery supply chains. While challenges remain in infrastructure, regulation, and capacity, the pathway toward a sustainable and inclusive blue economy is clear. By transforming marine waste into economic opportunity, Africa can chart a course toward food security, economic diversification, and environmental resilience—leveraging its oceans not just as sources of fish, but as engines of innovation and growth.

References

Abasubong KP, Gabriel NN, Adjoumani J-JY, Okon AO (2023) Waste to feed: an emerging technology to improve aquaculture in Africa. In: Emerging sustainable aquaculture innovations in Africa. Springer, pp 103–130

Abasubong KP, Amin AB, Uwem GU, Desouky HE (2025) A review of fish waste and byproducts in poultry production: effects on growth, egg quality, carcass traits, and broader knowledge beyond physiological response. Anim Adv 2(1):e021. https://doi.org/10.48130/animadv-0025-0027

Abu-Elala N, Marzouk M, Basuini M (2024) Chitosan for aquaculture: growth promotion, immune modulation, antimicrobial activity, bio-carrier utility, water quality management, and safety considerations. Ann Anim Sci 25(2):483–509. https://doi.org/10.2478/aoas-2024-0079

Adeleke B, Robertson-Andersson D, Moodley G, Taylor S (2020) Aquaculture in Africa: a comparative review of Egypt, Nigeria, and Uganda Vis-À-Vis South Africa. Rev Fish Sci Aquacult 29(2):167–197

Akbarian M, Khani A, Eghbalpour S, Uversky VN (2022) Bioactive peptides: synthesis, sources, applications, and proposed mechanisms of action. Int J Mol Sci 23(3)

Ameh VI, Ayeleru OO, Nomngongo PN, Ramatsa IM (2024) Bio-oil production from waste plant seeds biomass as pyrolytic lignocellulosic feedstock and its improvement for energy potential: a review. Waste Manag Bull 2(2):32–48

Amponsah SK (2025) Aquaculture in Africa: Production Systems, Challenges, and Opportunities. In IntechOpen. https://doi.org/10.5772/intechopen.1207770

Ashikin N, Redhwan A, Rashid Z, Nguang SI, Kamu A, Ferdinand R, Komilus C (2022) Crustacean shell waste as a potential feed material. Biosci Res 19:360–372

Athanasopoulou E, Mitsoula S, Dimitroglou A, Tsironi T (2025) Chitosan-based biodegradable coatings for securing nutrients in extruded fish feed pellets. Aquacult Fish. https://doi.org/10.1016/j.aaf.2025.04.001

Ayinla O (2007) Analysis of feeds and fertilizers for sustainable aquaculture development in Nigeria. FAO Fisheries Technical Paper

Balange AK, Layana P, Naidu BC (2023) Harnessing value and sustainability: fish waste valorization and the production of valuable byproducts. In: Capanoglu E, Navarro-Hortal MD, Forbes-Hernández TY, Battino M (eds) Advances in food and nutrition research, vol 107. Academic, pp 175–192

Balkhande DJ (2020) Devising of organic fertilizer from fish and crab wastes: Waste to best technology. Int J Fish Aquatic Stud 8(2):01–05

Bedzo OKK, Gottumukkala LD, Sasso GL, Kaminski K, Schlage W, Goffman F, Ivanov N, Hoeng J, Hayes DJ (2024) Process development for efficient pectin extraction from tobacco residues and its characterisation. Biomass Convers Biorefinery 14(23):29481–29501

Benjakul S, Sae-leaw T, Simpson B (2019) Byproducts from fish harvesting and processing. Delve Publishing, Ashland, pp 179–217

Bi M, Zhang Z, Guo X, Wan L (2023) Evaluation of sustainable utilization of African marine fishery resources. Fishes 8(1):4

Bilal M, Niu D, Wang Z-X (2024) Novel enzyme-fermentation process for bioconversion of restaurant food waste into isomaltooligosaccharide- and L-lactic acid-enriched animal feed. Front Sustain Food Syst 8:1326304. https://doi.org/10.3389/fsufs.2024.1326304

Cantillo J, Samdahl TM, Deshpande PC (2025) Promoting sustainable feed choices in Norwegian Salmon aquaculture by exploring sustainability awareness among fish feed manufacturers and fish farmers. Aquaculture 608:742698

Cattaneo A, Meneguz M, Dabbou S, Tambone F, Scaglia B (2024) Local circular economy: BSF insect rearing in the Italian Agri-Food Industry. Waste Manag 179:234–244

Chakravarty J, Edwards TA (2022) Innovation from waste with biomass-derived chitin and chitosan as green and sustainable polymer: a review. Energy Nexus 8:100149

Chandra P, Enespa, Singh R, Arora PK (2020) Microbial lipases and their industrial applications: a comprehensive review. Microb Cell Fact 19(1):169

Chaya S, Borah N, Borah BK, Rajkumari P, Gogoi K, Bhuyan PR (2025) Harnessing nature: Black soldier fly for sustainable organic waste management. Int J Adv Biochem Res 9(2):446–451. https://doi.org/10.33545/26174693.2025.v9.i2g.3818

Debrah J, Teye G, Dinis M (2022) Barriers and challenges to waste management hindering the circular economy in Sub-Saharan Africa. Urban Sci 6

DiNicolantonio JJ, O'Keefe JH (2020) The benefits of Omega-3 fats for stabilizing and remodeling atherosclerosis. Mo Med 117(1):65–69

El-Naggar M, El-Shabaka H, El-Rahman A, Khalil M, Suloma A (2021) Efficacy of dietary chitosan and chitosan nanoparticles supplementation on health status of Nile tilapia, Oreochromis niloticus (L.). Aquacult Rep 19

Enock A, Isaac A, Okyere E (2025) Impacts of eutrophication on freshwater biodiversity in the Afram river, Ghana: implications for ecosystem services. Int J Water Res 7:191

Falch E, Rustad T, Jonsdottir R, Shaw N, Dumay J, Berge J, Arason S, Kerry J, Sandbakk M, Aursand M (2006) Geographical and seasonal differences in lipid composition and relative weight of by-products from gadiform species. J Food Compos Anal 19(6–7):727–736

Ferraro V, Cruz I, Jorge R, Malcata F, Pintado M, Castro P (2010) Valorisation of natural extracts from marine source focused on marine by-products – a review. Food Res Int 43:2221–2233

Giri A, Ohshima T (2012) Bioactive marine peptides. Nutraceutical value and novel approaches. Adv Food Nutr Res 65:73–105

Gow RV, Hibbeln JR (2014) Omega-3 fatty acid and nutrient deficits in adverse neurodevelopment and childhood behaviors. Child Adolesc Psychiatr Clin N Am 23(3):555–590

Hamre K, Yufera M, Rønnestad I, Boglione C, Conceição L, Izquierdo M (2013) Fish larval nutrition and feed formulation: Knowledge gaps and bottlenecks for advances in larval rearing. Rev Aquacult 5(S26):–S58

Hasan B (2001) Microbial fermentation of fish waste (fish silage) for potential use in animal feed. Jurnal Peternakan dan Lingkungan

Hasimuna OJ, Maulu S, Nawanzi K, Lundu B, Mphande J, Phiri CJ, Kikamba E, Siankwilimba E, Siavwapa S, & Chibesa M (2023) Integrated agricultureaquaculture as an alternative to improving small-scale fish production in Zambia. Frontiers in Sustainable Food Systems, 7:1161121. https://doi.org/10.3389/fsufs.2023.1161121

Hassan N, Wahed N, Abdelhamid A, Ashraf M, Abdelfattah E (2023) Composting: an eco-friendly solution for organic waste management to mitigate the effects of climate change. Innovare J Soc Sci:1–7

Hassoun A, Rustad T, Bekhit A (2022) Bioconversion of marine by-products into edible protein. In: Alternative proteins. CRC Press, pp 297–327

He S, Franco C, & Zhang W (2013) Functions, applications and production of protein hydrolysates from fish processing co-products (FPCP). Food Research International, 50(1):289–297. https://doi.org/10.1016/j.foodres.2012.10.031

Idowu A, Igiehon O, Idowu S, Olatunde O, Benjakul S (2021) Bioactivity potentials and general applications of fish protein hydrolysates. Int J Peptide Res Ther 27:109

Irastorza A, Zarandona I, Andonegi M, Guerrero P, De la Caba K (2021) The versatility of collagen and chitosan: from food to biomedical applications. Food Hydrocoll 116:106633

Kaleem O, Bio Singou Sabi A-F (2020) Overview of aquaculture systems in Egypt and Nigeria, prospects, potentials, and constraints. Aquacult Fish 6:535–547

Kaleem O, & Bio Singou Sabi AF (2021) Overview of aquaculture systems in Egypt and Nigeria, prospects, potentials, and constraints. Aquaculture and Fisheries, 6(6):535–547. https://doi.org/10.1016/j.aaf.2020.07.017

Kals J, Opiyo MA, Rurangwa E, Soma K, Ndambi A, Vernooij A (2024) Use of black soldier fly larvae and freshwater shrimp to partly substitute commercial diet for Nile tilapia cultured in smallholder fish farms–a case study in Busia County, Kenya. Front Sustain Food Syst 8:1298551

Kamilya D, Khan MIR (2020) Chitin and chitosan as promising immunostimulant for aquaculture. In: Handbook of chitin and chitosan. Elsevier, pp 761–771

Khan N, Sudhakar K, Mamat R (2024) Eco-friendly nutrient from ocean: Exploring Ulva seaweed potential as a sustainable food source. J Agri Food Res 17:101239

Khiari Z (2024) Enzymes from fishery and aquaculture waste: research trends in the era of artificial intelligence and circular bio-economy. Marine Drugs 22:411

Kumar VK, Panwar NL (2024) Pyrolysis technologies for biochar production in waste management: a review. Clean Energy 8:61–78

Lin T-J, Chen S-W, Chang A-C (2006) Enrichment of n-3 PUFA contents on triglycerides of fish oil by lipase-catalyzed trans-esterification under supercritical conditions. Biochem Eng J 29:27–34

Ling Wen Xia F, Supri S, Djamaludin H, Nurdiani R, Leong Seng L, Wee Yin K, & Rovina K (2024) Turning waste into value: extraction and effective valorization strategies of seafood by-products. Waste Management Bulletin, 2:84–100. https://doi.org/10.1016/j.wmb.2024.06.008

Macwan S, Souza T, Dunshea F, Digiacomo K, Suleria H (2023) Black soldier fly larvae (*Hermetica illucens*) as a sustainable source of nutritive and bioactive compounds, and their consumption challenges. Anim Prod Sci 64

Magallanes LM, Tarditto LV, Grosso NR, Pramparo MC, Gayol MF (2019) Highly concentrated omega-3 fatty acid ethyl esters by urea complexation and molecular distillation. J Sci Food Agric 99(2):877–884
Mısır G (2022) Novel utilization of fish by-products and wastes: protein hydrolysates. Acta Aquatica Turcica 18
Mkadem H, Kaanane A (2019) Recovery and characterization of fish oil from by-products of sardine (Sardina pilchardus) in the canning process. J Aquatic Food Prod Technol 28(10):1037–1050
Moruf O, Aminu M, Abubakar A, Funke O (2022) Coastal shellfish as harvestable resources in Nigeria: a review
Nechita P (2017) Applications of chitosan in wastewater treatment
Nemati M, Shahosseini SR, Ariaii P (2024) Review of fish protein hydrolysates: production methods, antioxidant and antimicrobial activity and nanoencapsulation. Food Sci Biotechnol 33(8):1789–1803
Ndebele-Murisa MR, Mubaya CP, Dekesa CH, Samundengo A, Kapute F, & Yossa R (2024) Sustainability of Aqua Feeds in Africa: A Narrative Review. Sustainability, 16(23):10323. https://doi.org/10.3390/su162310323
Nguyen HT, Bao HND, Dang HTT, Tómasson T, Arason S, Gudjónsdóttir M (2022) Protein characteristics and bioactivity of fish protein hydrolysates from Tra Catfish (Pangasius hypophthalmus) side stream isolates. Foods 11(24)
Njinkoué J-M, Barnathan G, Miralles J, Gaydou E-M, Samb A (2002) Lipids and fatty acids in muscle, liver and skin of three edible fish from the Senegalese coast: Sardinella maderensis, Sardinella aurita and Cephalopholis taeniops. Comp Biochem Physiol B Biochem Mol Biol 131(3):395–402
Nur A, Hossain F, Hasan M, Zannat S, Chakroborty K, Rafiquzzaman SM (2020) Effect of selected seaweed powder as a fish feed on growth and immune system of tilapia (Oreochromis niloticus). Int J Fish Aquatic Stud 8:24–30
Obirikorang KA, Amisah S, Fialor SC, Skov PV (2015) Local agro-industrial by-products with potential use in Ghanaian aquaculture: a review. Aquacult Int 23:403–425
Olufemi-Phillips A, Igwe A, Toromade A, Eyo-Udo N, Adewale T (2024) Global trade dynamics' impact on food pricing and supply chain resilience: a quantitative model. World J Adv Res Rev 24:492–519
Olugbojo J, Akinyemi A, Obasa S, Dare E (2024a) Extraction and utilization of chitin and chitosan from waste yields of economically important Crustaceans and Molluscs for improved fish production. Asian J Fish Aquatic Res 26:95–114
Olugbojo JA, Akinyemi AA, Obasa SO, Dare EO (2024b) Extraction and utilization of chitin and chitosan from waste yields of economically important Crustaceans and Molluscs for improved fish production. Asian J Fish Aquatic Res 26(8):95–114
Ononogbo C, Ohwofadjeke P, Chukwu M, Nwawuike N, Obinduka F, Nwosu O, Ugenyi A, Innocent Chile N, Nwosu E, Nwakuba N, Osuagwu O, Echeta D, Eze V, Obodo R, Aniezi J, Eze C (2024) Agricultural and environmental sustainability in nigeria: a review of challenges and possible eco-friendly remedies. Environ Dev Sustain:1–47
Ore-Areche F, Carpio A, Flores D, Choque Rivera T, Ticsihua-Huamán J, Montalvo Otivo J, Malpartida Yapias RJ, Rojas J, Ccopa H, Mendoza P, Dominguez J (2024) Sustainable Seafood Processing: Reducing Waste and Environmental Impact in Aquatic Ecosystems. J Exp Biol Agri Sci 12:522–536
Otundo Richard M (2024) Strategic innovative community-based fisheries management practices for the success of the blue economy program in Kenya (July 30, 2024)
Oucho JA (2017) Effect of replacing shrimp shell meal with hydrolyzed feather meal on growth, diet digestibility and body composition of Oreochromis mossambicus (Peters, 1852), University of Eldoret
Paruğ Ş, Jabbr A, Lazrag F (2024) Sustainable development of fisheries and aquaculture in North Africa: current status and future prospects. Menba Kastamonu Üniversitesi Su Ürünleri Fakültesi Dergisi 10:90–105

Pateiro M, Domínguez R, Varzakas T, Munekata PES, Movilla Fierro E, Lorenzo JM (2021) Omega-3-Rich oils from marine side streams and their potential application in food. Mar Drugs 19(5)
Qin J, Kurt E, Bassi TL, Sa L, Xie D (2023) Biotechnological production of omega-3 fatty acids: current status and future perspectives. Front Microbiol 14:1280296
Ragasa C, Charo-Karisa H, Rurangwa E, Tran N, Shikuku K (2022) Sustainable aquaculture development in sub-Saharan Africa. Nat Food 3:1–3
Reppas CE, Behjat M, Peters G, Svanström M (2025) Creating circular pathways for nutrients in aquaculture using Biochar. Resour Conserv Recycl 227:108707
Rudovica V, Rotter A, Gaudêncio SP, Novoveská L, Akgül F, Akslen-Hoel LK, Alexandrino DA, Anne O, Arbidans L, Atanassova M (2021) Valorization of marine waste: use of industrial by-products and beach wrack towards the production of high added-value products. Front Mar Sci 8:723333
Rurangwa E, Agyakwah S, Boon H, Bolman B (2015) Development of Aquaculture in Ghana: analysis of the fish value chain and potential business cases
Salido M, Soto M, Seoane S (2024) Seaweed: nutritional and gastronomic perspective. A review. Algal Res 77:103357
Sapatinha M, Camacho C, Pais-Costa A, Fernando A, Marques A, Pires C (2024) Enzymatic hydrolysis systems enhance the efficiency and biological properties of hydrolysates from frozen fish processing co-products. Mar Drugs 23:14
Serra V, Pastorelli G, Tedesco DEA, Turin L, Guerrini A (2024) Alternative protein sources in aquafeed: current scenario and future perspectives. Vet Anim Sci 25:100381
Shahidi F, Varatharajan V, Peng H, Senadheera R (2019) Utilization of marine by-products for the recovery of value-added products. J Food Bioactives 6:10–61
Shi M, McHugh KJ (2023) Strategies for overcoming protein and peptide instability in biodegradable drug delivery systems. Adv Drug Deliv Rev 199:114904
Siddik S, Howieson J, Fotedar R, Partridge G (2021) Enzymatic fish protein hydrolysates in finfish aquaculture: a review. Rev Aquacult 13
Siddiqui S, Harahap I, osei-owusu J, Saikia T, Wu Y-S, Fernando I, Perestrelo R, Câmara J (2024a) Bioconversion of organic waste by insects – a comprehensive review. Process Saf Environ Prot 187:1–25
Siddiqui S, Süfer Ö, Çalışkan Koç G, Lutuf H, Rahayu T, Castro-Muñoz R, Fernando I (2024b) Enhancing the bioconversion rate and end products of black soldier fly (BSF) treatment – a comprehensive review. Environ Dev Sustain 27:9673–9741
Siddiqui SA, Singh S, Bahmid NA, Sasidharan A (2024c) Applying innovative technological interventions in the preservation and packaging of fresh seafood products to minimize spoilage – a systematic review and meta-analysis. Heliyon 10(8):e29066
Soma K, Vernooij A, Barbosa M, Opiyo M, Kals J, Cubero R, Komleh SP, Rurangwa E, Ndambi O, Schrijver R (2023) The protein transition: how to operationalise use of black soldier fly larvae (BSFL) and spirulina as feed ingrediencies in Kenya and Uganda? Wageningen Economic Research
Subasinghe R, Siriwardena S, Byrd K, Chan CY, Dizyee K, Shikuku K, Tran N, Adegoke A, Adeleke L, Anastasiou K, Beveridge M, Bogard J, Chu L, Fregene B, Ene-Obong H, Cheong KC, Nukpezah J, Olagunju OF, Powell A, Phillips M (2021) Nigeria fish futures aquaculture in Nigeria: increasing income, diversifying diets and empowering women report of the scoping study Citation Acknowledgments Creative Commons License Photo credits
Sun TM Xu, Z and Prinyawiwatkul W (2006) Fatty acid composition of the oil extracted from farmed Atlantic salmon (Salmo salar L.) viscera. J Am Oil Chem Soc 83(7):615–619
Suprijanto J, Widowati I (2024) Potential of Shells as a Source of Calcium. Buletin Oseanografi Marina 13:136–140
Tacon A, Metian M (2009) Fishing for aquaculture: non-food use of small pelagic forage fish—a global perspective. Rev Fish Sci 17:305–317

Thirukumaran R, Vijay AP, Krishnamoorthy S, Ramakrishnan P, Moses J, Chinnaswamy A (2022) Resource recovery from fish waste: prospects and the usage of intensified extraction technologies. Chemosphere 299:134361

Topić Popović N, Lorencin V, Strunjak-Perović I, Čož-Rakovac R (2023) Shell waste management and utilization: mitigating organic pollution and enhancing sustainability. Appl Sci 13(1):623

Torres-León C, Chávez ML, Hernández-Almanza A, Martinez Medina G, Ramírez Guzmán K, Londoño Hernández L, Aguilar C (2021) Recent advances on the microbiological and enzymatic processing for conversion of food wastes to valuable bioproducts. Curr Opin Food Sci 38:40–45

Tsoupras A, Brummell C, Kealy C, Vitkaitis K, Redfern S, Zabetakis I (2022) Cardio-protective properties and health benefits of fish lipid bioactives; the effects of thermal processing. Mar Drugs 20(3)

Tufail T, Bader Ul Ain H, Ashraf J, Mahmood S, Noreen S, Ijaz A, Ikram A, Arshad MT, Abdullahi MA (2025) Bioactive compounds in seafood: implications for health and nutrition. Food Sci Nutr 13(4):e70181

Wani A, Akhtar N, Mir TUG, Rahayu F, Suhara C, Anjli A, Chopra C, Singh R, Prakash A, El Messaoudi N, Dourado Fernandes C, Ferreira LF, Rather D-R, Américo-Pinheiro J (2023) Eco-friendly and safe alternatives for the valorization of shrimp farming waste. Environ Sci Pollut Res 31:38960–38989

Watanabe T, Murase T, Saito H (1995) Specificity of fatty acid composition of highly migratory fish. A comparison of docosahexaenoic acid content in total lipids extracted in various organs of bonito (Euthynnus pelamis). Comp Biochem Physiol B Biochem Mol Biol 111(4):691–695

Wong M, Mo W, Choi W-M, Cheng Z, Man Y (2016) Recycle food wastes into high quality fish feeds for safe and quality fish production. Environ Pollut 219:631

Wu W, Li H, Chen Y, Luo Y, Zeng J, Huang J, Gao T (2024) Recent advances in drying processing technologies for aquatic products. Processes 12:942

Younes I, Rinaudo M (2015) Chitin and chitosan preparation from marine sources. Structure, properties and applications. Mar Drugs 13(3):1133–1174

Yusoff MA, Mohammadi P, Ahmad F, Sanusi NA, Hosseinzadeh-Bandbafha H, Vatanparast H, Aghbashlo M, Tabatabaei M (2024) Valorization of seafood waste: a review of life cycle assessment studies in biorefinery applications. Sci Total Environ 952:175810

Chapter 15
Expanding Marine Fisheries and Aquaculture Supply Chains for Regional Food Security in Africa

Kenneth Prudence Abasubong, Mfon Timothy Udo, Ndakalimwe Naftal Gabriel, Uwem George Ubong, Aniefiokmkpong Okokon Okon, Michael Chukwuemeka Umehai, Umoh Uwana Emmanson, and Eyo Idorenyin Edem

Abstract Marine fisheries and aquaculture are central to enhancing regional food security in Africa, offering a sustainable source of nutrition, income, and economic growth. As land-based food systems face increasing pressure from climate change, population growth, and environmental degradation, aquatic food systems provide a resilient alternative. This chapter examines the strategic expansion of marine fisheries and aquaculture supply chains as a means to address Africa's food and nutritional challenges. It explores key stages of the value chain, harvesting, postharvest handling, storage, transportation, processing, marketing, and distribution, highlighting critical bottlenecks and opportunities for innovation. Emphasis is placed on inclusive approaches that empower small-scale fishers and women, strengthen cold chain logistics, improve access to finance, and promote environmentally responsible practices. This chapter also underscores the importance of policy harmonization, investment in infrastructure, and multi-stakeholder collaboration to build

K. P. Abasubong (✉) · U. G. Ubong · U. U. Emmanson · E. I. Edem
Department of Fisheries and Aquaculture, Faculty of Agriculture, Akwa Ibom State University (AKSU), PMB 1167, Obio Akpa Campus, Oruk Anam, Akwa Ibom State, Nigeria
e-mail: kennethabasubong@aksu.edu.ng

M. T. Udo
Department of Fisheries and Environmental Management, University of Uyo, Uyo, Akwa Ibom State, Nigeria

N. N. Gabriel
Department of Fisheries and Ocean Sciences, Faculty of Agriculture, Engineering, and Natural Sciences, Sam Nujoma Campus, University of Namibia, Henties Bay, Namibia

A. O. Okon
Department of Animal and Environmental Biology, University of Uyo, Uyo, Akwa Ibom State, Nigeria

M. C. Umehai
Department of Agricultural Extension, Delta State University, DELSU, Abraka, Nigeria

H. A. Mupambwa et al. (eds.), *Harnessing Coastal and Marine Resources for Improved Food Security in Africa*, Sustainability Sciences in Asia and Africa,
https://doi.org/10.1007/978-981-95-9725-3_15

efficient, equitable, and climate-resilient supply chains. Strengthening these systems is essential not only for improving access to nutritious food but also for advancing sustainable livelihoods, rural development, and regional economic integration across the African continent.

Keywords Supply chains · Food security · Marine fisheries · Aquaculture fisheries · Africa

15.1 Introduction

Marine fisheries and aquaculture are closely linked sectors essential to global and regional food systems (Béné et al. 2016). Marine fisheries involve harvesting fish and other marine organisms from oceans, seas, and coastal waters using traditional or industrial fishing methods (Mahboob and Shahzad 2024). In contrast, aquaculture, or fish farming, is the controlled cultivation of aquatic species—including fish, shellfish, and aquatic plants—in managed environments such as coastal ponds, cages, or tanks (Subasinghe et al. 2009). While marine fisheries rely on natural ecosystems, aquaculture enables planned, intensified production, allowing for better control over yield and sustainability.

In Africa, both sectors hold significant socioeconomic and nutritional value. Fisheries and aquaculture provide income, employment, and livelihoods for over 12 million people across the continent (N'Souvi et al. 2021). Artisanal and small-scale fisheries dominate marine fish production in many African countries, especially in West Africa, where they account for up to 80% of total catches (Nwafili and Gao 2007). Meanwhile, aquaculture is emerging as a vital solution to meet the continent's rising demand for animal protein. Countries like Egypt, Nigeria, Ghana, and Uganda have made significant progress, producing large quantities of tilapia, catfish, and other freshwater species (Babatunde et al. 2021). With Africa's population expected to double by 2050, these sectors are critical for ensuring food and nutrition security.

Fish also plays a crucial role in African diets by providing high-quality protein, omega-3 fatty acids, and essential micronutrients often lacking in other foods. In many coastal and island nations, fish constitutes more than half of the animal protein consumed (Viana et al. 2023). The accessibility of affordable fish directly impacts public health, especially in low-income communities where malnutrition is widespread. Additionally, marine fisheries and aquaculture are vital to regional trade, generating foreign exchange through exports of fresh, frozen, and processed seafood to global markets (Subasinghe et al. 2021).

Despite their promise, Africa's marine fisheries and aquaculture sectors face major structural and environmental challenges. These include overfishing, illegal, unreported, and unregulated (IUU) fishing, climate change effects, limited financing, inadequate infrastructure, and weak policy enforcement (Belhabib et al. 2016; Okafor-Yarwood 2019). Aquaculture, in particular, remains underdeveloped due to gaps in technology transfer, input supply, and investment. Therefore, sustainable development of these sectors is essential to Africa's goals of economic diversification, poverty reduction, and climate resilience. Strengthening governance, expanding research capacity, and fostering public–private partnerships are critical strategies to unlock the full potential of marine fisheries and aquaculture across the continent.

15.2 The Importance of Focusing on Marine Fisheries and Aquaculture

Population Growth Africa's population is projected to exceed 2.5 billion by 2050, intensifying pressure on food systems already strained by climate change, land degradation, and water scarcity (Simane et al. 2025). Traditional agriculture alone may struggle to meet this growing demand, and investing in sustainable, efficient alternatives such as marine fisheries and aquaculture is essential. Fish is a nutrient-dense, protein-rich food that requires fewer resources than land-based livestock. Expanding aquaculture, particularly near urban and coastal centers, can increase food supply, reduce postharvest losses, and address evolving consumer preferences. Strengthening these sectors will decrease dependence on imports, bolster national food sovereignty, and position Africa's blue economy as a key pillar of long-term food and nutrition security (Farmery et al. 2021).

Undernourishment Over 250 million people in Africa suffer from undernourishment, with vulnerable groups such as children, pregnant women, and rural poor disproportionately affected. This leads to stunting, wasting, and micronutrient deficiencies (Bain et al. 2013; Müller and Krawinkel 2005). Many traditional diets lack essential nutrients, causing "hidden hunger" despite sufficient calorie intake. Marine fisheries and aquaculture offer a nutrient-rich alternative, providing high-quality protein, omega-3 fatty acids, and critical micronutrients often absent from staple foods (Arshad et al. 2022; Bain et al. 2013). Expanding affordable fish production and supply can enhance dietary diversity and reduce malnutrition, particularly among low-income populations in high-risk regions.

Untapped Resources Africa holds vast, largely underutilized aquatic resources. Despite extensive coastlines, large lakes, and numerous inland water bodies, marine fisheries and aquaculture production remain far below their potential compared to other regions globally (Walakira et al. 2023). Limited infrastructure, technical capacity, and investment hinder full development. Aquaculture, especially, remains nascent, contributing only a small portion of the continent's fish supply (Obiero et al. 2019). This untapped potential presents a major opportunity to increase food production, generate income, and create jobs. Strategic investments in sustainable resource management, technology, and value chains can unlock this potential to improve food security and stimulate economic growth.

Economic Opportunity Marine fisheries and aquaculture are vital income and employment sources across Africa, supporting millions in coastal and inland communities. Beyond food production, these sectors offer diverse jobs in fishing, fish farming, processing, marketing, and transport—often engaging women and youth with limited alternatives. Expanding these sectors can invigorate local economies, reduce poverty, and promote entrepreneurship (Obiero et al. 2019). Furthermore, developing robust supply chains and value-added fish products opens access to regional and international markets, boosting foreign exchange earnings and enhancing economic resilience. Investing in fisheries and aquaculture is therefore a strategic driver for inclusive growth and improved livelihoods continent-wide (Muthuraman et al. 2023; Urugo et al. 2024).

15.3 Fish as Food—Role in Nutrition and Health

Fish plays a crucial role in global diets, especially in developing regions like Africa, where it serves as an affordable, accessible, and nutrient-dense food source (Bakhsh et al. 2023). As both wild-caught and farmed fish become increasingly integrated into food systems, their contribution to human health and nutrition continues to rise. Fish uniquely provides high-quality animal protein, essential fats, and a broad range of vitamins and minerals often lacking in plant-based diets (Byrd et al. 2022). This makes fish a strategic food in combating hunger, enhancing dietary diversity, and addressing micronutrient deficiencies in vulnerable populations.

Nutritionally, fish is an excellent source of easily digestible protein containing all nine essential amino acids (Abasubong et al. 2019). It is particularly valued for its high levels of omega-3 long-chain polyunsaturated fatty acids (LC-PUFAs), such as EPA and DHA, which support brain development, cardiovascular health, and immune function. Additionally, fish supply vital micronutrients including iodine, selenium, zinc, iron, calcium, and vitamins A, D, and B_{12} (Byrd et al. 2022). Small fish species consumed whole—bones and organs included—further enrich diets with bioavailable calcium and vitamin A, addressing critical nutrient gaps in many African communities.

Fish is essential in combating hunger and malnutrition, especially in regions where diets rely heavily on monotonous staple crops. In sub-Saharan Africa, where stunting and undernutrition affect many children under five, incorporating fish into diets can significantly improve outcomes. Research shows that even small amounts of fish enhance growth, cognitive development, and immune function in animals (Demelash Abera and Alefe Adimas 2024). For adults, regular fish consumption reduces the risk of noncommunicable diseases such as hypertension, heart disease, and type 2 diabetes. Thus, fish plays a dual role in addressing both undernutrition and overnutrition, tackling the "double burden" of malnutrition prevalent across Africa (Fanzo et al. 2022).

Moreover, fish is often the most affordable animal-source food for low-income households. In coastal, riverine, and lake communities, fish serve not only as a dietary staple but also as an economic lifeline. The availability of dried, smoked, or fermented fish ensures year-round access and wider distribution to inland and remote populations (Johnson et al. 2023). Expanding sustainable aquaculture and improving fisheries management are therefore central to food security strategies. By integrating fish into national nutrition policies, school feeding programs, and maternal-child health interventions, governments and development partners can strengthen efforts to combat hidden hunger and improve public health outcomes (Ahern et al. 2021).

15.4 The Current State of Marine Fisheries and Aquaculture in Africa

The current state of marine fisheries and aquaculture in Africa reflects a sector of high strategic importance, yet it remains constrained by structural and environmental challenges. Africa accounts for approximately 7% of global fish production, yielding around 13 million tonnes annually from both capture fisheries and aquaculture (Chan et al. 2021). Marine capture fisheries dominate, targeting key species such as sardines, sardinella, anchovies, tuna, tilapia, catfish, and shrimp. Inland fisheries, particularly around major water bodies like Lake Victoria, Lake Tanganyika, and the Niger and Congo basins, also contribute significantly to food security and employment (Njiru et al. 2008). Fishing activities are geographically widespread, with major marine fishing zones including the West African coast (Senegal, Ghana, Mauritania, Nigeria), the Southwest Indian Ocean (Mozambique, Tanzania, Madagascar), and North Africa (notably Morocco and Egypt). On the aquaculture side, growth is concentrated in Egypt, Nigeria, Ghana, Uganda, and Kenya, with Egypt alone responsible for over 70% of Africa's aquaculture output (Karani et al. 2022). These hubs primarily cultivate freshwater species, especially tilapia and catfish, due to their hardiness and strong market demand.

Despite this potential, the sector faces persistent challenges that limit its contribution to food security and economic growth. Postharvest losses—estimated between 20% and 30%—result from poor handling, inadequate cold chain infrastructure, outdated processing techniques, and limited access to electricity and clean water in rural and coastal areas (Abelti and Abera Teka 2024). Wider infrastructure deficiencies affect landing sites, transport logistics, hatcheries, and local feed production, constraining both capture fisheries and aquaculture development (Subasinghe et al. 2021). Governance weaknesses, including weak enforcement of regulations, rampant illegal, unreported, and unregulated (IUU) fishing, fragmented institutions, and limited cross-border coordination, further undermine sustainable resource management. Many countries operate with outdated or poorly implemented fisheries policies, while local communities often lack meaningful involvement in decision-making processes (Mozumder et al. 2023).

Nonetheless, promising opportunities point to a positive future for the sector. The growing emphasis on the blue economy—which advocates for sustainable ocean resource use to drive economic development—is gaining momentum in African policy circles, offering an integrated framework linking fisheries, aquaculture, coastal tourism, and marine conservation (Marwan 2023). Rising domestic and regional demand for affordable protein is spurring investments in aquaculture, value chain enhancement, and fish marketing (Ayoola 2010). Urbanization and changing dietary preferences continue to drive steady increases in fish consumption. Furthermore, the African Continental Free Trade Area (AfCFTA) presents a significant opportunity to boost regional trade in fish and aquaculture products by reducing tariffs, harmonizing standards, and integrating cross-border value chains (Timothy and Ejem 2025). If strategically harnessed through focused investments, policy reforms, and regional cooperation, these opportunities could transform Africa's fisheries and aquaculture sectors into key drivers of food and nutrition security, employment, and climate resilience.

15.5 Understanding the Supply Chains for Regional Food Security in Africa

Marine fisheries and aquaculture are critical to Africa's food security, nutrition, and economic development, yet their full potential remains underutilized due to fragmented supply chains, inefficiencies, and inadequate infrastructure. The journey of fish from water to table involves a complex supply chain—including harvesting, handling, processing, distribution, and retail—which varies depending on whether the source is marine capture or aquaculture, and whether the end market is local, regional, or international (Kruijssen et al. 2020). A clear understanding of this supply chain is essential for identifying bottlenecks, improving efficiency, reducing postharvest losses, ensuring food safety, and maximizing the nutritional and economic contributions of the sector. The Supply chain is presented in Fig. 15.1.

15.5.1 Production or Harvesting

Marine fisheries involve catching wild fish from ecosystems like continental shelves, estuaries, and open oceans. Methods vary by species, environment, and technology. Common techniques include bottom and mid-water trawling (for demersal and pelagic species), longlining (for tuna and swordfish), and artisanal fishing with gillnets, traps, handlines, and beach seines. Along West Africa's coast, artisanal fishers mainly target small pelagics like sardinella and anchovies using stationary gillnets, effective but prone to overfishing (Nunoo et al. 2015).

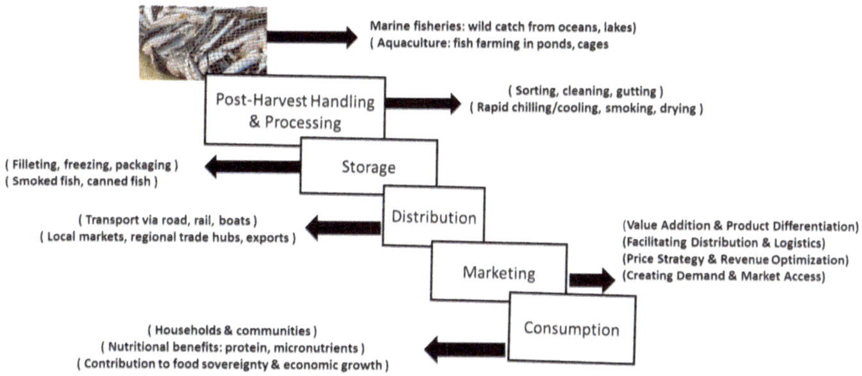

Fig. 15.1 Fisheries supply chain for regional food security in Africa. The illustration outlines the sequential stages from postharvest handling and processing to final consumption, emphasizing the importance of each step—storage, distribution, and marketing—in preserving fish quality, reducing losses, and ensuring availability and accessibility of nutritious fish products across the region

In aquaculture, fish are raised in controlled systems such as ponds, cages, raceways, and RAS. Nile tilapia and African catfish dominate due to their fast growth, adaptability, and market demand (Oboh 2022). Egypt leads in intensive tilapia cage farming along the Nile, while Nigeria, Ghana, and Uganda are rapidly expanding. However, challenges like poor seed quality, feed shortages, and disease outbreaks persist. Fish quality at harvest, impacted by stress, injury, and contamination, affects shelf life and market value. Proper handling and quick cooling help reduce spoilage (Getu and Misganaw 2015). Enhancing harvesting methods, infrastructure, and training is crucial for sustainable growth.

15.5.2 *Postharvest Handling and Processing*

After harvest, fish are sorted, cleaned, and stored to maintain quality and safety. Processing methods, such as drying, smoking, freezing, salting, filleting, and canning, preserve fish, improve taste, and boost market value (Siddiqui et al. 2024). In rural and coastal African areas, traditional techniques like sun-drying and smoking are widely used due to limited access to electricity and cold storage (Adeyeye 2016a). While low-cost and culturally accepted, these methods often lack proper hygiene controls, raising contamination and spoilage risks.

Poor handling, such as delays in chilling and inadequate packaging, can cause postharvest losses of up to 30% in sub-Saharan Africa (Ikbal et al. 2023), reducing food supply and income for fishers and traders. Solutions include investing in cold chain infrastructure (ice production, cold storage, refrigerated transport), improving hygiene training, and promoting better processing tools. Technologies like solar-powered cold rooms and mobile chillers are helping off-grid communities. Since women play key roles in processing and marketing, improving their access to equipment, training, and finance can boost incomes, promote gender equity, and support coastal livelihoods (Cliffe and Akinrotimi 2015).

15.5.3 Storage and Transport

Storage and transport are crucial for preserving fish quality and ensuring timely delivery to markets. Proper storage helps maintain freshness and nutrition, while efficient transport reduces delays, spoilage, and financial losses (Faisal et al. 2024). Because fish spoils quickly due to enzymes, microbes, and oxidation, maintaining a cold chain from harvest to sale is essential. Any break in this chain can compromise food safety and lead to major economic losses (Yu et al. 2021).

In many African regions, cold storage and transport systems are inadequate or unevenly distributed. Traditional preservation methods such as drying, smoking, and salting are still widely used, especially in rural areas without electricity (Ahmad et al. 2021). Although effective in prolonging shelf life, these methods can alter the fish's taste, texture, and marketability.

Modern cold chain infrastructure, like refrigerated trucks, cold rooms, and ice plants, is mostly found in urban or export-oriented zones. Rural producers often lack access to these facilities, which contributes to high postharvest losses (up to 30% in some cases) and limits market access for small-scale actors (Lisboa et al. 2024). Investing in mobile cold storage, solar-powered systems, and reliable transport infrastructure is essential to address these gaps.

15.5.4 The Distribution

Distribution connects fishers and fish farmers to consumers across local, regional, and international markets (Saini et al. 2023). Fish products, fresh, chilled, frozen, or processed, are moved from landing sites or farms based on demand and available preservation methods. In Africa, this process is largely managed by networks of traders, wholesalers, and transporters using trucks, boats, motorcycles, or bicycles. However, weak infrastructure and remote locations often disrupt logistics and cold chain systems, leading to delays and spoilage (Gligor et al. 2018). Women play a vital role in small-scale distribution, yet limited access to transport, storage, and finance restricts their earning potential. Addressing these challenges requires investment in cold chain logistics, reliable roads, and better coordination. The lack of refrigerated vehicles and adequate storage raises transport costs and reduces product quality. Innovative solutions, such as solar-powered cold rooms, insulated containers, mobile ice units, and digital route planning tools, are helping close these gaps (Nagy et al. 2023). Training in cold chain management can further enhance efficiency. Strengthening distribution systems is essential to improve access to fish, reduce losses, stabilize prices, and build resilience against climate and market shocks (Ojea et al. 2020).

15.5.5 Marketing

Marketing is a key stage in the fisheries and aquaculture supply chain. It involves promoting, selling, and delivering fish to consumers and determines how producers, processors, and traders earn value and access markets. In Africa, marketing is highly diverse, shaped by cultural preferences, income levels, urbanization, and retail systems (Okonkwo et al. 2023). Fish is sold through both informal outlets, like roadside stalls and open-air markets, and formal channels, such as supermarkets and urban fishmongers (Chipfupa 2023). High-value exports, such as tuna and shrimp, connect African producers to global markets.

However, effective marketing is often limited by poor digital access, low literacy, and fragmented supply chains. Better market information, through mobile tools, reports, and extension services, can improve decisions and profits (Ezeudu and Obimbua 2024). Certification schemes like MSC and ASC help access premium markets and promote sustainability, while local quality standards build consumer trust (Gulbrandsen 2009). Marketing also includes value addition, such as filleting, packaging, and ready-to-cook products. Women lead much of this work but face challenges like limited capital, tools, and training. Supporting women-led businesses and adapting to changing consumer demands, for safe, convenient, and nutritious fish, can boost sales, improve diets, and grow the sector's economic impact.

15.5.6 Consumption

Consumption is the final stage of the fish supply chain, where products reach households, restaurants, institutions, and other consumers (Tran et al. 2023). Access to fish depends on factors such as price, cultural preferences, seasonal availability, and proximity to markets (Arsil et al. 2018). While demand is growing across Africa due to urbanization, rising incomes, and greater awareness of fish's nutritional value (Abatekassa and Peterson 2011), access remains limited, especially in remote and low-income areas where infrastructure and affordability are lacking.

Improving the entire supply chain, from sustainable harvesting and aquaculture to processing and distribution, can boost food availability, enhance nutrition, and support livelihoods and economic growth across the continent.

15.6 Challenges in African Fisheries and Aquaculture Supply Chains

The fisheries and aquaculture sectors in Africa hold immense potential for improving food security, nutrition, and livelihoods. However, several systemic challenges limit the efficiency, sustainability, and inclusiveness of their supply chains. Among the most pressing issues are poor infrastructure, lack of storage and cold chain systems, and inadequate transport networks (March and Failler 2022). These obstacles contribute to high postharvest losses, reduced product quality, lower incomes for fishers and farmers, and limited access to markets, especially for small-scale operators.

15.6.1 Lack of Storage and Cold Chain Infrastructure

One of the critical weaknesses in African fisheries and aquaculture supply chains is the widespread absence of adequate cold storage facilities(Shoko et al. 2022). Many fish landing sites, aquaculture farms, and local markets lack ice-making machines, refrigerated storage units, or access to electricity. These are tools that are essential for preserving fish freshness from harvest to consumption. As a result, fish spoil quickly in high-temperature conditions, especially in tropical climates, leading to significant postharvest losses, sometimes exceeding 30% in certain regions (Mramba and Mkude 2022). These losses not only waste valuable protein resources but also reduce economic returns and discourage investment in fish production.

Furthermore, the limited availability of cold chain logistics affects the quality and safety of fish products (Hanafiah et al. 2024). Without reliable refrigeration, fish are prone to bacterial contamination, which poses public health risks and undermines consumer confidence. Small-scale fishers and aquaculture producers, who dominate the sector in many African countries, are particularly disadvantaged, as they often lack the capital or technical capacity to invest in cold storage or transportation technologies (Selamoglu 2021). As a result, their access to distant or higher value markets is constrained, and they are often forced to sell their fish quickly and at lower prices.

15.6.2 Inadequate Transport and Road Infrastructure

Transportation is another critical bottleneck in the fisheries and aquaculture value chains (Mgbomene 2025). In many coastal, rural, and inland areas, roads are either poorly constructed or seasonally impassable, making it difficult to move fish from production sites to processing centers or markets promptly. Transport delays not only reduce the freshness and value of the fish but also increase spoilage risks (Abera Teka 2024). For aquaculture farms located far from urban centers, the lack of feeder roads limits access to inputs such as quality fish seed, feed, and veterinary services, thereby affecting productivity and profitability (Maulu et al. 2020).

Additionally, poor road conditions increase transportation costs, which are often passed on to consumers in the form of higher fish prices. This reduces affordability and limits the nutritional impact of fish, particularly for low-income populations who rely on it as a vital source of animal protein (Eriegha et al. 2022). In many cases, the lack of reliable transport also discourages private sector investment and hampers the development of structured fish marketing systems (Alam and Yousuf 2024). Without improved infrastructure, efforts to expand and modernize the fisheries and aquaculture sectors will remain hamstrung by inefficiencies and missed economic opportunities.

15.6.3 *Other Infrastructure-Related Constraints*

Beyond storage and transport, the broader infrastructure landscape—including market facilities, electricity supply, water access, and digital connectivity—also presents significant challenges. Many fish markets lack hygienic conditions, sanitation services, and quality control mechanisms. This not only affects food safety but also reduces the competitiveness of African fish in both domestic and international markets (Kussaga et al. 2014). Moreover, the absence of modern processing facilities limits value addition opportunities, such as fish filleting, packaging, or freezing, which are essential for improving product shelf life and market reach.

Electricity deficits are particularly acute in rural and peri-urban regions, where many aquaculture operations and fish landing sites are based. Without reliable energy, operating cold rooms, ice plants, or aeration systems becomes nearly impossible. Likewise, weak digital infrastructure hinders the flow of market information, making it difficult for producers to respond to demand signals or access e-commerce platforms (Evareh et al. 2023). These infrastructure gaps collectively contribute to a fragmented and underperforming supply chain that fails to meet the growing demand for fish across Africa (Blessing et al. 2025).

15.7 Expanding Marine Fisheries and Aquaculture Supply Chains in Africa

As demand for fish continues to grow across Africa and globally, expanding and strengthening marine fisheries and aquaculture supply chains has become essential for improving food security, generating livelihoods, and supporting economic development. Both sectors offer immense potential to meet nutritional needs and reduce pressure on wild fish stocks. However, realizing this potential requires coordinated efforts to enhance production, reduce postharvest losses, improve infrastructure, and build inclusive market systems. By addressing gaps and inefficiencies along the entire value chain (from harvest to consumption), countries can unlock greater value, ensure sustainability, and build resilience in the face of climate and economic challenges.

15.7.1 Reducing Fish Loss and Waste in Africa

Fish loss and waste remain critical challenges in African fisheries and aquaculture systems, undermining food security, economic development, and the sustainability of aquatic resources. Postharvest losses, primarily due to spoilage, poor handling, and inadequate infrastructure, can account for 20–30% of total fish production in some regions (Muthuraman et al. 2023). Tackling these losses is essential to maximizing the benefits of fish production without necessarily increasing catch pressure or farming intensities. Fortunately, a combination of simple preservation techniques and community-based approaches offers practical, low-cost, and sustainable solutions to reduce spoilage and ensure that more fish reach consumers in good condition.

15.7.1.1 Simple, Low-Cost Preservation Methods

Traditional preservation methods such as smoking, salting, drying, and fermentation remain among the most widely used techniques in Africa to extend the shelf life of fish. These methods require minimal equipment and energy inputs, making them ideal for small-scale fisheries and aquaculture systems operating in rural or low-resource settings (Adeyeye 2016b). Smoking, for instance, not only preserves fish for several days or weeks but also adds market value through flavor enhancement. The introduction of improved smoking kilns, such as the FAO's Chorkor oven, has significantly reduced smoke exposure, increased batch size, and enhanced drying efficiency (Adeyeye 2016b).

Likewise, solar drying systems offer an environmentally friendly alternative to open-air drying. Covered solar dryers protect fish from contamination by dust, flies, and rain, while speeding up the drying process and improving product quality (Olokor et al. 2009). Salting and brining techniques are also effective in inhibiting bacterial growth, especially when combined with drying. In addition, the use of simple ice storage, using insulated containers or locally produced ice blocks, can significantly reduce microbial spoilage during transport and at landing sites (Indiarto et al. 2021). Providing fishers with access to affordable ice is a key strategy in reducing loss, particularly in hot climates.

15.7.1.2 Community-Based Cold Chain and Processing Initiatives

Local, community-driven cold chain systems have emerged as effective models for minimizing postharvest fish spoilage. Cooperatives and women-led processing groups often lead these efforts, pooling resources to invest in shared cold rooms, solar-powered freezers, or ice production units (Kitinoja 2014). These facilities improve the freshness and marketability of fish and enable communities to access more distant or urban markets where fish fetch higher prices (Brummett 2000). Solar-powered cold storage has shown promising results in off-grid communities, offering sustainable and cost-effective refrigeration while reducing dependence on fossil fuels.

Community-based fish processing centers also enhance efficiency and reduce waste by offering standardized processing, packaging, and quality control. These centers often act as hubs for skill development, where fishers, especially women, are trained in postharvest handling, hygiene practices, and value addition (e.g., producing fish powder or vacuum-packed products) (Subasinghe et al. 2021). Moreover, partnerships with local governments or NGOs can provide technical and financial support to scale these initiatives.

15.7.1.3 Strengthening Education and Extension Services

A critical component in reducing fish loss is improving knowledge and capacity among stakeholders throughout the value chain (Türkten et al. 2024). Many losses result from poor handling practices, such as rough transportation, delayed processing, or improper storage. Training programs targeting fishers, vendors, processors, and transporters can raise awareness about the economic and health costs of fish spoilage and demonstrate best practices for preservation, hygiene, and timely market access.

Extension services can introduce low-cost innovations such as clean handling mats, insulated fish boxes, or mobile fish processing units, which allow for immediate preservation in remote fishing villages. Education also plays a key role in promoting dietary diversification, ensuring that undervalued or bycatch species are processed and consumed rather than discarded (Priestley et al. 2022). Targeted campaigns can also educate consumers and vendors about food safety, storage, and the responsible handling of fish products.

15.7.1.4 Policy and Institutional Support

Supportive policy frameworks are essential to scale and sustain loss-reduction strategies (Lambin et al. 2020). Governments can facilitate access to credit for investing in cold chain equipment, subsidize renewable energy-based preservation technologies, and provide incentives for community-led fish processing facilities. National standards for postharvest handling and quality control should be developed and enforced to ensure safe and marketable fish products. Partnerships with development agencies, NGOs, and the private sector are also critical to bridge gaps in financing, innovation, and infrastructure.

Investment in infrastructure, such as roads, clean water supply, and energy systems, will further reduce spoilage during transport and enable wider adoption of preservation technologies. Additionally, integrating fish loss and waste reduction into national fisheries development plans can help align local and regional efforts under a common goal of enhancing food security and economic resilience (Olaifa et al. 2022).

15.7.2 Improving Markets and Trade for Fish Products in Africa

Markets and trade are fundamental to the success of fisheries and aquaculture in Africa. While fish plays a vital role in food and nutrition security, its economic potential is often underutilized due to inefficiencies in marketing systems, poor infrastructure, and limited access to regional and international markets (Maulu et al. 2024). Strengthening the fish trade, both locally and across African borders, can unlock greater economic opportunities for producers, reduce postharvest losses, and improve food distribution (Ibengwe et al. 2022). To ensure this, inclusive market systems and equitable pricing mechanisms must be developed to support small-scale fishers, aquaculture farmers, and fish traders—particularly women and youth, who form a major part of the postharvest value chain.

15.7.2.1 Local Fish Sales: Strengthening Community-Based Trade

Selling fish within local markets remains the backbone of the fish trade in many African countries. Fresh, dried, smoked, or salted fish are commonly sold in village markets, roadside stalls, or urban centers. However, challenges such as a lack of cold storage, inadequate transportation, and fluctuating prices limit the profitability of local sales. Improving local market infrastructure, such as modern landing sites, cold storage facilities, and hygienic processing centers, can greatly increase the shelf life and value of fish products (Zego and Husny 2023).

Empowering local traders through access to microfinance and training on market literacy can further improve incomes and reduce exploitation. Equally important is building trust and transparency in price negotiations. Digital tools and mobile apps that share real-time market prices can prevent unfair pricing and ensure that fishers and farmers are compensated fairly for their products (Kamal and Bablu 2023). Strengthening local demand also requires public awareness campaigns to promote the nutritional value of locally sourced fish.

15.7.2.2 Regional Trade: Expanding Intra-African Fish Markets

Africa's fish trade is often underdeveloped across borders due to weak regional integration, trade barriers, and poor logistics. However, with the implementation of the African Continental Free Trade Area (AfCFTA), opportunities for intra-African trade are expanding (Udombana 2020). Countries such as Uganda, Egypt, Nigeria, and Ghana already serve as major fish producers with the capacity to supply neighboring markets with fresh, frozen, or processed fish products.

Expanding cross-border fish trade requires harmonized standards for food safety and quality, streamlined customs procedures, and improved road and port infrastructure. Building regional market hubs and strengthening transport corridors will facilitate the movement of goods and reduce postharvest loss during transit. Investing in regional value chains, where fish produced in one country is processed and packaged in another, can create jobs and improve trade efficiency across the continent (Kaminski et al. 2018).

15.7.2.3 Fair Prices and Inclusive Market Access

One of the most pressing concerns for small-scale fishers and aquaculture producers is the issue of unfair prices and middlemen dominance (Kimani et al. 2020). Many producers lack bargaining power and are forced to sell their fish at low prices immediately after harvest. Creating cooperatives or producer organizations can strengthen collective bargaining and improve access to better-paying markets. Additionally, inclusive certification schemes and labeling (such as "Sustainably Sourced" or "Women-Farmed Fish") can boost product value and attract premium buyers.

Gender equity is particularly important in market access. Women dominate fish processing and retailing in many African countries but face systemic barriers in accessing finance, market stalls, and training (Elias et al. 2024). Targeted programs that support women traders with credit, storage, and legal protections can significantly enhance their economic security and market participation.

15.7.2.4 Building Stronger Fish Markets for a Food-Secure Africa

Improving markets and trade for fish products is a cornerstone of sustainable fisheries and aquaculture development in Africa. Strengthening local trade systems, facilitating regional integration, ensuring fair pricing, and supporting inclusive access can enhance both livelihoods and food security. Policy reforms, infrastructure investments, and regional cooperation under frameworks like AfCFTA are essential to transform the continent's fish value chains (Udombana 2020). Ultimately, an efficient and equitable market system will not only improve the economic well-being of producers and traders but also ensure that nutritious fish reaches more African households in affordable and sustainable ways.

15.7.3 Technology to Improve the Fish Supply Chain in Africa

The integration of technology into the fish supply chain is becoming increasingly critical in Africa as the continent strives to enhance food security, reduce postharvest losses, and increase the profitability of fisheries and aquaculture (Chan et al. 2019). From catch to consumer, innovative tools such as mobile applications, solar-powered drying systems, GPS tracking, and digital finance solutions are revolutionizing how fish are harvested, processed, stored, and marketed. These technologies not only improve efficiency and traceability but also empower small-scale producers and traders by improving access to markets, information, and services.

15.7.3.1 Mobile Applications: Connecting Producers to Markets and Knowledge

Mobile apps are among the most impactful technologies in the fisheries and aquaculture sector. They serve as platforms for market information, digital payments, training content, and fisheries management tools (Verma et al. 2024). Apps like eFishery, Tulaa, and Farm Crowdy allow fish farmers to monitor feeding schedules, track inventory, access real-time price updates, and connect with buyers or extension officers. In Kenya, the M-Farm app enables fish producers to get up-to-date market prices, eliminating the reliance on middlemen and allowing farmers to negotiate better deals (Baumüller 2013). Similarly, in Nigeria, the AgUnity platform supports transparency in transactions and builds trust among stakeholders in the fish value chain (Akisanmi and Ogunderu 2024). These digital platforms increase income potential and reduce postharvest losses caused by delays in sales.

15.7.3.2 Solar Dryers and Cold Storage: Enhancing PostHarvest Handling

Traditional drying methods often expose fish to contamination, spoilage, and poor hygiene. Solar drying technology offers a sustainable and cost-effective alternative. Solar fish dryers use enclosed structures with transparent roofs to dry fish hygienically, reducing spoilage, improving product quality, and extending shelf life (Regupathi and Balasubramanian 2018). In Uganda, community-based solar dryers introduced through public–private partnerships have significantly reduced spoilage rates and improved the income of women fish processors along Lake Victoria (Al-Ismaili 2021; Whiston et al. 2022). Additionally, solar-powered cold storage units, such as those introduced by ColdHubs in Nigeria, help small-scale fishers store unsold catch safely, cutting postharvest losses by over 50%. These technologies are particularly vital in remote areas lacking reliable electricity and infrastructure.

15.7.3.3 GPS and Remote Sensing: Improving Fisheries Monitoring and Logistics

GPS technology and remote sensing tools are transforming how fisheries are managed and how fish are distributed. Fisherfolk use GPS tracking devices for safer navigation, optimized fishing routes, and real-time location tracking of boats, thus improving safety and efficiency. In Ghana, the introduction of GPS-enabled fishing logbooks has helped artisanal fishers optimize their trips and reduce fuel costs.

Remote sensing, combined with drones or satellite imagery, supports environmental monitoring and resource mapping (Srivastava et al. 2022). Governments and NGOs can use these tools to track fish stocks, detect illegal fishing activities, and make informed decisions on fishery closures or restocking. On the logistics side, digital tracking of supply chains ensures traceability, helping to meet quality standards for export markets.

15.7.3.4 Building a Smart and Inclusive Fish Supply Chain

Technology has the potential to transform Africa's fish supply chains by enhancing productivity, ensuring food safety, and reducing postharvest losses (Tawakalitu et al. 2025). Mobile applications, solar technologies, GPS tracking, and remote sensing tools offer practical and scalable solutions to long-standing challenges in the fisheries and aquaculture sectors. However, successful implementation depends on user-friendly design, affordability, access to training, and inclusive policies that ensure women and small-scale actors are not left behind (Samuel & Yeboah 2023). To maximize impact, governments and development partners must invest in digital infrastructure, support innovation ecosystems, and integrate technology into national fisheries strategies. With the right approach, technology can help build resilient, profitable, and equitable fish supply chains across Africa.

15.8 Empowering Small-Scale Fishers and Women

Recognizing women's vital roles in the fisheries supply chain (fish processing, marketing, and distribution) is essential, as their contributions ensure fish availability, particularly in rural and peri-urban areas. However, social norms, limited resource access, and legal barriers often restrict their participation and advancement (Breckwoldt et al. 2020). Addressing these gender disparities through targeted policies and programs enhances equity and strengthens community resilience.

Limited capital access significantly hinders small-scale fishers, especially women, from investing in better fishing gear, cold storage, transportation, and processing equipment due to social norms and collateral constraints. Microfinance schemes, cooperative savings groups, and tailored credit facilities have proven effective in enabling women to upgrade operations, gain financial independence, and improve incomes. Supporting access to critical inputs such as quality seed and feed for aquaculture further boosts productivity and sustainability (Avadí et al. 2022). Capacity building through training in sustainable fishing, postharvest handling, business management, and marketing empowers women with skills to increase operational efficiency and market competitiveness. Their active participation in cooperatives and fisher associations fosters collective bargaining, resource sharing, and improved market access, while advocating for women's representation in policy dialogues ensures their voices influence fisheries governance (Chambon et al. 2023). Finally, facilitating access to appropriate technologies, like improved gear, solar-powered cold storage, mobile market information, and digital payments, further enhances productivity, reduces losses, and strengthens women's roles across the value chain.

15.8.1 Policies and Partnerships for Better Fish Supply Chains

Africa's fish supply chains play a crucial role in ensuring food security, nutrition, employment, and economic development. With over 12 million people engaged in fisheries and aquaculture across the continent, effective management and coordination are vital to support livelihoods while maintaining ecological sustainability (Subasinghe et al. 2009). However, challenges such as overfishing, weak governance, limited infrastructure, and climate vulnerability hinder the sector's growth (Kapapa et al. 2024). To address these issues, African governments and organizations have increasingly focused on implementing structured policies and forging partnerships aimed at building more resilient and equitable fish supply chains.

15.8.1.1 Strengthening Fisheries Governance and Resource Management

Governments across Africa are recognizing the need for sound fisheries governance as a foundation for sustainable supply chains. Policies that regulate fishing effort, prevent overexploitation, and protect marine ecosystems are critical. For example, Senegal's National Strategy for the Sustainable Development of Fisheries and Aquaculture emphasizes ecosystem-based management and involves fishing communities in decision-making through Local Artisanal Fisheries Councils (Corral and Romero Manrique de Lara 2017). These councils play a key role in comanaging marine resources and have improved compliance with fishing regulations while empowering small-scale fishers.

15.8.1.2 Combatting Illegal, Unreported, and Unregulated (IUU) Fishing

IUU fishing is a major threat to Africa's marine resources and undermines legal supply chains. Governments have enacted laws and participated in regional initiatives to monitor and control fishing activities. Ghana, for instance, has taken steps to curb IUU fishing through its Fisheries Enforcement Unit, which conducts patrols and uses Vessel Monitoring Systems (VMS) and Automatic Identification Systems (AIS) to track vessels (Kurekin et al. 2019). Additionally, the FISH-i Africa initiative, a regional partnership involving countries like Kenya, Mozambique, and Tanzania, uses shared intelligence, regional cooperation, and digital tools to combat illegal fishing in the Western Indian Ocean. These efforts help secure legal supply chains and improve transparency.

15.8.1.3 Supporting Small-Scale Fishers through Inclusive Policies

Small-scale fisheries provide up to 80% of the fish consumed in sub-Saharan Africa, yet fishers often face marginalization, lack of market access, and insufficient infrastructure (Muzari 2016). Governments have begun to address these gaps by developing inclusive fisheries policies. For example, The Gambia's Fisheries Policy includes provisions for improved landing sites, cold storage, and training for small-scale fishers, particularly women who dominate postharvest activities (Alam 2013). These investments strengthen local supply chains by reducing postharvest losses and improving quality control, enabling fishers to access better markets.

15.8.1.4 Promoting Aquaculture Development for Supply Chain Resilience

Aquaculture is emerging as a strategic response to overfishing and rising demand for fish in Africa. National policies are increasingly supporting aquaculture as a means to boost domestic fish production and reduce import dependence. Nigeria, the largest aquaculture producer in sub-Saharan Africa, has developed the National Aquaculture Strategy focusing on public–private partnerships, access to finance, and development of hatcheries and feed industries (Murekezi et al. 2024). These policies have led to increased production and job creation along the aquaculture supply chain, from fingerling production to processing and distribution.

15.8.1.5 Enhancing Cold Chain and Infrastructure Development

Inadequate cold storage, processing facilities, and transport systems are major barriers to efficient fish supply chains in Africa (Ejairu and Researcher 2022). To address this, governments and development partners have launched infrastructure projects aimed at reducing spoilage and improving value addition. For instance, in Tanzania, the government, in partnership with the Food and Agriculture Organization (FAO), established cold storage facilities and modernized landing sites in Lake Victoria fishing communities (Odongkara et al. 2010). These efforts have improved fish quality, increased earnings for fishers, and facilitated access to regional and international markets.

15.8.1.6 Women's Empowerment and Gender-Inclusive Policies

Women play a central role in African fish supply chains, particularly in processing and trade, yet they face barriers in access to finance, training, and decision-making. Governments and organizations have begun to integrate gender perspectives into fisheries policy. For instance, in Sierra Leone, the National Strategy for Gender Mainstreaming in Fisheries was developed with support from the Environmental

Justice Foundation (EJF). It promotes women's access to resources, participation in cooperatives, and leadership roles, thereby strengthening the social and economic fabric of the fish value chain (Sesay and Kargbo 2022).

15.8.1.7 Certification and Market Access Initiatives

To access premium markets, African fisheries are increasingly pursuing certification schemes and value chain upgrades. In Namibia, the government has supported hake fisheries in obtaining Marine Stewardship Council (MSC) certification by investing in monitoring and traceability systems (Butterworth 2016). This has improved sustainability, enabled access to European markets, and enhanced Namibia's reputation as a sustainable seafood producer. Such certification initiatives help align supply chains with global sustainability standards while providing economic incentives for responsible practices.

15.8.1.8 NGO and Community-Based Organization (CBO) Partnerships

Nongovernmental organizations have partnered with governments and communities to pilot innovative fisheries comanagement models and promote sustainability. In Mozambique, the NGO Rare has worked with local fishing communities to establish Managed Access Areas with No-Take Zones, empowering fishers to manage their resources while protecting biodiversity (Paul et al. 2016). These initiatives have resulted in increased fish biomass, better livelihoods, and improved trust between communities and authorities, making them scalable models for other African coastal regions.

15.8.1.9 Research and Innovation for Policy and Supply Chain Development

African universities and research institutes, often in collaboration with international partners, play a key role in generating data and innovations that inform policy and improve supply chain efficiency (Fu et al. 2022). Institutions like the WorldFish Center and the University of Ghana's Centre for Coastal Management conduct studies on fish stock dynamics, postharvest technologies, and climate impacts. Their findings help shape national policies and support the adoption of innovative practices, such as solar drying, improved fish feed, and low-cost cold storage technologies.

15.9 Marine Fisheries and Aquaculture Supply Chains' Contribution to Food Security in Africa

Marine fisheries and aquaculture contribute to food security across four key pillars: availability, accessibility, utilization, and stability. Their supply chains play a critical role in Africa by increasing the supply and affordability of nutrient-dense fish. These sectors support livelihoods, minimize postharvest losses, and enhance dietary diversity. Efficient value chains—from harvesting and processing to distribution and marketing—ensure a consistent flow of fish to both urban and rural areas, directly addressing hunger and malnutrition across the continent. The pillar of the food security framework is presented in Fig. 15.2.

15.9.1 Availability

In Africa, the availability of aquatic foods is vital for food security, especially where fish is a primary source of high-quality protein and essential micronutrients like omega-3 fatty acids, vitamins A and D, zinc, and iron. Fish plays a crucial role in preventing malnutrition, supporting child development, and reducing chronic disease risks (Byrd et al. 2022). However, rising global demand and pressures from overfishing, habitat loss, and climate change threaten the consistent supply of wild

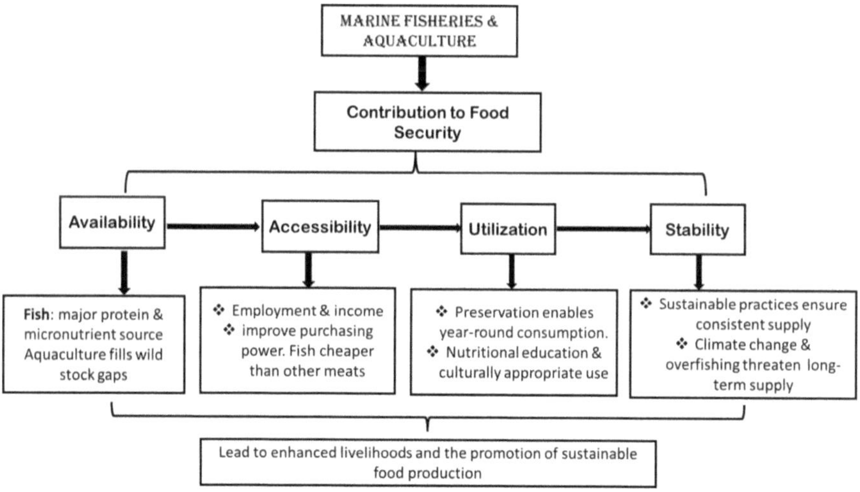

Fig. 15.2 Adapted from the FAO's four pillars of food security framework—Availability, Accessibility, Utilization, and Stability, this diagram illustrates how marine fisheries and aquaculture contribute to food and nutrition security in Africa. It highlights the sectors' role in providing nutrient-rich food, supporting livelihoods, enabling proper dietary use, and promoting sustainability amid environmental challenges

fish. Aquaculture, the continent's fastest growing food sector, is increasingly filling this gap by producing fish year-round under controlled conditions. It enables selective breeding for local resilience and improved nutrition, while innovations in sustainable feed, disease control, and water-efficient systems boost productivity (Lal et al. 2024). These advances help ensure a stable, scalable, and nutritious fish supply, securing a vital nutritional lifeline for millions across Africa now and into the future.

15.9.2 Accessibility

Access to fish and aquatic foods in Africa is closely tied to the livelihoods of millions relying on marine and inland fisheries. The fisheries and aquaculture sectors employ over 12 million people, many small-scale fishers, processors, and traders—especially women who lead postharvest activities (Elias et al. 2024). This employment supports household incomes and boosts purchasing power in vulnerable communities, improving access to diverse, nutritious foods. In coastal and lakeside areas, fish remains one of the most affordable animal proteins compared to beef, poultry, or imports. Local species like tilapia, catfish, and sardines are widely consumed for their cultural significance, nutrition, and low cost (Hasselberg et al. 2020). Informal markets and roadside vendors further lower access barriers by offering small, affordable portions suited to daily budgets. However, seasonal supply fluctuations, poor infrastructure, and gender inequities limit consistent access (Roever 2014). Still, investments in cold chains, digital marketplaces, inclusive governance, and microfinance for small-scale actors are strengthening accessibility as a resilient and equitable pillar of food security across Africa.

15.9.3 Utilization

Utilization of fish in Africa is shaped by traditional knowledge, cultural preferences, and changing nutritional needs (Mosepele et al. 2007). Preservation methods like drying, salting, smoking, and fermenting have long extended fish shelf life, especially where cold storage is limited. These low-cost techniques reduce postharvest losses and enable year-round consumption, particularly in rural and landlocked areas (Rabiepour et al. 2024). Preserved fish is a dietary staple in many communities, central to culturally important dishes that support both identity and nutrition. While fish is rich in omega-3s, iron, zinc, and vitamins A and D, nutritional education remains crucial to maximize its health benefits. Culturally relevant nutrition messaging through school feeding, health campaigns, and women's training promotes balanced diets and fights malnutrition (Xu et al. 2021). Combining traditional practices with modern food safety enhances health outcomes and acceptance. By blending indigenous knowledge with education, Africa is transforming fish from just food to a key driver of health and food security.

15.9.4 Stability

The stability of fish supply in Africa depends on balancing sustainable management with growing pressures from climate change and overfishing (Lam et al. 2012). Millions in coastal and inland communities rely on fisheries for food and livelihoods, making a steady supply vital for food security and economic resilience. Sustainable methods, like community management, marine protected areas, and responsible aquaculture, restore stocks, protect breeding grounds, and reduce harmful fishing, securing stable yields (Koch 2018). However, climate change, via rising ocean temperatures, shifting currents, and acidification, threatens fish distribution and productivity. Overfishing worsens these issues, depleting stocks and damaging ecosystems, especially where enforcement is weak (Tao 2023). Regional cooperation, improved monitoring, and adaptive management are building resilience. By combining traditional knowledge, science, and policy, Africa aims to safeguard fisheries and aquaculture stability, ensuring future generations benefit from this vital nutrition and income source amid climate change.

15.10 Future Pathways: Scaling Impact Across the Continent

As Africa confronts growing challenges such as food insecurity, environmental degradation, and socioeconomic inequality, the strategic expansion of marine fisheries and aquaculture supply chains offers a viable solution to enhance nutrition and livelihoods. Moving beyond localized interventions, the next phase must focus on scaling impact across regions through inclusive, coordinated, and innovative approaches. Central to this is mainstreaming technology, e.g., recirculating aquaculture systems (RAS), solar-powered cold storage, digital market platforms, and mobile-based extension services. These can improve productivity, traceability, and supply chain efficiency. Ensuring these tools are accessible and affordable, particularly for women and small-scale producers, is essential to maximizing their benefits.

In parallel, policy harmonization and regional integration will be key to enabling scale. Collaboration through frameworks led by the African Union, NEPAD, and regional economic communities can help align regulations, standardize quality and safety protocols, and facilitate trade across borders. Investment in rural infrastructure, such as transport networks, processing hubs, and cold chain logistics, will reduce geographic disparities and support broader market access. This must be accompanied by financial inclusion strategies, including microcredit, insurance schemes, and targeted subsidies to empower smallholders, especially women and youth, to engage competitively in aquatic food systems.

Equipping the sector for long-term impact also requires strengthening education and capacity development. Training programs and national fisheries institutions must be empowered to build a skilled workforce with knowledge in sustainable practices, aquaculture management, climate adaptation, and business development. Environmental sustainability must remain central, through ecosystem-based management, enforcement against illegal fishing, and climate-smart practices. Together, these interventions can transform marine fisheries and aquaculture into resilient, inclusive engines of food security and economic transformation across Africa.

15.11 Conclusion

Expanding marine fisheries and aquaculture supply chains is essential for achieving regional food security in Africa. As the continent faces growing populations, climate variability, and increasing pressure on land-based food systems, sustainably harnessing aquatic resources offers a viable solution to reduce hunger, improve nutrition, and support economic development. Strategic investments in sustainable fisheries management, aquaculture development, infrastructure, value addition, and inclusive capacity-building, especially for youth and women, are central to transforming these supply chains.

Success stories from countries like Egypt, Nigeria, Kenya, and Ghana demonstrate that coordinated policies, public–private partnerships, and community involvement can lead to substantial gains in productivity, sustainability, and market access. However, for such progress to be replicated across the continent, regional cooperation, harmonized regulations, and knowledge sharing must be strengthened. By expanding marine fisheries and aquaculture in a responsible, inclusive, and integrated manner, Africa can secure its food future while preserving marine ecosystems and creating livelihoods for millions.

References

Abasubong KP, Liu WB, Adjoumani YJJ, Xia SL, Xu C, & Li XF (2019) Xylooligosaccharides benefits the growth, digestive functions and TOR signaling in Megalobrama amblycephala fed diets with fish meal replaced by rice protein concentrate. Aquaculture, 500:417–428. https://doi.org/10.1016/j.aq

Abatekassa G, Peterson C (2011) Market access for local food through the conventional food supply chain. Int Food Agribus Manag Rev 14(2):135–160

Abelti A, Abera Teka T (2024) Intervening fish post-harvest losses to narrow the gap between demand and supply: a review on the magnitude of fish post-harvest losses in some Sub-Saharan African countries. Aquacult Fish Fish 4(1):15. https://doi.org/10.1111/aff.12345

Abera Teka T (2024) Fish handling and postharvest losses around the Fincha'a, Amarti, and Nashe reservoirs in Oromiya, Ethiopia. Sci Rep 14:Article 12345. https://doi.org/10.1038/s41598-024-12345-x
Adeyeye S (2016a) An overview of traditional fish smoking in Africa. J Culin Sci Technol 14(4):311–323. https://doi.org/10.1080/15428052.2016.1170929
Adeyeye S (2016b) Traditional fish processing in Nigeria: a critical review. Nutr Food Sci 46:3. https://doi.org/10.1108/NFS-11-2015-0148
Ahern M, Thilsted S, Kjellevold M, Overå R, Toppe J, Doura M, Kalaluka E, Wismen B, Vargas M, Franz N (2021) Locally-procured fish is essential in school feeding programmes in Sub-Saharan Africa. Foods 10:2080
Ahmad J, Ali M, Arif M, Iftikhar S, Ramzan I, Hussain M, Javed S, Adnan S, Bilal H, Hayat S (2021) Review article on: traditional and modern techniques for food preservation. 2021
Akisanmi D, Ogunderu O (2024) Using telegram web-3 mini-apps to connect smallholder farmers with markets for sustainable food chains in Lagos. Open J Manag Sci 5:45–57. (ISSN: 2734-2107)
Alam AN (2013) Advanced training for post-harvest fisheries management in The Gambia for minimizing loss and maximizing profit
Alam MS, Yousuf A (2024) Fishermen's community livelihood and socio-economic constraints in coastal areas: an exploratory analysis. Environ Chall 14:100810
Al-Ismaili A (2021) A review on solar drying of fish. J Agric Mar Sci 26:1–9
Arshad N, Samat N, Lee LK (2022) Insight into the relation between nutritional benefits of aquaculture products and its consumption hazards: a global viewpoint. Front Mar Sci 9:925463
Arsil P, Sularso K, Mulyani A (2018) Factors influencing consumer preferences for locally produced food: a comparison between traditional and modern markets. Int J Eng Technol Sci 7:535–538
Avadí A, Cole SM, Kruijssen F, Dabat M-H, Mungule CM (2022) How to enhance the sustainability and inclusiveness of smallholder aquaculture production systems in Zambia? Aquaculture 547:737494
Ayoola SO (2010) Sustainable fish production in Africa. Afr J Food Agric Nutr Dev 10:1–9
Babatunde A, Deborah R-A, Gan M, Simon T (2021) A quantitative SWOT analyses of key aquaculture players in Africa. Aquac Int 29(4):1753–1770
Bain LE, Awah PK, Geraldine N, Kindong NP, Sigal Y, Bernard N, Tanjeko AT (2013) Malnutrition in Sub-Saharan Africa: burden, causes and prospects. Pan Afr Med J 15:120. https://doi.org/10.11604/pamj.2013.15.120.2535
Bakhsh M, Ghazali MH, Kashif M, Channo A, Ijaz M, Subhan A, Randhawa N, Waqar Z, Shehzad A, Kaka S (2023) The role of fish in global food and nutrition security: current aspects and future prospects. Univ Sindh J Anim Sci 4:23–35
Baumüller H (2013) Enhancing smallholder market participation through mobile phone-enabled services: the case of M-farm in Kenya. Electron J Inf Syst Dev Countries 56(1):1–17. https://doi.org/10.1002/j.1681-4835.2013.tb00448.x
Belhabib D, Lam V, Cheung W (2016) Overview of west African fisheries under climate change: impacts, vulnerabilities and adaptive responses of the artisanal and industrial sectors. Mar Policy 71:15–28
Béné C, Arthur R, Norbury H, Allison E, Beveridge M, Bush S, Campling L, Leschen W, Little D, Squires D, Thilsted S, Troell M, Williams M (2016) Contribution of fisheries and aquaculture to food security and poverty reduction: assessing the current evidence. World Dev 79:177–196. https://doi.org/10.1016/j.worlddev.2015.11.007
Blessing S, Femi J, Christopher G (2025) Impact of E-commerce technologies on supply chain management in retail
Breckwoldt A, Ameyaw A, Reuter H, Aheto DW (2020) From fish to cash: analyzing the role of women in fisheries in the Western region of Ghana. Mar Policy 113:103800. https://doi.org/10.1016/j.marpol.2020.103800
Brummett R (2000) Factors influencing fish prices in southern Malawi. Aquaculture 186:243–251

Butterworth DS (2016) The South African experience with MSC certification: a perspective. Fish Res 182:169–177. https://doi.org/10.1016/j.fishres.2016.06.004

Byrd KA, Shieh J, Mork S, Pincus L, O'Meara L, Atkins M, Thilsted SH (2022) Fish and fish-based products for nutrition and health in the first 1000 days: a systematic review of the evidence from low and middle-income countries. Adv Nutr 13(6):2458–2487

Chambon M, Miñarro S, Alvarez-Fernandez S, Porcher V, Reyes-García V, Drouet HT, Ziveri P (2023) A synthesis of women's participation in small-scale fisheries management: why women's voices matter. Rev Fish Biol Fish 34:1–21

Chan CY, Tran N, Pethiyagoda S, Crissman CC, Sulser TB, Phillips MJ (2019) Prospects and challenges of fish for food security in Africa. Glob Food Sec 20:17–25

Chan CY, Tran N, Cheong KC, Sulser TB, Cohen PJ, Wiebe K, Nasr-Allah AM (2021) The future of fish in Africa: employment and investment opportunities. PLoS One 16(12):e0261615

Chipfupa U (2023) Formal and informal agricultural markets in Sub-Saharan Africa, pp 78–96

Cliffe PT, Akinrotimi O (2015) Role of women in fishery activities in some coastal communities of Rivers State, Nigeria. Int J Agric Res 10:24–32

Corral S, Romero Manrique de Lara D (2017) Local community-based approach for sustainable management of artisanal fisheries on small islands. Ocean Coast Manag 142:162–171. https://doi.org/10.1016/j.ocecoaman.2017.03.011

Demelash Abera B, Alefe Adimas M (2024) Health benefits and health risks of contaminated fish consumption: current research outputs, research approaches, and perspectives. Heliyon 10(13):e33905

Ejairu E, Researcher i, I (2022) Analyzing the critical failure points and economic losses in the cold chain logistics for perishable agricultural produce in Nigeria. Int J Supply Chain Manag 1:1–21

Elias M, Zaremba H, Tavenner K, Ragasa C, Paez Valencia AM, Choudhury A, de Haan N (2024) Towards gender equality in forestry, livestock, fisheries and aquaculture. Glob Food Sec 41:100761

Eriegha O, Ekelemu J, Eyo V (2022) The drivers and impacts of inflation on fish production in Nigeria chapter. In: Research trends in fisheries and aquatic sciences, vol 15. AkiNik Publications, pp 105–132

Evareh J, Effiom P-C, Effiom S, Odu P, Uket I, Nwankwo S, Ojah O (2023) The transformative role of artificial intelligence in smart energy transition for unprecedented energy sustainability in Nigeria. In: Conference: international conference on engineering, science, & technology, University of Cross River State

Ezeudu T, Obimbua E (2024) Enhancing rural market access and value chain integration for sustainable agricultural development in Nigeria: a study of constraints, strategies, and implications. Int J Res Innov Soc Sci 8:528–550

Faisal S, Pandey Y, Muzamil M, Ali M (2024) Food processing technology, storage and post-harvest management. In: Essentials of food technology and nutritional science. AkiNik Publications

Fanzo J, Rudie C, Sigman I, Grinspoon S, Benton TG, Brown ME, Covic N, Fitch K, Golden CD, Grace D, Hivert MF, Huybers P, Jaacks LM, Masters WA, Nisbett N, Richardson RA, Singleton CR, Webb P, Willett WC (2022) Sustainable food systems and nutrition in the 21st century: a report from the 22nd annual Harvard nutrition obesity symposium. Am J Clin Nutr 115(1):18–33

Farmery AK, Allison EH, Andrew NL, Troell M, Voyer M, Campbell B, Eriksson H, Fabinyi M, Song AM, Steenbergen D (2021) Blind spots in visions of a "blue economy" could undermine the ocean's contribution to eliminating hunger and malnutrition. One Earth 4(1):28–38

Fu X, Fu X, Ghauri P, Hou J (2022) International collaboration and innovation: evidence from a leading Chinese multinational enterprise. J World Bus 57(4):101329

Getu A, Misganaw K (2015) Post-harvesting and major related problems of fish production. Fish Aquac J 06:123. https://doi.org/10.4172/2150-3508.1000123

Gligor D, Tan A, Trang N (2018) The obstacles to cold chain implementation in developing countries: insights from Vietnam. Int J Logist Manag 29

Gulbrandsen L (2009) The emergence and effectiveness of the marine stewardship council. Mar Policy 33:654–660

Hanafiah H, Azami E, Fitriyandi D, Rahadi D (2024) Logistic analysis of fishery cold supply chain. J Econ Finance Manag Stud 07. https://doi.org/10.47191/jefms/v7-i6-04

Hasselberg AE, Aakre I, Scholtens J, Overå R, Kolding J, Bank MS, Atter A, Kjellevold M (2020) Fish for food and nutrition security in Ghana: challenges and opportunities. Glob Food Sec 26:100380

Ibengwe LJ, Onyango PO, Hepelwa AS, Chegere MJ (2022) Regional trade integration and its relation to income and inequalities among Tanzanian marine dagaa fishers, processors and traders. Mar Policy 137:104975

Ikbal A, Bhattacharya A, Ray B (2023) Post harvest losses in fisheries and its mitigation measures: an overview. Vigyan Varta Int E-Magazine Sci Enthus 4:141–145

Indiarto R, Jeanette G, Mulkya H, Yusra N, Subroto E (2021) A mini-review of salting techniques to improve food quality. Int J Sci Technol Res 10:285–289

Johnson D, Campion B, Chuenpagdee R, Varquez J, Ward A, Arias Schreiber M (2023) Chapter 5. The promise of dried and fermented small fish processing to enhance food security and nutrition, p 26

Kamal M, Bablu T (2023) Mobile applications empowering smallholder farmers: an analysis of the impact on agricultural development. Int J Soc Anal 8:36–52

Kaminski AM, Genschick S, Kefi AS, Kruijssen F (2018) Commercialization and upgrading in the aquaculture value chain in Zambia. Aquaculture 493:355–364

Kapapa L, Onyango P, Bwanthondi P, Mfilinge P (2024) The vulnerability of fisheries-based livelihoods to climate variability and change in coastal small pelagic fishing communities in Tanzania. Mar Policy 169:106344

Karani P, Failler P, Gilau AM, Ndende M, Diop ST (2022) Africa blue economy strategies integrated in planning to achieve sustainable development at national and regional economic communities (RECs). J Sustain Res 4(3):e220011

Kimani P, Wamukota A, Manyala J, Mlewa C (2020) Analysis of constraints and opportunities in marine small-scale fisheries value chain: a multi-criteria decision approach. Ocean Coast Manag 189:105151

Kitinoja L (2014) Use of cold chains in reducing food losses in developing countries. In: Proceedings of the international conference on postharvest technology for developing countries. World Food Preservation Center, pp 1–12

Koch V (2018) Solutions in focus: sustainable fisheries and aquaculture. World Wildlife Fund (WWF)

Kruijssen F, Tedesco I, Ward A, Pincus L, Love D, Thorne-Lyman AL (2020) Loss and waste in fish value chains: a review of the evidence from low and middle-income countries. Glob Food Sec 26:100434

Kurekin AA, Loveday BR, Clements O, Quartly GD, Miller PI, Wiafe G, Adu Agyekum K (2019) Operational monitoring of illegal fishing in Ghana through exploitation of satellite earth observation and AIS data. Remote Sens 11(3):293. https://doi.org/10.3390/rs11030293

Kussaga JB, Luning PA, Tiisekwa BPM, Jacxsens L (2014) Challenges in performance of food safety management systems: a case of fish processing companies in Tanzania. J Food Prot 77(4):621–630

Lal J, Verma DK, Jana A, Jayaswal R, Chakraborty A, Kumar S, Devati M, Pavankalyan S (2024) Emerging innovations in aquaculture: navigating towards sustainable solutions. Int J Environ Climate Chang 14:83–96

Lam V, Cheung W, Swartz W, Sumaila R (2012) Climate change impacts on fisheries in West Africa: implications for economic, food and nutritional security. Afr J Mar Sci 34:103–117

Lambin EF, Kim H, Leape J, Lee K (2020) Scaling up solutions for a sustainability transition. One Earth 3(1):89–96

Lisboa HM, Pasquali MB, dos Anjos AI, Sarinho AM, de Melo ED, Andrade R, Batista L, Lima J, Diniz Y, Barros A (2024) Innovative and sustainable food preservation techniques: enhancing food quality, safety, and environmental sustainability. Sustainability 16(18):8223

Mahboob K, Shahzad SM (2024) Marine fisheries and its impact on community: an overview. Biol Clin Sci Res J 2024(1):1085. https://doi.org/10.54112/bcsrj.v2024i1.1085

March A, Failler P (2022) Small-scale fisheries development in Africa: lessons learned and best practices for enhancing food security and livelihoods. Mar Policy 136:104925

Marwan Y (2023) Blue economy literature review. Int J Bus Manag 18:12–12

Maulu S, Hasimuna OJ, Monde C, Mweemba M (2020) An assessment of post-harvest fish losses and preservation practices in Siavonga district, Southern Zambia. Fish Aquatic Sci 23(1):25. https://doi.org/10.1186/s41240-020-00170-x

Maulu S, Musuka CG, Molefe M, Ngoepe TK, Gabriel NN, Mphande J, Phiri M, Muhala V, Macuiane MA, Ndebele-Murisa MR, Hasimuna OJ, Bokhutlo T, Mulumpwa M, Erasmus VN, Jere W, Dekesa CH, Mubaya CP, Baloi MF, Iitembu JA, Siankwilimba E, Zhang L (2024) Contribution of fish to food and nutrition security in southern Africa: challenges and opportunities in fish production. Front Nutr 11:1424740

Mgbomene C (2025) Harnessing alternative transportation system for efficient food supply chain in nigeria: role of the blue economy. Jalingo J Soc Manag Sci 6:67–78

Mosepele K, Mmopelwa G, Mosepele B, Kgathi DL (2007) Indigenous knowledge and fish utilization in the Okavango Delta, Botswana: implications for food security. In: Ahmed A (ed) Managing knowledge, technology and development in the era of information revolution. World Association for Sustainable Development; Griffith University, pp 292–302

Mozumder MMH, Uddin MM, Schneider P, Deb D, Hasan M, Saif SB, Nur A-A (2023) Governance of illegal, unreported, and unregulated (IUU) fishing in Bangladesh: status, challenges, and potentials. Front Mar Sci 10:Article 1150213. https://doi.org/10.3389/fmars.2023.1150213

Mramba RP, Mkude KE (2022) Determinants of fish catch and post-harvest fish spoilage in small-scale marine fisheries in the Bagamoyo district, Tanzania. Heliyon 8(6):e09574

Müller O, Krawinkel M (2005) Malnutrition and health in developing countries. CMAJ 173(3):279–286

Murekezi P, Habimana O, Menezes A (2024) Public–private partnership innovations for aquaculture development with a focus on sub-Saharan Africa, FAO fisheries and aquaculture technical paper no. 713. Food and Agriculture Organization of the United Nations. https://doi.org/10.4060/cc9252en

Muthuraman S, Al-Haziazi M, Subramanian K, Al Husaini Y (2023) Exploring the opportunities to promote value-added products in the fisheries sector in the Sultanate of Oman. Soc Sci J 13:1716–1723

Muzari W (2016) Small scale fisheries and fish farming, processing and Marketing in Sub-Saharan Africa: implications for poverty alleviation, food security and nutrition. Int J Sci Res 5:1740–1749

N'Souvi K, Sun C, Zhang H, Broohm DA, Okey MKN (2021) Fisheries and aquaculture in Togo: overview, performance, fisheries policy, challenges and comparative study with Ghana, Mali, Niger and Senegal fisheries and aquaculture. Mar Policy 132:104681

Nagy G, Bányai Á, Illés B, Varga A (2023) The impact of increasing digitalization on the logistics sector and logistics services providers. Multidiszciplináris Tudományok 13:19–29

Njiru M, Kazungu J, Ngugi C, Gichuki J, Muhoozi L (2008) An overview of the current status of Lake Victoria fishery: opportunities, challenges and management strategies. Lakes Reserv Res Manag 13:1–12

Nunoo FKE, Asiedu B, Kombat EO, Samey B (2015) Sardinella and other small pelagic value and supply chain of the fishery sector, Ghana. USAID/Ghana Sustainable Fisheries Management Project (SFMP). Coastal Resources Center, Graduate School of Oceanography, University of Rhode Island and Netherlands Development Organisation. GH2014_ACT044_SNV

Nwafili S, Gao T (2007) Structure and dynamics of fisheries in Nigeria. J Ocean Univ China 6:281–291

Obiero K, Meulenbroek P, Drexler S, Dagne A, Akoll P, Odong R, Kaunda-Arara B, Waidbacher H (2019) The contribution of fish to food and nutrition security in Eastern Africa: emerging trends and future outlooks. Sustainability 11(6):1636

Oboh A (2022) Diversification of farmed fish species: a means to increase aquaculture production in Nigeria. Rev Aquac 14(4):2089–2098. https://doi.org/10.1111/raq.12690
Odongkara OK, Abila OR, Luomba J (2010) The contribution of Lake Victoria fisheries to National Economies. Afr J Trop Hydrobiol Fish 12:47–51
Ojea E, Lester SE, Salgueiro-Otero D (2020) Adaptation of fishing communities to climate-driven shifts in target species. One Earth 2(6):544–556
Okafor-Yarwood I (2019) Illegal, unreported and unregulated fishing, and the complexities of the sustainable development goals (SDGs) for countries in the Gulf of Guinea. Mar Policy 99:414–422. https://doi.org/10.1016/j.marpol.2017.09.016
Okonkwo I, Mujinga J, Namkoisse E, Francisco A (2023) Localization and global marketing: adapting digital strategies for diverse audiences. J Digit Market Commun 3:66–80
Olaifa E, Osabuohien E, Haruna I (2022) Enhancing fish production for food security in Nigeria. Mater Today Proc 65:2208–2214
Olokor J, Omojowo F, Samuel S (2009) Adaptation and improvement of a simple solar tent dryer to enhance fish drying. Nat Sci 7(10):1–4
Paul SAL, Wilson AMW, Cachimo R, Riddell MA (2016) Piloting participatory smartphone mapping of intertidal fishing grounds and resources in northern Mozambique: opportunities and future directions. Ocean Coast Manag 134:79–92
Priestley M, Williams J, Dissanayake S (2022) Teaching fisheries bycatch: exploring economic and behavioral drivers of bycatch through a classroom game. Agric Resour Econ Rev 51:1–32
Rabiepour A, Zahmatkesh F, Babakhani A (2024) Preservation techniques to increase the shelf life of seafood products: an overview. J Food Eng Technol 13:1–24
Regupathi R, Balasubramanian S (2018) A study on solar drying system for fish preservation. Asia Pac J Res 1(88):137–142
Roever S (2014) Informal economy monitoring study sector report: street vendors. WIEGO
Saini VP, Paul T, Bhutia TP, Abhiman, Meharoof M, Pai M, Acharya A (2023) Farmer friendly innovations in fisheries: rural poverty to prosperity. College of Fisheries, Kishanganj. ISBN 978-81-963832-2-0
Selamoglu M (2021) Importance of the cold chain logistics in the marketing process of aquatic products: an update study. J Surv Fish Sci 8(1):25–29. https://doi.org/10.18331/SFS2021.8.1.2
Sesay V, Kargbo IR (2022) Women's participation in the aquaculture value chain in selective regions of Sierra Leone. Int J Rec Innov Acad Res 6:12–33
Shoko AP, Limbu SM, Ulotu EE, Shayo SD, Silas MO, Chimatiro SK, Madalla NA, Tamatamah RA (2022) Fish seed and feed value chains analysis and their critical factors for aquaculture development in Tanzania. Aquacult Fish Fish 3(1). https://doi.org/10.1002/aff2.84
Siddiqui SA, Singh S, Bahmid NA, Sasidharan A (2024) Applying innovative technological interventions in the preservation and packaging of fresh seafood products to minimize spoilage – a systematic review and meta-analysis. Heliyon 10(8):e29066
Simane B, Kapwata T, Naidoo N, Cissé G, Wright C, Berhane K (2025) Ensuring Africa's food security by 2050: the role of population growth, climate-resilient strategies, and putative pathways to resilience. Foods 14:262
Srivastava S, Seng K, Ang L, Pachas A, Lewis T (2022) Drone-based environmental monitoring and image processing approaches for resource estimates of private native Forest. Sensors 22:7872
Subasinghe R, Soto D, Jia J (2009) Global aquaculture and its role in sustainable development. Rev Aquac 1:2–9
Subasinghe RP, Siriwardena SN, Byrd KA, Chan CY, Dizyee K, Shikuku KM, Tran N, Adegoke A, Adeleke L, Anastasiou K, Beveridge M, Bogard J, Chu LH, Fregene BT, Ene-Obong H, Cheong KC, Nukpezah JA, Olagunju OF, Powell A, Steensma JT, Williams G, Shelley CC, Phillips MJ (2021) Nigeria fish futures. Aquaculture in Nigeria: Increasing income, diversifying diets and empowering women: Report of the scoping study (Program Report No. 2021-16). WorldFish, Penang
Tao Y (2023) Ocean overfishing under the "tragedy of the commons" theory. Adv Econ Manag Polit Sci 14:188–193

Tawakalitu A, Adediran B, Onyegbula A, Olowolaju E, Atanda S, Damilola A (2025) Post-harvest management in Africa: a review of innovative technologies and multidisciplinary approaches for reducing food losses. J Appl Sci 6:150–158

Timothy A, Ejem E (2025) African continental free trade agreement (AfCFTA) and supply chain issues in Africa: a qualitative exploration. Preprint. SSRN. https://doi.org/10.13140/RG.2.2.29822.29769

Tran N, Rodriguez UP, Chan CY, Aung YM, Chu L, Saiful Islam AHM, Barman BK, Phillips MJ (2023) Future scenarios of fish supply and demand for food and nutrition security in Bangladesh: an analysis with the AsiaFish model. Aquaculture 568:739288

Türkten H, Yıldırım Ç, Yüceer S (2024) Food loss, waste, and environmental impacts in the fish value chain. Osmaniye Korkut Ata Univ J Instit Sci Technol 7(2):782–801. https://doi.org/10.47495/okufbed.1399800

Udombana NJ (2020) A step closer: economic integration and the African continental free trade area. Duke J Comp Int Law 31:1–89. SSRN Electronic Journal. https://doi.org/10.2139/ssrn.3797090

Urugo MM, Yohannis E, Teka TA, Gemede HF, Tola YB, Forsido SF, Tessema A, Suraj M, Abdu J (2024) Addressing post-harvest losses through agro-processing for sustainable development in Ethiopia. J Agric Food Res 18:101316

Verma DK, Monika, Barad RR, Singh S, Chandra I, Maurya NK, Ranjan D (2024) Digitalization as innovative development in aquaculture and fisheries as future importance. In: Futuristic trends in agriculture engineering & food sciences, Vol. 3, Book 15, pp 689–708). IIP Series. e-ISBN 978-93-5747-931-8. https://doi.org/10.58532/V3BCAG15P6CH1

Viana DF, Zamborain-Mason J, Gaines SD, Schmidhuber J, Golden CD (2023) Nutrient supply from marine small-scale fisheries. Sci Rep 13(1):11357

Walakira JK, Hinrichsen E, Tarus V, Langi S, Ibrahim NA, Badmus O, Aziz A, Baumüller H (2023) Scaling aquaculture for food security and employment in Africa: insights from Egypt, Kenya and Nigeria. (ZEF Working Paper No. 223). Center for Development Research, University of Bonn. https://doi.org/10.48565/bonndoc-130

Whiston A, Rost A, Mangeni Sande R, Bruce C, Brierley A (2022) Drying fish (Rastrineobola argentea) on the bank of Lake Victoria in a prototype solar-heated enclosure: using renewable energy to reduce drying-time and improve product quality and value. Afr J Food Sci 16:289–300

Xu YY, Sawadogo-Lewis T, King SE, Mitchell A, Roberton T (2021) Integrating nutrition into the education sector in low- and middle-income countries: A framework for a win-win collaboration. Matern Child Nutr 17(3):e13156

Yeboah SA (2023) Breaking barriers: bridging the gender gap in entrepreneurial finance in developing nations. J Surv Fish Sci 8(1):25–29. https://doi.org/10.18331/SFS2023.8.1.2

Yu Z, Liu Y, Wang Q, Sun L, Sun S (2021) Research on food safety and security of cold chain logistics. IOP Conf Ser: Earth Environ Sci 647:012176

Zego S, Husny Z (2023) Performance of cold chain logistics service providers in the fast-moving consumer goods industry in Nigeria: a systematic review. IOP Conf Ser Earth Environ Sci 1274:012021

Chapter 16
Soil Fertility Status of the Lake Liambezi Maize Farm and Implication for Sustainable Cultivation

V. Hepute, J. Abah, G. Kampungu, G. M. Mulonda, and N. Mwazi

Abstract The Lake Liambezi area in the Zambezi Region of Namibia has a very high potential for maize farming. However, the intensive maize production and poor cultivation practices in this area could lead to soil degradation due to nutrient loss and changes in soil physiochemical properties, which have adverse implications for sustainable cultivation. Thus, this study determined the soil fertility status of the Lake Liambezi maize farms in relation to maize crop nutrient requirements, and, hence, estimated the potential for sustainable cultivation. Soil samples were randomly collected from a total of 21 maize farms in the area. An area of one hectare (1 ha) was marked in each farm and divided into two sampling grids (0.5 ha each) per farm, giving rise to 42 sampling grids. Then, two soil samples were collected within each sampling grid at depths of 0–15 cm (topsoil) and 15–30 cm (subsoil). Thus, a total of 84 soil samples were collected and analyzed using standard laboratory procedures for soil texture, pH, organic matter content, nitrogen, phosphorus, potassium, calcium, magnesium, sodium, and electrical conductivity, respectively. Descriptive statistics were used to analyze the data generated from replicate analyses of the soil parameters and the mean levels were compared with standard maize crop growth nutrients requirement and the soil parameters. Additionally, one-way Analysis of Variance (ANOVA, $p < 0.05$) was used to determine the significance of mean data variations between the maize farms. Within the two soil depths sampled, the mean results obtained across the maize farms revealed that the farms' soils have

V. Hepute (✉)
The Namibian Agronomic Board, Windhoek, Namibia

Department of Agricultural Sciences and Agribusiness, Faculty of Health, Natural Resources and Applied Sciences, Namibia University of Science and Technology, Windhoek, Namibia
e-mail: vhepute@nust.na

J. Abah
Department of Applied Educational Sciences, School of Education, University of Namibia, Katima Mulilo Campus, Katima Mulilo, Namibia

G. Kampungu · G. M. Mulonda · N. Mwazi
The Namibian Agronomic Board, Windhoek, Namibia

H. A. Mupambwa et al. (eds.), *Harnessing Coastal and Marine Resources for Improved Food Security in Africa*, Sustainability Sciences in Asia and Africa,
https://doi.org/10.1007/978-981-95-9725-3_16

heterogeneous soil characterized by loamy sand, sandy loam and loamy textural classes, neutral pH (7.08 and 7.27), low electrical conductivity (513.01 μS/cm and 481.83 uS/cm), low organic matter (1.63% and 1. 59%), low nitrogen (0.062% and 0.060%), low phosphorus (8.29 ppm and 5.59 ppm), very high potassium (453.86 ppm and 541.38 ppm), very high magnesium (222.19 ppm and 218.19 ppm), very high calcium (3579.57 ppm and 3329.31 ppm), and high sodium (202.19 ppm and 169.05 ppm). The low levels of nitrogen and phosphorus across the farm soils post risk for sustainable maize cultivation in the study area.

Keywords Soil fertility • Maize production • Sustainable cultivation • Lake Liambezi

16.1 Background of the Study

Maize crop has been recognized as an important crop worldwide due to its extensive utilization in human and animal nutrition, as well as in industry (Schaefer et al. 2025). Maize crop possesses a high yield potential and can respond positively to various agro-management practices, with a reported global average productivity of 5.75 metric tons per hectare (Biswas et al. 2023). Despite its relevance, however, maize productivity faces a myriad of challenges, especially concerning soil management and water availability (Schaefer et al. 2025). Maize production and traditional cultivation activities are associated with nutrient uptake from the soil and adverse physicochemical changes, which may promote soil degradation, nutrient depletion, and soil pH alteration. These practices may eventually create bigger problems, leading to a gradual decline in soil productivity and low maize production (Shah and Wu 2019). It has been observed that large agricultural areas in sub-Saharan Africa (SSA) are covered by inherently poor soils that have been subjected to soil fertility depletion and land degradation for many decades due to poor management, including low nutrient and organic matter application (Van der Velde et al. 2014). In an earlier study, Erenstein et al. (2022) noted that the declining farm soil productivity, soil fertility, and nutrient availability for crop uptake have been a major concern to crop producers worldwide. Thus, efficient soil fertility management and sustainable agronomic activities have become expedient and a precondition for guaranteeing a sustainable food production system, especially white maize production in the Lake Liambezi farming area. This is very crucial for meeting both current and future food demands without compromising the farm soil functionality and productivity.

Sustainable agriculture entails utilizing and maintaining the capability of agricultural natural resources, ecosystem productivity, and its usefulness in meeting the human food needs over the long run. Maize production primarily depends on sustainable agronomic practices such as effective and efficient fertilizer application to promote balanced soil nutrients. Effective soil fertilization, proper water utilization, as well as weed and pest management have been reported to increase maize yield and productivity (Shah and Wu 2019). Thus, sustainable maize farming and proper soil fertilization are very crucial in the Lake Liambezi maize farms, not only for the production of staple food crops but also for food security, poverty alleviation, and rural socioeconomic development of the area.

Maize, wheat, and rice have been identified as the world's leading staple cereals, accounting for an estimated 42% of the world's food calories and 37% of protein intake (Erenstein et al. 2022). Maize, especially the white variety, is exclusively produced in Namibia for human consumption and is one of Namibia's staple food crops and a controlled product. At the Lake Liambezi maize farms, farmers cultivate maize without any proper fertilizer application program, under the assumption that the soil is naturally fertile. Hence, fertilizer application by the farmers was mainly guided by visual assessment of the maize crop appearance, in which a yellowish appearance is taken as a sign of nutrient deficiency. The farmers often attempt to remedy this perceived nutrient deficiency by applying nitrogen fertilizer to the specific spot. Despite the potential threat of this traditional agronomic activity to both maize production and soil productivity in the long term, there is currently no documented study to track the effect of the current practices on the soil health and nutrient status of the Lake Liambezi maize farms. Thus, there is a need to determine the soil fertility status and crop nutrient availability levels at the Lake Liambezi maize farms in relation to maize crop nutrient requirements. This is critical for providing recommendations on the potential fertility management approaches that could optimize the productive capability of the farm soil. This study outcome, therefore, creates baseline information for farmers and future studies.

16.1.1 Problem Statement

The Lake Liambezi area in the Zambezi Region has a very high potential for maize farming, and a significant number of small-scale farmers are currently cultivating maize in the area. However, the cultivation of maize crops without prior knowledge of the nutrient status of the soil to inform relevant soil fertility management over an extended period of time could adversely affect the productive capability of the soil and lead to loss of soil productivity value. In the study area, it is perceived by the small-scale farmers that the natural soil nutrients are adequate for maize cultivation, and, hence, there is no need for full field fertilizer application. Most often, the farmers only apply nitrogen fertilizer to a particular area of the field where there is perceived nutrient deficiency as a result of the yellowish appearance of maize leaves. This practice might lead to soil nutrient imbalance and threaten sustainable maize crop productivity in the Lake Liambezi maize farms. Over time, this results in nutrient mining leading to soil fertility degradation and soil organic matter depletion (Nezomba et al. 2015). Thus, scientific soil fertility management practices through appropriate soil nutrient tests and management programs are crucial for sustainable crop production (Subedi and Ma 2009).

Moreover, long-term experiments in Africa have provided evidence on the benefits of integrating organic and inorganic fertilizers to replenish nutrient and carbon reserves in poor soils (Munera-Echeverri et al. 2020; Oldfield et al. 2019). Therefore, the main objective of the current study was to determine the soil nutrient status of the Lake Liambezi maize farms in relation to standard maize crop nutrient requirements.

16.2 Literature Review

Unsustainable practices in crop farming could threaten soil functionality and the ability to sufficiently supply essential nutrients for long-term and sustainable cultivation. Thus, understanding the soil fertility status is fundamental for sustainable cultivation, especially in maize-dominated farming systems, where nutrient depletion and poor management can easily limit crop yield potentials and overall performance (MacCarthy et al. 2023). This literature review synthesizes contemporary research on soil fertility parameters relevant to the current study on the nutrient status of the Lake Liambezi maize farms. These include soil pH, soil organic matter, electrical conductivity, soil nitrogen, phosphorus, potassium, calcium, magnesium, and sodium.

16.2.1 Importance of Soil Fertility

The sustainable capability and capacity of the soil to supply essential nutrients for optimum crop growth and yield performance are crucial factors influencing agricultural productivity, which requires continuous soil fertility enhancement (Kwari et al. 2011). Soil fertility reflects the availability of essential macro- and micronutrients, soil physiochemical status, organic matter content, pH, and biological activity. According to Goshu (2015), the depletion of essential soil nutrients such as nitrogen, phosphorus, and potassium is a common problem in sub-Saharan Africa maize production that requires targeted interventions. This is very crucial for the Lake Liambezi maize farms—the study area, which has been under intensive and continuous cultivation. Moreover, maintaining and enhancing soil fertility ensures that crop nutrient requirements are met to support optimal growth and yield, promoting long-term utilization of limited arable fields.

16.2.2 Sustainable Cultivation Practices

According to Wang et al. (2024), sustainable agricultural practices generally focus on transforming the use of fertilizers, implementing integrated pest management, adopting crop rotation, utilizing precision irrigation, practicing cover cropping, and employing conservation tillage, among other techniques, with the goal to enhance agricultural productivity and efficiency. Additionally, Integrated Soil Fertility Management (ISFM), combining organic and inorganic fertilizer sources, has proven to be very effective in improving nutrient availability in smallholder farmers (MacCarthy et al. 2023). While practices such as minimal tillage, organic amendments, and maintaining soil cover are known to protect farm soils against erosion and nutrient losses, Sanchez (2019) emphasized that continuous soil fertility status analysis is crucial for maintaining soil health and enhancing productivity. At the Lake Liambezi maize farms, however, the obvious lack of good agronomic practices and lack of soil fertility monitoring data could threaten long-term and sustainable productivity of the farm soils.

16.2.3 Soil Texture Impacts on Fertility

According to Upadhyay and Raghubanshi (2020), soil texture significantly impacts soil fertility through its effects on water retention, aeration, and nutrient-holding capacity. Generally, clay and loam soils have higher cation exchange capacity, water and nutrient holding capacity compared with sandy soils, which leach nutrients faster (Kowalenko and Ivarson 1978). Therefore, establishing knowledge on soil textural classes at the Lake Liambezi maize farm will enable understanding of site-specific fertility challenges and creation of appropriate recommendations on sustainable soil fertility management practices (Nhunda et al. 2024).

16.2.4 Soil pH and Its Effects

Soil pH plays a fundamental role in determining nutrient availability and plant health, making it a crucial factor in agricultural productivity and sustainability (Khaled and Sayed 2023). Measured on a scale of 0–14, soil pH provides quantitative information about the acidity and alkalinity of farm fields. It regulates nutrient solubility and microbial activity critical for nutrient cycling (Nwite 2022). According to Khaled and Sayed (2023), most soil nutrients are optimally available to plants in a slightly acidic to neutral pH range between 6.0 and 7.0, noting that deviations from this range can lead to nutrient deficiencies or toxicities with implications for plant growth and crop yields. In addition, soil pH controls microbial community composition and activities, defines soil organic matter stabilization and destabilization mechanisms, and regulates intensities of many biogeochemical processes (Wang and Kuzyakov 2024), and these have implications for nutrient cycling in the soil, and hence crop productivity. Khaled and Sayed (2023) emphasized that the impact of soil pH on nutrient availability underscores the importance of regular soil testing and monitoring. They noted that by understanding the current pH levels and nutrient status of their soils, farmers can make informed decisions about the necessary amendments and management practices to optimize soil conditions. In response to attaining favorable soil pH levels at the Lake Liambezi maize farms, site-specific soil assessment and amendment practices that enhance availability and improve uptake of essential nutrients are required.

16.2.5 Electrical Conductivity

Electrical Conductivity (EC) serves as a gauge for measuring soil salinity and water-soluble salts, which are pivotal indicators of mineral nutrients in the topsoil readily available for plant absorption (Ahmad et al. 2025). EC measurements from soil extracts can serve as reliable indicators for predicting soil nutrient content (Kim and Park 2024; Heiniger et al. 2003). Nutrients such as nitrogen and potassium directly influence the electrical conductivity of removed soil solutions (Ouyang et al. 1998). However, Ahmad et al. (2025) argued that soil salinity is a key factor measured by EC, and crops vary in their tolerance to saline conditions. The ideal EC value for optimal plant growth falls within the range of 0.8–1.8 mS/cm, with a maximum threshold of 2.5 mS/cm (Liyan 2022). Thus, monitoring the EC range is crucial for selecting appropriate crops for the given soil salinity levels, especially in maize fields. This is critical for establishing effective and sustainable EC amendment needs at Lake Liambezi maize fields.

16.2.6 Soil Cation Exchange Capacity

Cation Exchange Capacity (CEC) is a soil property that describes its ability to provide nutrients in the soil solution, making them available for plant uptake (Monroy et al. 2024). CEC determines the soil's capacity to retain and supply cations essential for healthy plant growth and development, and hence it is a key indicator for assessing soil fertility (García et al. 2012). Soil exchangeable cations (such as Ca^{2+}, Mg^{2+}, K^+, and Na^+) measure the soil's cation exchange capacity and indicate the ability of a soil to hold exchangeable cations, which are important for retaining nutrients for plant use. Soils with higher CEC, such as clay-rich soils, retain nutrients better than sandy soils (Goshu et al. 2015). Thus, analyzing the exchangeable cations becomes imperative in informing fertilizer application decisions for sustainable maize nutrient management (Nhunda et al. 2024).

16.2.7 Soil Organic Matter (SOM)

The importance of soil organic matter (SOM) in agriculture cannot be overemphasized. It is often regarded as one of the most vital constituents of the soil, forming the backbone of soil health and fertility (Lal 2016). Healthy soils, characterized by good structure, optimal nutrient availability, and a diverse microbial community, tend to have higher levels of SOM (Kibblewhite et al. 2008). MacCarthy et al. (2023) noted that SOM enhances soil fertility by improving water retention, nutrient supply, and microbial populations, eventually fostering soil structure and buffering pH fluctuation. Apart from improving soil structure, enhancing soil nutrient, and water retention, SOM is critical for supporting a rich and diverse soil biota (Postma-Blaauw et al. 2010). This biodiversity is an integral part of the soil ecosystem, contributing to soil fertility, suppressing soil-borne diseases, decomposing organic matter and residues, and enhancing crop nutrient availability (Bender et al. 2016). Thus, SOM plays a key role in numerous soil functions, contributing to soil structure, nutrient cycling, and water retention, among other things (Janzen 2006). Given the vital role that SOM plays in soil health and crop production, understanding its dynamics is crucial for sustainable agricultural practices and food security. This can contribute substantially to higher productivity and resilience in agricultural systems.

16.3 Materials and Methods

16.3.1 Study Area

The Lake Liambezi area comprises about 300 km^2, of which 100 km^2 is open water when the lake is full. The Lake is located (Latitude 17.9110°S and longitude 24.3728°E (Peel et al., 2015), between the Linyanti and Chobe rivers, about 60 km South of Katima Mulilo in the Zambezi region. The area receives precipitation up to 600 mm per annum. Despite a fishery that dominates the livelihood activities of the people of Lake Liambezi area, the Lake is surrounded by several maize crop farms, which contribute substantially to the food security and economy of the area. Figure 16.1 shows the map of the Lake Liambezi maize farms.

16.3.2 Soil Sample Collection

As seen in Table 16.1, a total of 21 maize farms from the Lake Liambezi area were randomly sampled for their soil nutrient status analysis. In each of the farms, an area of 1 ha was marked and divided into two sampling grids (0.5 ha each). According to the International Plant Nutrition Institute (IPNI) sampling grid/segment minimizes field selection bias and offers systematic and structured samples with proper field representation. Then, two soil samples (consisting of one sample within topsoil [0–15 cm] and another one sample within subsoil [15–30 cm]) were randomly collected from each sampling grid using an auger soil sampler at an interval of about 10 m between sampling points. Thus, a total of 84 soil samples were collected for laboratory analysis.

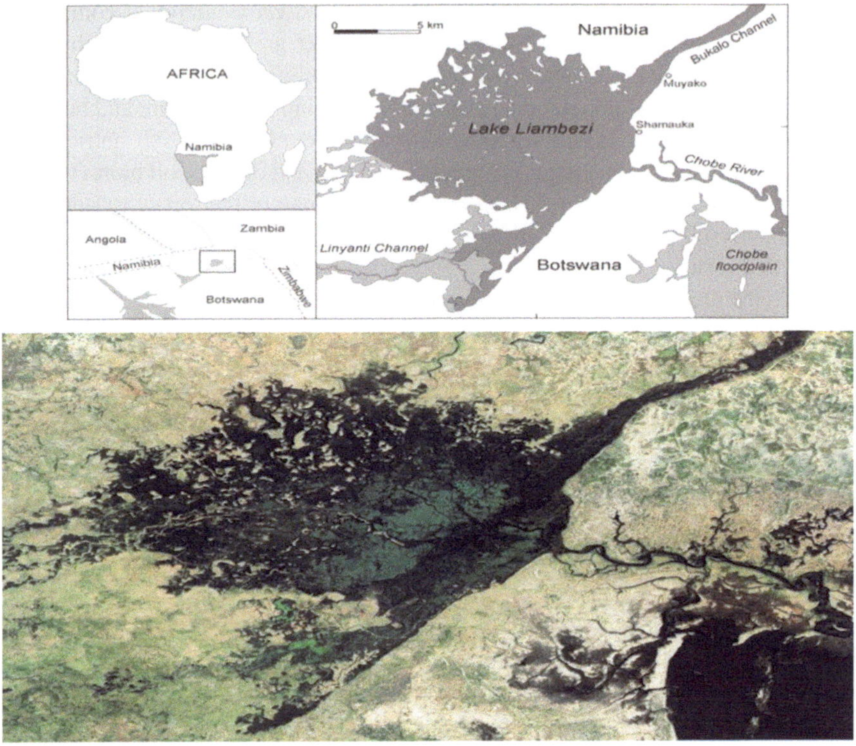

Fig. 16.1 Lake Liambezi Map. (Source: Peel et al., 2015)

Table 16.1 Soil sample size

Farm Zone	Farm location	Sample size			Total sample
		No. of farms sampled	No. of segments	No. of sampling depths	
Zambezi	Lake Liambezi	21	2	2	84

16.3.3 Sample Preparation and Analysis

The soil samples were dried at room temperature for 14 consecutive days. Sample weight was taken periodically at 08:30 and 16:30 each day until a constant weight was obtained at three consecutive times. Then, each sample was crushed and sieved using a 2 mm sieve mesh, and laboratory analyses of the soil parameters were then carried out on less than 2 mm soil fractions using standard laboratory procedures (Kettler et al. 2001). The part with bigger soil particle size (more than 2 mm) was regarded as gravel or stones and was discarded. The laboratory analysis was carried out using standard procedures at the Ministry of Agriculture, Water and Forestry's

Analytical Laboratory in Windhoek, Namibia. Each soil sample was analyzed for the following soil parameters:

16.3.3.1 Soil Texture

The soil texture was determined by particle size groups or classes after determining the sand, silt, and clay fractions using the pipette method (Beretta et al. 2014; Kettler et al. 2001). The pipette method gives precise and accurate result in the determination of silt and clay (Beretta et al. 2014). A soil textural triangle (USDA classification system) was used to determine soil texture (Beretta et al. 2014).

16.3.3.2 pH

The soil pH was measured in 1:2.5 suspensions in water, following a two-point calibration of the pH meter using fresh buffer solutions of pH 4.0 and 7.0. Between sample measurements, the pH electrode was washed thoroughly with deionized water (Öhlinger 1996).

16.3.3.3 Electrical Conductivity (EC)

A 1:5 soil:water suspension of the soil sample was prepared by weighing 10 g air-dry soil (<2 mm) into a bottle, followed by the addition of 50 mL deionized water. The solution was shaken vigorously until the soil dissolved completely. Then, the conductivity meter was calibrated according to the manufacturer's instructions using the KCl reference solution (0.01 M KCl) to obtain the cell constant (Rayment and Higginson 1992). The cell was rinsed thoroughly with deionized water and gently dipped into the soil suspension to measure the EC. The cell was rinsed thoroughly after each measurement. All the soil EC measurements were done at room temperature (25 °C).

16.3.3.4 Organic Matter

The Walkley–black method was used to determine the soil organic matter (De Vos et al. 2007).

16.3.3.5 Total Nitrogen

The soil total nitrogen was determined following the Kjeldahl digestion method, and the nitrogen content of the sample was multiplied by the conversion factor, N × 6.25 (Bremner 1960).

16.3.3.6 Available Phosphorus

The soil available phosphorus was determined as soil phosphate following the Olsen method and measured spectrophotometrically using the phosphomolybdate blue method (Koralage et al. 2015).

16.3.3.7 Exchangeable Cations

The soil exchangeable cations like calcium (Ca), magnesium (Mg), and potassium (K) were extracted using 1 N ammonium acetate at pH 7.0 (Normandin et al. 1998), and their levels were determined using Inductively Coupled Plasma–Optical Emission Spectrophotometer (ICP–OES)—iCAP 6000 series. The results obtained in ppm were converted to meq/100 g soil equivalent by dividing each by the corresponding element's conversion factor (Ca = 10.0, Mg = 2.0, K = 0.26, and Na = 0.09) (Öhlinger 1996).

16.3.4 Data Analysis

Using descriptive statistics, the data generated from replicate analyses of the soil parameters was computed as the mean of each parameter. Furthermore, analysis of variance (ANOVA, $p < 0.05$) was calculated to determine the significance of mean data variations in the soil parameters across the sampled farms. In the event of a significant difference, the Welch's t-test post hoc analysis was conducted using pairwise t-tests: two-sample assuming unequal variances with Excel. When an ANOVA's p-value is significant, a series of post hoc tests, using t-tests, can be performed to determine which specific pairs of groups have statistically significant differences in their means. According to Riffenburgh (2012), the tests are robust against differing variances, provided the sizes of the two samples are about the same. In our study, the sample sizes are the same ($n = 42$).

16.4 Result and Discussion

16.4.1 Soil Texture

Based on the United States Department of Agriculture (USDA)'s soil texture triangle (Fig. 16.2) for the classification of soil texture, the laboratory analysis results of the soil particle sizes largely revealed heterogeneous soil characterized by loamy sand, sandy loam, and loamy textural classes, with the mean value revealing sandy loam (Table 16.2). The negative kurtosis indicates fewer outliers in the data set

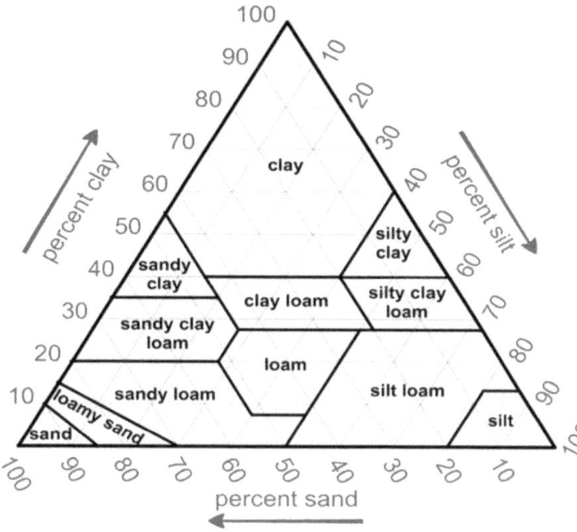

Fig. 16.2 USDA soil texture triangle

Table 16.2 Descriptive statistics of the Lake Liambezi maize farm soil particle sizes

	Soil Particle Sizes			
Descriptive stat	Sand %	Silt %	Clay %	Mean textural class
Mean	50.90	26.26	22.86	Sandy loam
Median	44.80	24.35	20.75	
Mode	44.80	35.20	33.30	
Standard deviation	22.77	16.39	15.74	
Kurtosis	−1.14	−0.97	−0.17	
Skewness	0.47	0.26	0.78	
Range	71.70	59	59.20	
Minimum	18.90	1.20	3.80	
Maximum	90.60	60.20	63.00	
Count	42	42	42	
Confidence level (95%)	7.09	5.11	4.90	

($n = 42$). Moreover, the data skewness (0.26–0.78) falls within the excellent classification −1 to +1 as espoused by Hair (2014). The recorded soil textures have good drainage and good nutrient retention capability. Soil texture affects the water and nutrient-holding ability of a particular farm soil (Huntley 2023). It determines soil suitability and affects the water and fertilizer management approach to be applied to a specific farm (Abd-Elmabod et al. 2019). Thus, with good tillage, proper fertilization, and addition of organic matter through mulching and cover crop, the Lake Liambezi maize farm soils can be maintained for long-term productivity.

Table 16.3 Descriptive statistics of the top soil (0–15 cm) properties of Lake Liambezi maize farms

Descriptive statistics	Ph	EC	OM	N	P	K	Ca	Mg	Na
Mean	7.08*	513.01*	1.63*	0.06^{NS}	8.29*	453.86*	3579.57*	222.19*	202.19*
Median	7.20	321.50	1.62	0.06	1.20	467.50	3882.00	226.00	123.50
Mode	7.16	256.00	1.60	0.06	0.00	64.00	4194.00	96.00	0.00
Standard deviation	1.02	465.09	0.48	0.03	21.94	243.35	1035.96	114.18	228.37
Kurtosis	0.56	1.01	0.60	9.49	28.43	−1.23	2.23	−0.63	0.83
Skewness	−0.86	1.49	0.00	2.54	5.05	−0.14	−1.38	−0.08	1.25
Range	4.34	1610.00	2.27	0.18	135.00	796.00	4763.00	446.00	795.00
Minimum	4.45	70.00	0.55	0.02	0.00	64.00	211.00	3.00	0.00
Maximum	8.79	1680.00	2.82	0.20	135.00	860.00	4974.00	449.00	795.00
Count	42	42	42	42	42	42	42	42	42
Confidence level (95%)	0.32	144.93	0.15	0.01	6.84	75.83	322.83	35.58	71.16

Key: Data with asterisk (*) varied significantly between farms ($p < 0.05$)
NS not significantly different

16.4.2 Soil pH

Soil pH is a critical factor that influences soil health and environmental sustainability (Dewangana et al. 2023). It is the most important parameter that influences the chemical, biological, and physiological processes in the soil, as well as the nutrients available for the plant's uptake (Xia et al. 2024). The results of the farm soil pH (Table 16.3) ranged from 4.45 to 8.79, with an average of 7.08 within the top soil (0–15 cm), and 4.42–8.74 with an average of 7.27 within the subsoil (15–30 cm) (Table 16.4). The result of analysis of variance of the soil pH was significantly different across the farms ($p < 0.05$). The present results largely revealed heterogeneous pH characteristics with slightly acidic, slightly alkaline, and neutral soil. The desirable pH range for the maize crop is 6.0–7.0 (Wakwoya et al. 2022). According to Dewangana et al. (2023), nutrients that are essential for plant growth, such as nitrogen, phosphorus, and potassium, are mostly nutrients decreases, and plants may suffer from nutrient deficiencies (Barrow and Hartemink 2023). Thus, regular pH assay of the farm soils is crucial to guide the soil management practice and optimize nutrient availability. This will not only optimize the availability to plants at a soil pH of between 6.0 and 7.0—when the soil pH is too low or too high, the availability of these sustainable maize crop production—but also enhance the soil microbial activities, organic matter, and trace nutrients balance (Subedi and Ma 2009).

Table 16.4 Descriptive statistics of the subsoil (15–30 cm) properties of Lake Liambezi maize farms

Descriptive statistics	Ph	EC	OM	N	P	K	Ca	Mg	Na
Mean	7.27*	481.83*	1.59*	0.06^{NS}	5.59*	541.38*	3329.31*	218.19*	169.05*
Median	7.46	310.00	1.61	0.06	0.00	549.00	3469.50	235.50	137.50
Mode	7.44	335.00	1.71	0.06	0.00	–	–	301.00	0.00
Standard deviation	1.01	371.82	0.42	0.02	8.66	370.64	1291.37	110.19	167.29
Kurtosis	1.55	−0.15	0.15	16.07	7.28	9.59	1.32	−1.05	2.08
Skewness	−1.25	1.02	−0.53	3.09	2.31	2.25	−1.30	0.03	1.44
Range	4.32	1320.40	1.89	0.17	42.90	2198.00	5070.00	404.00	724.00
Minimum	4.42	96.60	0.47	0.02	0.00	32.00	44.00	27.00	0.00
Maximum	8.74	1417.00	2.36	0.19	42.90	2230.00	5114.00	431.00	724.00
Count	42	42	42	42	42	42	42	42	42
Confidence level (95%)	0.31	115.87	0.13	0.01	2.70	115.50	402.42	34.34	52.13

Key: Data with asterisk (*) varied significantly between farms ($p < 0.05$)
NS not significantly different

16.4.3 Soil Organic Matter

As shown in the Tables 16.3 and 16.4, the levels of the soil organic matter (SOM) in the Lake Liambezi maize farm ranged from 0.55% to 2.82% with an average of 1.62% within the top soil (0–15 cm), and 0.47%–2.36% with an average of 1.59 within the subsoil (15–30 cm). The result of analysis of variance (ANOVA) of the SOM differed statistically across the farms ($p < 0.05$). The status of SOM varies substantially in cropping fields, driven by differences in management practices, soil types, and landscape position (Tittonell et al. 2008; Zingore and Njoroge 2022). Due to its strong influence on soil biological, chemical, and physical properties and crucial soil functions, SOM is an essential indicator of soil quality with direct implications for crop productivity, food security, and human livelihood (Zingore and Njoroge 2022). According to Biernbaum (2012), the amount of organic matter in mineral soils (sand, loam, or clay) ranging from 1% and below is at minimal to very low levels for optimum crop production and sustainable soil productivity. However, organic matter content at 2–4% is at a moderate to an optimum level, and at 5% and greater is high (Biernbaum 2012). Thus, the results of the current study revealed a low level of soil organic matter, which may require the application of compost, animal manure, and/or organic materials like leaves, straw, hay, bark, and wood shavings to boost the soil organic matter content and optimize maize production at Lake Liambezi farming area. The introduction of cover crops or green manure may also significantly provide food for soil microorganisms and ultimately increase microbial activities and soil organic matter.

16.4.4 Nitrogen

The soil total nitrogen content of the Lake Liambezi maize farms (Tables 16.3 and 16.4) revealed low soil nitrogen levels, ranging from 0.019% to 0.195% with an average of 0.063% within the top soil (0–15 cm), and 0.017–0.185% with an average of 0.060% within the subsoil (15–30 cm). Using the conversion factor of 4.43, these values correspond to 0.084–0.864 mg/kg NO_3-N (top soil) and 0.075–0.820 mg/kg NO_3-N (subsoil). The result of ANOVA of the soil total nitrogen was not significantly different across the farms ($p > 0.05$). Being a critical macronutrient for plant function, nitrogen is often regarded as the most essential nutrient in crop production. According to Heckman (2002), soil nitrate-N (NO_3-N) concentrations in the range of 25–30 mg/kg indicate sufficiency for most crops, but nitrogen fertilizer practice should be adjusted based on local extension recommendations. This recommendation is necessary because several other factors such as initial soil fertility, crop variety, and yield target may influence the nitrogen requirement of a specific field to optimize maize crop productivity. Thus, effective nitrogen fertilizer management and soil content monitoring is crucial for optimization of soil nitrogen and maize cultivation in the Lake Liambezi maize farms.

16.4.5 Phosphorus

The levels of the soil available phosphorus (Tables 16.3 and 16.4) of the Lake Liambezi maize farms revealed an extremely low mean phosphorus level of 8.29 ppm in the topsoil and 5.59 ppm within the subsoil. The result of ANOVA of the soil available phosphorus differed statistically across the farms ($p < 0.05$). The critical levels of soil phosphorus for optimal maize yields range from 11 $mg \cdot kg^{-1}$ to 12.5 $mg \cdot kg^{-1}$ (Bai et al. 2013; Sucunza et al. 2018). Generally, the present levels of the soil available phosphorus of the Lake Liambezi maize farm are unable to sustainably support maize crop production for optimum yield output in the long term. Hence, there is a need for effective and efficient application of phosphorus fertilizer to the soil of the Lake Liambezi maize farms through various means of phosphorus fertilization programs. However, the exact amount of phosphorus to apply will depend on other various factors such as the potential and target grain yield of each farmer, economic analysis, climate, production practices at each farm during any specific cropping season, as well as the type and form of fertilizer.

However, Sharmin et al. (2024) observed that the low phosphorus-use-efficiency (PUE) of plants in the field condition (15–20%) highlights that most of the phosphorus applied to the farm soil remains unavailable to plants, and the excess of it causes ground and surface water contamination (i.e., eutrophication) through leaching and runoff, which ultimately results in environmental pollution. Therefore, it is crucial to apply phosphorus fertilizers based on the soil test value and PUE to protect the environment from contamination and sustainable management of phosphorus resources (Sharmin et al. 2024). Ultimately, the negative impacts of agricultural practices such as phosphate fertilizer applications must be understood when managing soils in agricultural production (Derpsch et al. 2024). Thus, organic soil amendment methods should be incorporated into the soil fertility management system at the Lake Liambezi maize crop fields. This involves incorporating locally produced kraal manure, chicken manure, and compost that slowly release fertilizers into the soil. This approach enhances soil structure and nutrient retention. Furthermore, riparian buffer strips should be constructed as natural filters to prevent nutrient flow into adjacent water bodies. This is an environmentally conscious approach that not only supports agricultural productivity but also protects aquatic ecosystems from eutrophication, ensuring the sustainability of both terrestrial and freshwater resources.

In addition, other sustainable fertilization methods that address environmental concerns near sensitive ecosystems like Lake Liambezi involve integrated Good Agricultural Practices (GAP) and Best Management Practices (BMPs). Loss of fertilizers from leaching may be addressed by splitting fertilizer applications (Sitthaphanit et al. 2009). Splitting applications allow for the strategic timing of nutrient delivery, thereby matching fertilizer applications to crop demand to improve efficiency. Furthermore, when applications are split, the risk of losing nutrients to intense rainfall and flooding soon after the application is reduced (Philip et al. 2019). This has the ability to reduce nutrient runoff and leaching while holistically promoting sustainable fertility management at Lake Liambezi maize crop fields.

16.4.6 Potassium

The results (also in Tables 16.3 and 16.4) depict the soil potassium content of the Lake Liambezi maize farms. The results revealed wide variation, ranging from 64 to 860 ppm with an average of 452.26 ppm in the top soil (0–15 cm), and 32–2230 ppm with an average of 549 ppm in the subsoil (15–30 cm). The result of ANOVA of the soil potassium contents differed statistically across the farms ($p < 0.05$). Potassium is a critical nutrient required in large amounts for proper maize growth. According to Pettigrew (2008), potassium is required for enzymes and many chemical processes within plant tissues and the production of starch as well as protein in maize crops. Potassium enhances drought resistance in maize crops by regulating stomatal movement, improving water uptake through enhanced root growth, and maintaining cell turgor pressure (Ul-Allah et al., 2020). The optimum potassium values for sustainable maize crop production range from 40 to 120 ppm (Muthaura et al. 2017). In many of the maize farms, the recorded soil potassium in the current study is above the recommended limit for the maize crop. This might promote nutrient imbalances and pose a toxicity threat to the maize crop. Therefore, it is imperative to develop an effective potassium fertilizer management program and monitor potassium levels in the soils of the Lake Liambezi maize farms.

16.4.7 Calcium

Tables 16.3 and 16.4 also showed that the soil calcium levels of the Lake Liambezi maize farms ranged from 211 to 4974 ppm with an average of 3583.58 ppm within the top soil and 44–5114 ppm with an average of 199.26 ppm in the subsoil. The result of ANOVA of the soil calcium contents differed statistically across the farms ($p < 0.05$). As a crucial element for plants, calcium is involved in photosynthesis and nutrient absorption, and affects the growth of plants (Weng et al. 2022). Calcium is crucial for root development as it is necessary for the translocation and storage of carbohydrates and proteins (do Moraes Gatti et al. 2023). It has been reported that plants with calcium deprivation had shorter roots and a dark brown color, and displayed initial symptoms of chlorosis in their young leaves, which eventually led to necrosis and tipping (do Moraes Gatti et al. 2023). The reported optimum levels of soil calcium for crop growth ranged from 600 to 1000 ppm (Sahota 2018). While the current results of the soil calcium contents showed wide variation across the maize farms, there is a need for a proper soil management regime to optimize the soil calcium content, and hence guarantee a sustainable maize productivity.

16.4.8 Magnesium

The soil magnesium levels of the Lake Liambezi maize farms (also seen in Tables 16.3 and 16.4) showed wide variation of the soil magnesium content, ranging from 3 to 449 ppm with an average of 224.30 ppm within the top soil and 27–431 ppm with an average of 218.19 ppm in the subsoil. The result of ANOVA of the soil magnesium contents differed statistically across the farms ($p < 0.05$). An ideal level of magnesium in the soil should range from 25 to 45 ppm (do Moraes Gatti et al. 2023). Thus, the recorded magnesium mean values are high and require remedial practices to lower the levels to the optimum range for sustainable maize crop production. However, soil undergoing intensive crop forage and harvest are not being replenished with magnesium fertilizers, resulting in depletion of indigenous magnesium from the soil and large-scale deficiency (Wang et al. 2020). This may be a future concern in the Lake Liambezi maize farm area, despite the current high level of soil magnesium, as the farms are under intensive and continuous maize cultivation without any soil fertility management regime in place. Moreover, it has been observed that the deficiency of soil magnesium affects photosynthesis and carbohydrate partitioning in crops (Farhat et al. 2016) and reduces the sustainability of agricultural production and development (Rude and Gruber 2004; de Baaij et al. 2015).

16.4.9 Sodium

The soil sodium levels of the Lake Liambezi maize farm area (Tables 16.3 and 16.4) recorded in the current study revealed wide variation of the soil sodium content, ranging from 0 to 795 ppm with an average of 207.44 ppm within the top soil and 0–724 ppm with an average of 169.04 ppm in the subsoil. The result of ANOVA of the soil sodium contents differed statistically across the farms ($p < 0.05$). Determining the level of soil sodium is critical to knowing the salinity levels in the soil and applying appropriate practices to flush the salt through the soil profile and out of the root system. Sodium is not considered a plant nutrient; therefore, high soil levels can cause problems of soil salinity, disturbance to soil structure, and poor nutrient uptake by plants. The optimum sodium level of agricultural soils should not exceed 40 ppm (do Moraes Gatti et al. 2023). Therefore, the recorded mean soil sodium of the maize farms is high and require remedial practices to lower the levels to the optimum range for sustainable maize crop production.

16.4.10 Electrical Conductivity

The results displayed in Tables 16.3 and 16.4 show that the soil electrical conductivity (EC) levels of the Lake Liambezi maize farms ranged from 70 ppm to 1680 μS/cm with an average of 524.03 μS/cm within the top soil and 96.6 ppm to 1417/cm with an average of 481.83 μS/cm in the subsoil. The result of ANOVA of the soil EC levels differed statistically across the farms ($p < 0.05$). EC measures nutrient availability for crop uptake, and it is an important indication of soil fertility as well as a determinant of salinity level in the soil (Ehtaiwesh 2022). EC levels in the soil have a positive correlation with plant nutrients such as nitrogen, phosphorus, potassium, calcium, and magnesium. The optimum soil EC levels ranged from 800 to 2500 μS/cm (do Moraes Gatti et al. 2023). Thus, the recorded average EC values in this study are low and require improvement practices to boost the EC levels to the optimum range for sustainable maize crop production, as this might severely affect crop performance. Within a specific range, EC can be used to indicate the status of nutrient availability for plants, with the low end indicating nutrient-poor soil that is structurally unstable and dispersed, while high EC values often reflect the potential for salinity problems (Fard and Sadikhani 2014).

16.5 Conclusion

The results of soil fertility status of the Lake Liambezi maize farms varied widely across the 21 farms sampled. The results revealed low mean levels of total nitrogen, available phosphorus, soil organic matter, and electrical conductivity, while high mean levels of potassium, magnesium, and calcium were recorded. These conditions may create concerns about nutrient imbalance, which is a limiting factor for sustainable crop production. Despite sodium not being considered an important plant nutrient, the high mean soil levels recorded in the Lake Liambezi maize farms might cause concern about potential salinity problems in many of the maize farms. Interestingly, the recorded soil textures largely fell within the loamy sand, sandy loam, and loamy categories, with the mean value revealing a sandy loam textural class. These soil textures have good drainage and good nutrient retention capabilities. However, unsustainable agronomic practice, such as the current state of continuous maize cultivation observed at Lake Liambezi maize farms without any appropriate fertility management program, poses a threat to the soil's productive capacity in the long run. Thus, the results of this study show the need to adopt a sustainable nutrient management program to achieve long-term productivity of the maize farms.

16.6 Implication of the Current Study

As observed by Goshu et al. (2015), agricultural land use without appropriate soil management can deplete soil nutrient contents, posing challenges to food security, which is already creating greater challenges to most developing countries to feed the ever-increasing human population. Thus, the cultivation of maize crops without prior knowledge of the nutrient status of the soil to inform relevant soil fertility management, which characterized our study area, could adversely affect the productive capability of the soil and lead to loss of soil value over a long time. In the study area, it is perceived by the small-scale farmers that the natural soil nutrients are adequate for maize cultivation, and hence there is no need for field fertilizer application. Most often, the farmers only apply nitrogen fertilizer to a particular area of the field where there is perceived nutrient deficiency as a result of the yellowish appearance of maize leaves. This practice might pose severe negative implications for soil fertility and threaten sustainable crop productivity in the Lake Liambezi maize farms. Thus, there is a need to adopt best practices and scientific approaches to sustainable soil fertility management and farmers' education in the Lake Liambezi maize farm areas of the Zambezi Region.

16.7 Lesson Learned

The results of this study revealed low mean levels of soil total nitrogen, available phosphorus, organic matter, and electrical conductivity, while high mean levels of potassium, magnesium, and calcium across the 21 sampled farms. These conditions may create concerns about nutrient imbalance, which is a serious limiting factor for sustainable crop production. The recorded soil nutrient status is contrary to the small-scale farmers' perception that the natural soil nutrients were adequate for continuous maize cultivation, and hence saw no need for field fertilizer treatment. Moreover, the fact that some farmers are already carrying out isolated treatment of their maize farms using the yellowish appearance of maize leaves as an indicator of nutrient deficiency partly confirms our findings of the low soil nutrient status. Thus, there is a need for governmental intervention through its Department of Agricultural and Rural Extension Services to carry out farmers' education on the best practice of soils fertility management, as well as the provision of fertilizer subsidies to the maize farmers in the Lake Liambezi farm areas.

References

Abd-Elmabod SK, Bakr N, Muñoz-Rojas M, Pereira P, Zhang Z, Cerdà A, Jordán A, Mansour H, De la Rosa D, Jones L (2019) Assessment of soil suitability for improvement of soil factors and agricultural management. Sustainability 11(6):1588

Ahmad MN, Anuar MI, Nordiana Abd Aziz, Murdi AA (2025) Function and application of soil electrical conductivity (EC) sensor in agriculture: a review. Adv Agric Food Res J 6(1). https://doi.org/10.36877/aafrj.a0000552

Aslam M, Zamir MSI, Afzal I, Yaseen M (2013) Morphological and physiological response of maize hybrids to potassium application under drought stress. J Agric Res 51(4):443–454

Bai Z, Li H, Yang X, Zhou B, Shi X, Wang B, Li D, Shen J, Chen Q, Qin W (2013) The critical soil P levels for crop yield, soil fertility and environmental safety in different soil types. Plant Soil 372(1):27–37

Barrow NJ, Hartemink AE (2023) The effects of pH on nutrient availability depend on both soils and plants. Plant Soil 487(1):21–37

Bender SF, Wagg C, van der Heijden MG (2016) An underground revolution biodiversity and soil ecological engineering for agricultural sustainability. Trends Ecol Evol 31(6):440–452. https://doi.org/10.1016/j.tree.2016.02.016

Beretta AN, Silbermann AV, Paladino L, Torres D, Kassahun D, Musselli R, Lamohte AG (2014) Soil texture analyses using a hydrometer: modification of the Bouyoucos method. Ciencia e Investigaci6n Agraria: Revista Latinoamericana de Ciencias de La Agricultura 41(2):263–271

Biernbaum J (2012) Organic matters: feeding the soil and building soil quality. Department of Horticulture, Michigan State University

Biswas S, Saha A, Debnath S (2023) Effect of potassium level on growth, yield, nutrient uptake and economics of hybrid maize. Environ Ecol 41(3D):2155–2159

Bremner JM (1960) Determination of nitrogen in soil by the Kjeldahl method. J Agric Sci 55(1):11–33

de Baaij JHF, Hoenderop JGJ, Bindels RJM (2015) Magnesium in man: implications for health and disease. Physiol Rev 95:1–46. https://doi.org/10.1152/physrev.00012.2014

do Moraes Gatti VC, da Silva Barata H, Silva VFA, da Cunha FF, de Oliveira RA, de Oliveira JT, Silva PA (2023) Influence of calcium on the development of corn plants grown in hydroponics. AgriEngineering 5(1):623–630

De Vos B, Lettens S, Muys B, Deckers JA (2007) Walkley–Black analysis of forest soil organic carbon: recovery, limitations and uncertainty. Soil Use Manag 23(3):221–229

Dewangana SK, Minj D, Paul AC, Shrivastava SK (2023) Evaluation of physicochemical characteristics of water sources in Dawana Odgi area, Surajpur, Chhattisgarh. Int J Sci Res Eng Dev 6(6)

Ehtaiwesh AF (2022) The effect of salinity on nutrient availability and uptake in crop plants. Sci J Appl Sci Sabratha Univ 9(9):55–73

Erenstein O, Jaleta M, Sonder K, Mottaleb K, Prasanna BM (2022) Global maize production, consumption and trade: trends and R&D implications. Food Secur 14(5):1295–1319
Fard FG, Sadikhani MR (2014) The role of soil quality in today's world. Indian J Fundam Appl Life Sci 4(S4):1249–1257
Farhat N, Elkhouni A, Zorrig W, Smaoui A, Abdelly C, Rabhi M (2016) Effects of magnesium deficiency on photosynthesis and carbohydrate partitioning. Acta Physiol Plant 38(6):145
Hair JF (2014) A primer on partial least squares structural equation modeling (PLS-SEM). Sage
Heckman JR (2002) In-season soil nitrate testing as a guide to nitrogen management for annual crops. HortTechnology 12(4):706–710
Heiniger RW, McBride RG, Clay DE (2003) Using soil electrical conductivity to improve nutrient management. Agron J 95(3):508–519
Huntley BJ (2023) Soil, water and nutrients. In: Ecology of Angola: terrestrial biomes and ecoregions. Springer, pp 127–147
Janzen HH (2006) The soil carbon dilemma: shall we hoard it or use it? Soil Biol Biochem 38(3):419–424. https://doi.org/10.1016/j.soilbio.2005.10.008
Kettler TA, Doran JW, Gilbert TL (2001) Simplified method for soil particle-size determination to accompany soil-quality analyses. Soil Sci Soc Am J 65(3):849–852
Khaled F, Sayed A (2023) Soil pH and its influence on nutrient availability and plant health. Int J Adv Chem Res 5(2):61–63. https://doi.org/10.33545/26646781.2023.v5.i2a.204
Kibblewhite MG, Ritz K, Swift MJ (2008) Soil health in agricultural systems. Philos Trans R Soc B Biol Sci 363(1492):685–701. https://doi.org/10.1098/rstb.2007.2178
Kim HN, Park JH (2024) Monitoring of soil EC for the prediction of soil nutrient regime under different soil water and organic matter contents. Appl Biol Chem 67:1. https://doi.org/10.1186/s13765-023-00849-4
Koralage ISA, Silva NRN, De Silva CS (2015) The determination of available phosphorus in soil: a quick and simple method. OUSL J 8:1
Kwari JD, Kamara AY, Ekeleme F, Omoigui L (2011) Soil fertility variability in relation to the yields of maize and soybean under intensifying cropping systems in the tropical savannas of northeastern Nigeria. In: Innovations as key to the green revolution in Africa: exploring the scientific facts. Springer Netherlands, Dordrecht, pp 457–464
Lal R (2016) Soil health and carbon management. Food Energy Secur 5(4):212–222. https://doi.org/10.1002/fes3.96
Liyan G (2022) Soil electrical conductivity (EC): what's it, why important, how to measure & more. https://www.seeedstudio.com/blog/2022/07/15/soil-electricalconductivity/# sidr-nav (Megtekintve: 2024.05. 01.)
MacCarthy DS, Adamtey N, Freduah BS, Fosu-Mensah BY, Ofosu-Budu GK, Fliessbach A (2023) Modeling the effect of soil fertility management options on maize yield stability under variable climate in a sub-humid zone in Ghana. Front Sustain Food Syst 7:1132732
Monroy, R. T., Monsalve, M.O., Galeano-Vasco, L.F., Medina-Sierra, M., & Ceron-Munoz, M.F. (2024). Cation exchange capacity in grazing systems and a case study for quantification by hyperspectral imaging. IntechOpen. https://doi.org/10.5772/intechopen.112991
Munera-Echeverri JL, Martinsen V, Strand LT, Cornelissen G, Mulder J (2020) Effect of conservation farming and biochar addition on soil organic carbon quality, nitrogen mineralization, and crop productivity in a light textured Acrisol in the sub-humid tropics. PLoS One 15(2):e0228717
Muthaura C, Mucheru-Muna M, Zingore S, Kihara J, Muthamia J (2017) Effect of application of different nutrients on growth and yield parameters of maize (Zea mays), case of Kandara Murang'a County. ARPN J Agric Biol Sci 12(1):19–33
Nezomba H, Mtambanengwe F, Tittonell P, Mapfumo P (2015) Point of no return? Rehabilitating degraded soils for increased crop productivity on smallholder farms in eastern Zimbabwe. Geoderma 239:143–155
Nhunda D, Semoka JM, Tindwa HJ (2024) Assessment of soil fertility status in selected fields under maize production in Kongwa District, Dodoma region, Tanzania. J Agric Ecol Res Int 25(1):32–47

Normandin V, Kotuby-Amacher J, Miller RO (1998) Modification of the ammonium acetate extractant for the determination of exchangeable cations in calcareous soils. Commun Soil Sci Plant Anal 29(11–14):1785–1791

Nwite JN (2022) Effect of pH on soil nutrient availability and maize performance. Soil Chem Lett 14(4):210–218

Oldfield EE, Bradford MA, Wood SA (2019) Global meta-analysis of the relationship between soil organic matter and crop yields. Soil 5(1):15–32

Ouyang D, MacKenzie AF, Fan M (1998) Phytotoxicity of banded urea amended with triple superphosphate and potassium chloride. Agron J 90(6):734–739

Öhlinger R (1996) Soil sampling and sample preparation. In: Methods in soil biology. Springer, pp 7–11

Pettigrew WT (2008) Potassium influences on yield and quality production for maize, wheat, soybean and cotton. Physiol Plant 133(4):670–681

Postma-Blaauw MB, de Goede RGM, Bloem J, Faber JH, Brussaard L (2010) Soil biota community structure and abundance under agricultural intensification and extensification. Ecology 91(2):460–473. https://doi.org/10.1890/09-0666.1

Rayment GE, Higginson FR (1992) Australian laboratory handbook of soil and water chemical methods. Inkata Press, Melbourne

Riffenburgh RH (2012) Chapter 12 – Tests on means of continuous data. In: Riffenburgh RH (ed) Statistics in medicine, 3rd edn. Academic, pp 249–274. https://doi.org/10.1016/B978-0-12-384864-2.00012-3

Rude RK, Gruber HE (2004) Magnesium deficiency and osteoporosis: animal and human observations. J Nutr Biochem 15(12):710–716

Sahota TS (2018) What does a soil test report indicate? Northwest Link:3–5

Schaefer PE, Tabaldi LA, Müller TM, Ribeiro LP, Martin TN (2025) Development of maize under different pH values, humidity and presence of Azospirillum brasilense. Braz J Biol 84:e287643

Shah F, Wu W (2019) Soil and crop management strategies to ensure higher crop productivity within sustainable environments. Sustainability 11(5):1485

Subedi KD, Ma BL (2009) Assessment of some major yield-limiting factors on maize production in a humid temperate environment. Field Crop Res 110(1):21–26

Sucunza FA, Boem FHG, Garcia FO, Boxler M, Rubio G (2018) Long-term phosphorus fertilization of wheat, soybean and maize on mollisols: soil test trends, critical levels and balances. Eur J Agron 96:87–95

Tittonell P, Vanlauwe B, Corbeels M, Giller KE (2008) Yield gaps, nutrient use efficiencies and response to fertilisers by maize across heterogeneous smallholder farms of western Kenya. Plant Soil 313(1):19–37

Ul-Allah S, Ijaz M, Nawaz A, Sattar A, Sher A, Naeem M, Shahzad U, Farooq U, Nawaz F & Mahmood K (2020) Potassium application improves grain yield and alleviates drought susceptibility in diverse maize hybrids. Plants, 9(1):75. https://doi.org/10.3390/plants9010075

Van der Velde M, Folberth C, Balkovič J, Ciais P, Fritz S, Janssens IA, Obersteiner M, See L, Skalský R, Xiong W (2014) African crop yield reductions due to increasingly unbalanced nitrogen and phosphorus consumption. Glob Chang Biol 20(4):1278–1288

Wakwoya MB, Woldeyohannis WH, Yimamu FK (2022) Effects of minimum tillage and liming on maize (Zea mays L.) yield components and selected properties of acid soils in Assosa Zone, West Ethiopia. J Agric Food Res 8:100301

Wang C, Kuzyakov Y (2024) Soil organic matter priming: the pH effects. Glob Change Biol 30:e17349, | 1–21. https://doi.org/10.1111/gcb.17349

Wang Z, Hassan MU, Nadeem F, Wu L, Zhang F, Li X (2020) Magnesium fertilization improves crop yield in most production systems: a meta-analysis. Front Plant Sci 10:495191

Wang X, Qi J, Kan Z (2024) Sustainable management and tillage practice in agriculture. Agronomy 14(12):2891. https://doi.org/10.3390/agronomy14122891

Weng X, Li H, Ren C, Zhou Y, Zhu W, Zhang S, Liu L (2022) Calcium regulates growth and nutrient absorption in poplar seedlings. Front Plant Sci 13:887098

Xia Y, Feng J, Zhang H, Xiong D, Kong L, Seviour R, Kong Y (2024) Effects of soil pH on the growth, soil nutrient composition, and rhizosphere microbiome of Ageratina adenophora. PeerJ 12:e17231

Zingore S, Njoroge S (2022) Soil organic matter regulates maize productivity and fertilizer response in maize production. Growing Africa 1(1):8–11

García-Ruiz R, Ochoa MV, Hinojosa MB & Gómez-Muñoz B (2012) Improved soil quality after 16 years of olive mill pomace application in olive oil groves. Agron Sustain Dev 32(3):803–810

Sanchez PA (2019) Properties and Management of Soils in the Tropics. Cambridge University Press

Phillips HR, Guerra CA, Bartz ML, Briones MJ, Brown G, Crowther TW & Eisenhauer N (2019) Global distribution of earthworm diversity. Sci 366(6464):480–485

Peel RA, Tweddle D, Simasiku EK, Martin GD, Lubanda J, Hay CJ, & Weyl OLF (2015) Ecology, fish and fishery of Lake Liambezi, a recently refilled floodplain lake in the Zambezi Region, Namibia, Afr J Aquat Sci DOI: 10.2989/16085914.2015.1105779

Goshu D (2015) The dynamics of food price convergence in Ethiopia. Ethiop J Econ 24(1):31–58

Sitthaphanit S, Limpinuntana V, Toomsan B, Panchaban S & Bell RW (2009) Fertiliser strategies for improved nutrient use efficiency on sandy soils in high rainfall regimes. Nutr Cycl Agroecosystems 85(2):123–139

Upadhyay S & Raghubanshi AS (2020) Determinants of soil carbon dynamics in urban ecosystems. In Urban Ecol (pp. 299–314). Elsevier

Sharmin S, Wang Q, Islam MR, Wang W & Enyoh CE (2024) Microplastic contamination of non-mulched agricultural soils in Bangladesh: Detection, characterization, source apportionment and probabilistic health risk assessment. J Xenobiotics 14(2):812–826

Derpsch R, Kassam A, Reicosky D, Friedrich T, Calegari A, Basch G & dos Santos DR (2024) Nature's laws of declining soil productivity and Conservation Agriculture. Soil Secur 14:100127

Kowalenko CG & Ivarson KC (1978) Effect of moisture content, temperature and nitrogen fertilization on carbon dioxide evolution from field soils. Soil Biol Biochem 10(5):417–423

MIX
Papier aus verantwortungsvollen Quellen
Paper from responsible sources
FSC® C105338
FSC
www.fsc.org

If you have any concerns about our products,
you can contact us on
ProductSafety@springernature.com

In case Publisher is established outside the EU,
the EU authorized representative is:
Springer Nature Customer Service Center GmbH
Europaplatz 3, 69115 Heidelberg, Germany

Printed by Libri Plureos GmbH
in Hamburg, Germany